Fundamentals of
MOLECULAR EVOLUTION

Fundamentals of

MOLECULAR

EVOLUTION

WEN-HSIUNG LI
The University of Texas, Houston

DAN GRAUR
Tel Aviv University

SINAUER ASSOCIATES, INC. · PUBLISHERS
Sunderland, Massachusetts

THE COVER

The nine ubiquitin-coding repeats in the polyubiquitin C locus in humans exhibit evidence of concerted evolution (Chapter 6). Regions that may have been recently subjected to gene conversion are shown in a single color. Ambiguous relationships are indicated by split shading. Substitutions that have occurred subsequent to the conversion event are denoted by X. Modified from P. M. Sharp and W.-H. Li, 1987, Molecular evolution of ubiquitin genes. Trends in Ecology & Evolution 2: 328-332.

FUNDAMENTALS OF MOLECULAR EVOLUTION

Library of Congress Cataloging-in-Publication Data

Li, Wen-Hsiung, 1942–
Fundamentals of molecular evolution / Wen-Hsiung Li and Dan Graur.
p. cm.
Includes bibliographical references and index.
ISBN 0-87893-452-9
1. Chemical evolution. I. Graur, Dan, 1953– . II. Title.
QH325.L65 1991
575.1—dc20 90-43581
CIP

Printed in U.S.A.

5 4 3 2 1

To our families,

Sue Jean, Vivian, Herman, and Joyce
Mina, Or, and Inar

CONTENTS

PREFACE

In 1869, Johann Friedrich Miescher discovered DNA. For a while he entertained the "wild" notion that DNA might have something to do with heredity, but he abandoned this "absurd" idea and spent more than 30 years studying protamines, a group of very basic proteins in sperm cells, which, according to Miescher, were the key to heredity. As it turned out, DNA molecules are not only the key to heredity but, as one of the pioneers of molecular evolution, Emile Zuckerkandl, phrased it, they are "documents of evolutionary history." In fact, the DNA of every living organism is an accumulation of historical records. However, the information contained in these records is in a disorderly state, scattered and at times fragmentary. Some of it is hidden or camouflaged beyond recognition, and parts of it are lost without a trace. The purpose of molecular evolution is to unravel these historical records, put the information in order, read it, and decipher its meaning.

Since evolutionary processes leave their distinctive marks on the genetic material, it is possible to use molecular data not only to reconstruct the chronology of evolution but also to identify the driving forces behind the evolutionary process. Spectacular achievements in molecular biology, such as gene cloning, DNA sequencing, and restriction endonuclease fragment analysis have, in a sense, placed scientists in a new and privileged position. We can peer into a previously unseen world where genes evolve by such processes as gene duplication, shuffling of DNA, nucleotide substitution, transposition, and gene conversion. In this world genomes appear both static and fluid, sometimes experiencing little change over long periods of time and at other times changing dramatically during the geological equivalent of the blink of an eye.

By studying the genetic material we can also attempt to build a classification of the living world that is based not so much on taxonomic convenience but on phylogenetic facts. As opposed to the traditional fields of evolutionary inquiry, such as comparative anatomy, morphology, and paleontology, which out of necessity restrict themselves to the study of evolutionary relationships among closely related organisms, we can now build gigantic family trees connecting vertebrates, insects, plants, fungi, and bacteria, and can trace their common ancestry to times truly and geologically immemorial.

The purpose of this book is to describe the dynamics of evolutionary change at the molecular level, the driving forces behind the evolutionary process, and the effects of various molecular mechanisms on the long-term evolution of genomes, genes, and their products. In addition, the book provides basic methodological tools for comparative and phylogenetic analyses of molecular data from an evolutionary perspective. We emphasize that, although individuals are the entities affected by natural selection and other processes, it is populations and genes that change over evolutionary time. To make the connection between these seemingly disparate levels, we work through basic concepts of population genetics.

We have set out to write a book for "beginners" in molecular evolution. At the same time, we have tried to maintain the standards of the scientific method and to include quantitative treatments of the issues at hand. Therefore, in describing evolutionary phenomena and mechanisms at the molecular level, both mathematical and intuitive explanations are provided. Neither is meant to be at the expense of the other; rather, the two approaches are intended to complement each other and to help the reader achieve a better grasp of the issues. We have not attempted to attain encyclopedic completeness, but have provided a large number of examples to support and clarify the many theoretical arguments and discussions.

For many years, biochemists and molecular biologists have regarded evolutionary studies as an aggregate of wild speculations, unwarranted assumptions, and undisciplined methodology. While this assessment has never been accurate, the introduction of molecular methods has undoubtedly turned evolution into a "hard" science in which relevant parameters can be measured, counted, or computed from empirical data, and theories can be tested against reality. Conjectures in evolutionary studies today serve the same purpose as in physics; they are quantitative working hypotheses meant to encourage experimental work so that the theory can be verified, refined, or refuted. One of the main aims of this book is to show that by strengthening the factual basis of the field, evolutionary studies have achieved what Sir William Herschel in 1831 called the true goal of all natural sciences: namely to phrase its propositions "not vaguely and generally, but with all possible precision in place, weight, and measure."

We are indebted to many colleagues, students, and friends for helping us put this book together. Their comments, suggestions, corrections, and discussions greatly improved this work and saved us many embarrassments. For favors great and small, we thank Sara Barton, Adina Breiman, David Cutler, David Hewett-Emmett, Winston Hide, Austin Hughes, Li Jin, Margaret Kidwell, Amanda Ko, Giddy Landan, William S. Lewis, Volker

Loeschcke, Ora Manheim, David Mindell, Tatsuya Ota, Lori Sadler, Paul Sharp, Yuval Shuali, Jürgen Tomiuk, David Wool, and Chung-I Wu. Especially, we thank C. William Birky Jr. and Bruce Walsh for thoroughly reviewing the entire manuscript and for their valuable suggestions. We would also like to express our gratitude to Dr. Masatoshi Nei for his inspiration and guidance in our research. Support from the National Institutes of Health and the U.S.–Israel Binational Science Foundation have made this cooperative endeavor possible.

W.-H. Li
D. Graur

Fundamentals of
MOLECULAR EVOLUTION

INTRODUCTION

WHAT IS MOLECULAR EVOLUTION?

Molecular evolution encompasses two areas of study: (1) the evolution of macromolecules and (2) the reconstruction of the evolutionary history of genes and organisms. By "evolution of macromolecules" we refer to the rates and patterns of change occurring in the genetic material (e.g., DNA sequences) and its products (e.g., proteins) during evolutionary time and to the mechanisms responsible for such changes. The second area, also known as "molecular phylogeny," deals with the evolutionary history of organisms and macromolecules, as inferred from molecular data.

It might appear that the two areas of study constitute independent fields of inquiry, for the object of the first is to elucidate the causes and effects of evolutionary changes in molecules, while the second uses molecules merely as a tool to reconstruct the biological history of organisms and their genetic constituents. In practice, however, the two disciplines are intimately interrelated, and progress in one area facilitates studies in the other. For instance, phylogenetic knowledge is essential for determining the order of changes in the molecular characters under study. And conversely, knowledge of the pattern and rate of change of a given molecule is crucial in attempts to reconstruct the evolutionary history of a group of organisms.

Traditionally, a third area of study, prebiotic evolution or the "origin of life," is also included within the framework of molecular evolution. This subject, however, involves a great deal of speculation and is less amenable to quantitative treatments. Moreover, the rules that govern the process of information transfer in prebiotic systems (i.e., systems devoid of replicable genes) are not known at the present time. Therefore, this book will not deal with the origin of life. Interested readers may consult Oparin (1957), Cairns-Smith (1982), Dyson (1985), and Loomis (1988).

The study of molecular evolution has its roots in two disparate disciplines: population genetics and molecular biology. Population genetics provides the theoretical foundation for the study of evolutionary processes, while molecular biology provides the empirical data. Thus, to understand molecular evolution it is essential to acquire some basic knowledge of both the theory of population genetics and molecular biology.

1

GENE STRUCTURE AND MUTATION

This chapter provides some basic background in molecular biology that is required for studying evolutionary processes at the DNA level. The most essential parts are the genomic structure of a typical eukaryotic gene and the various types of mutation. Further molecular background will be provided in the relevant chapters. Some terms are defined in the Glossary.

DNA SEQUENCES

The hereditary information of all living organisms, with the exception of some viruses, is carried by **deoxyribonucleic acid** (**DNA**) molecules. DNA usually consists of two complementary chains twisted around each other to form a right-handed helix. Each chain is a linear polynucleotide consisting of four nucleotides. There are two **purines**: **adenine** (**A**) and **guanine** (**G**), and two **pyrimidines**: **thymine** (**T**) and **cytosine** (**C**). The two chains are joined together by hydrogen bonds between pairs of nucleotides. Adenine pairs with thymine by means of two hydrogen bonds, also referred to as the **weak bond**, and guanine pairs with cytosine by means of three hydrogen bonds, the **strong bond** (Figure 1).

Each nucleotide in a DNA sequence contains a pentose sugar (deoxyri-

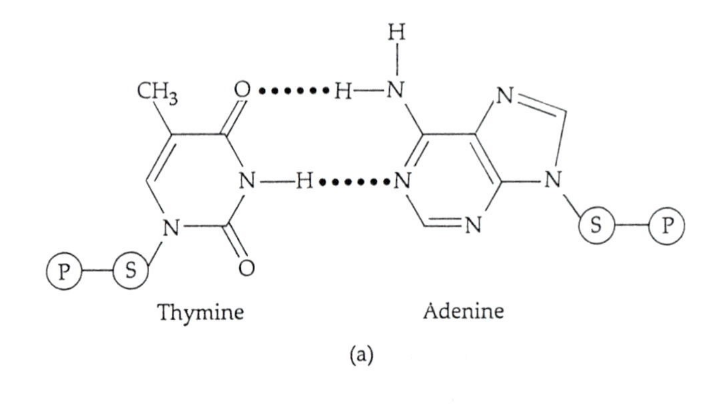

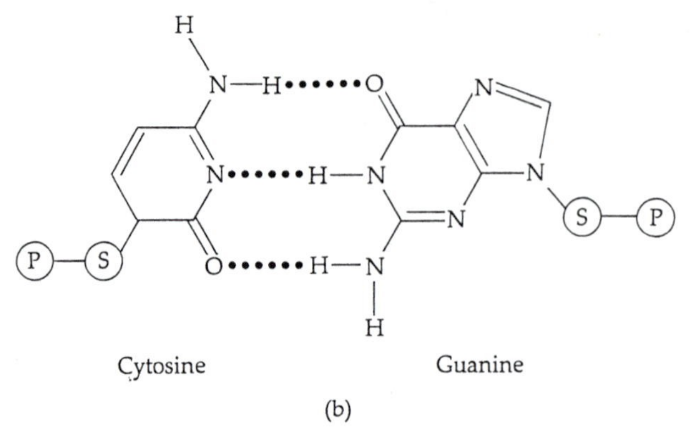

Figure 1. Complementary base pairing by means of hydrogen bonds (dotted lines) between (a) thymine and adenine (weak bond), and (b) cytosine and guanine (strong bond). (P), phosphate; (S), sugar.

bose), a phosphate group, and a purine or a pyrimidine base. The backbone of the DNA molecule consists of sugar and phosphate moieties, which are covalently linked together by asymmetrical 5′–3′ phosphodiester bonds. Consequently, the DNA molecule is polarized, one end having a phosphoryl radical (—P) on the 5′ carbon of the terminal nucleotide, the other possessing a free hydroxyl (—OH) on the 3′ carbon of the terminal nucleotide. The direction of the phosphodiester bonds determines the molecule's character; thus, for instance, the sequence 5′—G—C—A—A—T—3′ is different from the sequence 3′—G—C—A—A—T—5′. By convention, DNA sequences are written in the order they are transcribed, i.e., from the 5′ end to the 3′ end, also referred to as the **upstream** and **downstream** directions, respectively. The double helical form of DNA has two strands in antiparallel array (Figure 2).

Ribonucleic acid (RNA) is found as either a double- or a single-stranded molecule. RNA differs from DNA by having ribose, instead of deoxyribose,

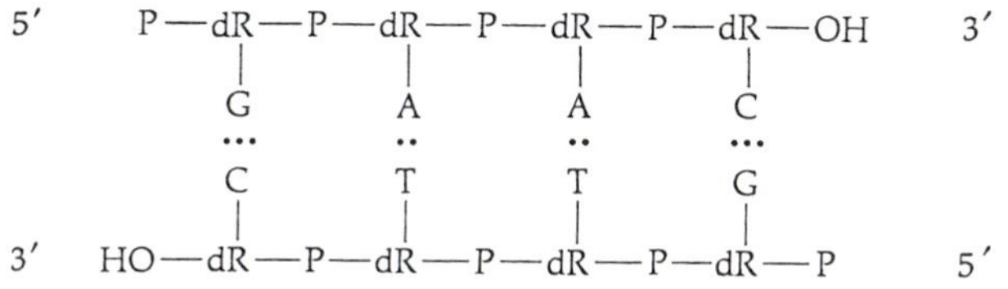

```
5'   P—dR—P—dR—P—dR—P—dR—OH   3'
       |     |     |     |
       G     A     A     C
      ···    ··    ··   ···
       C     T     T     G
       |     |     |     |
3'  HO—dR—P—dR—P—dR—P—dR—P     5'
```

Figure 2. Schematic representation of the antiparallel structure of double-stranded DNA. P, phosphate; dR, deoxyribose; OH, hydroxyl; A, adenine; G, guanine; C, cytosine; T, thymine; —, covalent bond; ··, weak bond; ···, strong bond.

as its backbone sugar moiety, and by using the nucleotide uracil instead of thymine. Adenine, cytosine, guanine, and thymine/uracil are referred to as the standard nucleotides. Some functional RNA molecules, most notably tRNAs, contain nonstandard nucleotides, i.e., chemical modifications of standard nucleotides that have been introduced into the RNA after its transcription.

GENE STRUCTURE

Traditionally, a **gene** was defined as a segment of DNA that codes for a polypeptide chain or specifies a functional RNA molecule. Recent molecular studies, however, have radically altered our perception of genes, and we shall adopt a somewhat vaguer definition. Accordingly, a gene is a sequence of genomic DNA or RNA that is essential for a specific function. Performing the function may not require the gene to be translated or even transcribed.

At present, three types of genes are recognized: (1) **protein-coding genes**, which are transcribed into RNA and subsequently translated into proteins, (2) **RNA-specifying genes**, which are only transcribed, and (3) **regulatory genes**. According to a narrow definition, the third category includes only untranscribed sequences. Transcribed regulatory genes essentially belong to one of the first two categories. Protein-coding genes and RNA-specifying genes are also referred to as **structural genes**. Note that some authors restrict the definition of structural genes to include only protein-coding genes.

The transcription of structural genes in bacteria is carried out by only one type of DNA-dependent RNA polymerase. In eukaryotes, on the other hand, three types of RNA polymerases are employed. Ribosomal RNA (rRNA) genes are transcribed by RNA polymerase I, protein-coding genes are transcribed by RNA polymerase II, and small cytoplasmic RNA (scRNA) genes, such as genes specifying transfer RNAs (tRNAs), are transcribed by RNA polymerase III. Some small nuclear RNA (snRNA) genes are transcribed by polymerase II, others by polymerase III. One snRNA gene, U6, may be transcribed by either polymerase II or III.

Protein-coding genes

A standard eukaryotic protein-coding gene consists of transcribed and nontranscribed parts (Figure 3a). The nontranscribed parts are designated according to their location relative to the protein-coding regions as 5′ and 3′ **flanking sequences**. The 5′ flanking sequence contains several signals (specific sequences) that determine the initiation, tempo, and timing of the transcription process. Because these regulatory sequences promote the transcription process, they are also referred to as **promoters**, and the region in which they reside is called the **promoter region**. The promoter region consists of the following signals: the **TATA box** located 19–27 base pairs (bp) upstream of the startpoint of transcription, the **CAAT box** farther upstream, and one or more copies of the **GC box**, consisting of the sequence GGGCGG or its variants and surrounding the CAAT box (Figure 3a). The CAAT and GC boxes, which may function in either orientation, control the initial binding of the RNA polymerase, while the TATA box controls the choice of the startpoint of transcription. We note, however, that none of the above signals is uniquely essential for promoter function. Some genes do not possess a TATA box and thus do not have a unique startpoint of transcription. Other genes possess neither a CAAT box nor a GC box, and their transcription initiation is controlled by other elements in the 5′ flanking region. The 3′ flanking sequence contains signals for the termination of the transcription process and poly(A)-addition. At the present time, it is impossible to delineate with precision the points at which a gene begins and ends.

The transcription of protein-coding genes in eukaryotes starts at the **transcription-initiation site** (the **cap site** in the RNA transcript), and ends at the **termination site**, which may or may not be identical with the **polyadenylation** or **poly(A)-addition site** of the mature messenger RNA (mRNA) molecule. In other words, termination of transcription may occur farther downstream from the poly(A)-addition site. The transcribed RNA, also referred to as **pre-messenger RNA (pre-mRNA)**, contains 5′ and 3′ **untranslated regions**, **exons,** and **introns**. Introns, or intervening sequences, are those transcribed sequences that are excised during the processing of the pre-mRNA molecule. All genomic sequences that remain in the mature mRNA following splicing are referred to as exons. Exons or parts of exons that are translated are referred to as protein-coding exons or **coding regions**.

There are several types of introns, which are characterized by the specific mechanism with which the intron is cleaved out of the pre-mRNA (see Lewin 1990). Here we are concerned only with the introns in the nuclear genes that

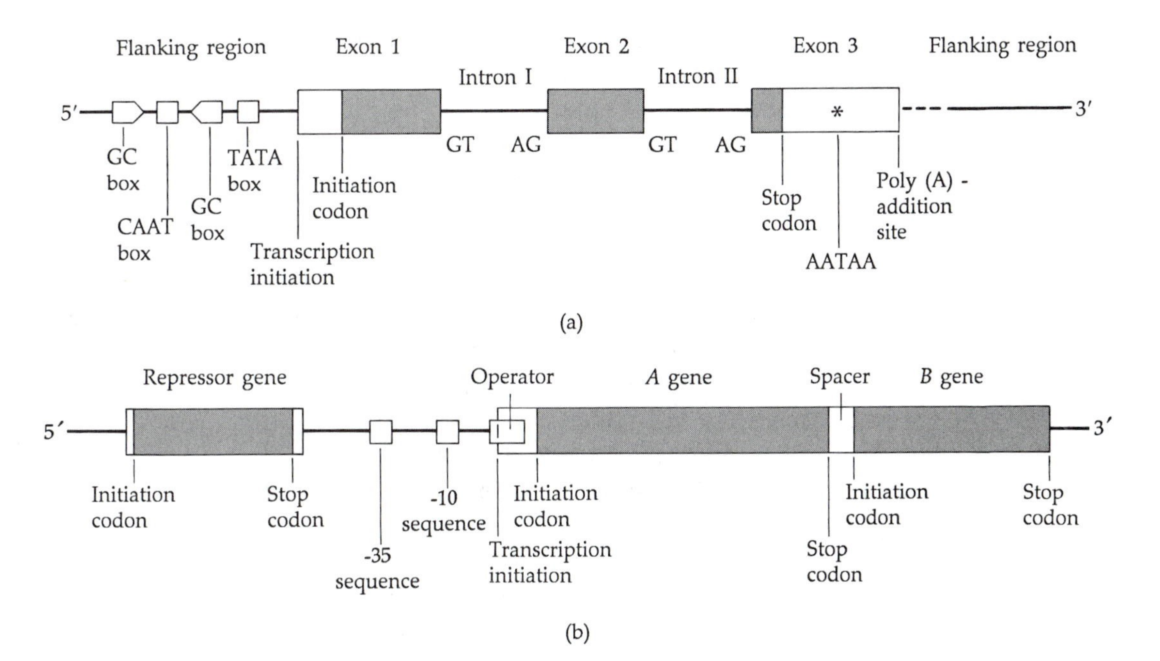

Figure 3. (a) Schematic structure of a typical eukaryotic protein-coding gene. Note that, by convention, the 5′ end is at the left. (b) Schematic structure of an induced prokaryotic operon. Genes *A* and *B* are protein-coding genes and are transcribed into a single messenger RNA. The repressor gene encodes a repressor protein, which binds to the operator and prevents the transcription of the structural genes by blocking the movement of the RNA polymerase. The operator is a DNA region with at least 10 bases, which may overlap the transcribed region of the genes in the operon. By binding to an inducer (a small molecule), the repressor is converted to a form that cannot bind to the operator. The RNA polymerase can then initiate the transcription of the genes *A* and *B* in the operon (see Lewin 1990). In both (a) and (b), the regions are not drawn according to scale.

are transcribed by RNA polymerase II. These introns are enzymatically cleaved out of the pre-mRNA during the maturation of the molecule. The **splicing sites** or **junctions** of these introns are probably determined by nucleotides at the 5′ and 3′ ends of each intron, known as the **donor** and **acceptor** sites, respectively. For instance, all eukaryotic nuclear introns begin with **GT** and end in **AG** (the **GT–AG rule**), and these sequences have been shown to be essential for correct excision and splicing (Figure 3a). Exon sequences adjacent to introns may also contribute to the determination of the splicing site. In addition, each intron contains a specific sequence, called the **TACTAAC box**, located about 30 bases upstream of the 3′ end of the intron. This sequence is well conserved among yeast nuclear genes, though it is more variable in the genes of higher eukaryotes. The splicing involves the cleavage of the 5′ splice junction and the creation of a phosphodiester bond between the G at the 5′ end of the intron and the A in the sixth position of the TACTAAC box. Subsequently, the 3′ splice junction is cleaved, and the two exons are ligated.

The number of introns varies from gene to gene. Some genes possess dozens of introns, some of which may be thousands of nucleotides long. Others (e.g., most histone genes) are devoid of introns altogether. Exons are not distributed evenly over the length of the gene. Some exons are clustered; others are located at great distances from neighboring exons. One such example is shown in Figure 4. Note that the vast majority of nuclear protein-coding genes consist mostly of introns. Not all introns interrupt coding regions. Some occur in untranslated regions, mainly in the region between the transcription-initiation site and the initiation codon.

Protein-coding genes in eubacteria are different from those in eukaryotes in several respects. Most importantly, they do not contain introns (Figure 3b). Promoters in eubacteria contain a **−10 sequence** and a **−35 sequence**, so named because they are located, respectively, 10 bp and 35 bp upstream of the initiation site of transcription. The former, also known as the **Pribnow box**, has the sequence TATAAT or its variant, while the latter has the sequence TTGACA or its variant. A prokaryotic promoter may also contain other specific sequences farther upstream from the −35 sequence.

Several structural genes in prokaryotes may be arranged consecutively to form a unit of gene expression which is transcribed into one molecule of mRNA and subsequently translated into different proteins. Such a unit usually contains genetic elements that control the coordinated expression of the genes belonging to the unit. The entire arrangement of genes is called an **operon** (Figure 3b).

RNA-specifying genes

The structure of RNA-specifying genes is usually similar in eukaryotes and prokaryotes. These genes generally contain no introns. However, in some

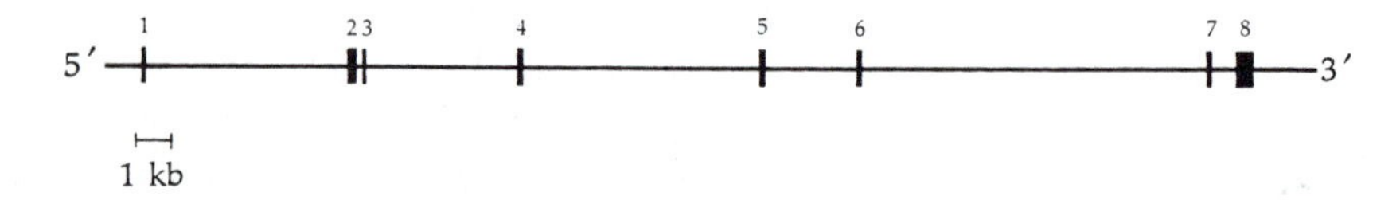

Figure 4. The localization of the eight exons in the human factor-IX gene. The vertical bars represent the eight exons. Only the transcribed region is shown. The exons and introns are drawn to scale. Note that the 5′ untranslated region is shorter than the 3′ untranslated region. Data from Yoshitake et al. (1985).

organisms, such as ciliates, slime molds, and bacteria, RNA-specifying genes may contain introns that must be spliced out before the RNA molecule becomes functional. Sequence elements involved in the regulation of transcription of some RNA-specifying genes are sometimes included within the sequence specifying the functional end product. In particular, all tRNA-specifying genes contain an internal transcriptional start recognized by RNA polymerase III.

Many RNA molecules are modified following transcription. Such modifications include the incorporation of standard and nonstandard nucleotides, modification of standard nucleotides into nonstandard ones, and the enzymatic addition of terminal sequences of ribonucleotides to either the 5′ or the 3′ end.

Regulatory genes

Our knowledge of regulatory genes is less advanced than that of the other types of genes. Several such genes or families of genes have been tentatively identified. They are: (1) **replicator genes**, which specify the sites for initiation and termination of DNA replication, (2) **recombinator genes**, which provide specific recognition sites for recombination enzymes, (3) **segregator genes**, which provide specific sites for attachment of the chromosomes to the segregation machinery during meiosis and mitosis, and (4) **attachment sites** for proteins, hormones, or other molecules. Many more regulatory genes may exist, some of which could be independent of structural genes in both function and location and could be involved in complex regulatory functions, such as the ontogenetic development of multicellular organisms.

GENETIC CODES

The synthesis of proteins involves a process of decoding, whereby the genetic information carried by an mRNA molecule is translated into amino acids through the use of transfer RNA (tRNA) mediators. A list of the 20 primary amino acids and their abbreviations is given in Table 1. Translation starts at the translation-initiation site and proceeds to a stop signal. Translation involves the sequential recognition of adjacent nonoverlapping triplets of nucleotides, called **codons.** The phase at which a sequence is translated is determined by the initiation codon and is referred to as the **reading frame**. In the translational machinery at the interphase between the ribosome and the mRNA molecule, each codon is translated into a specific amino acid, which is subsequently added to the elongating polypeptide. The correspondence between the codons and the amino acids is determined by a set of

Table 1. Primary amino acids and their three- and one-letter abbreviations.

Name	Three-letter abbreviation	One-letter abbreviation
Alanine	Ala	A
Arginine	Arg	R
Asparagine	Asn	N
Aspartic acid	Asp	D
Cysteine	Cys	C
Glutamic acid	Glu	E
Glutamine	Gln	Q
Glycine	Gly	G
Histidine	His	H
Isoleucine	Ile	I
Leucine	Leu	L
Lysine	Lys	K
Methionine	Met	M
Phenylalanine	Phe	F
Proline	Pro	P
Serine	Ser	S
Threonine	Thr	T
Tryptophan	Trp	W
Tyrosine	Tyr	Y
Valine	Val	V

rules called the **genetic code**. With few exceptions (see later), the genetic code for nuclear protein-coding genes is "**universal**," i.e., the translation of almost all eukaryotic nuclear genes and prokaryotic genes is determined by the same set of rules.

The universal genetic code is given in Table 2. Since a codon consists of three nucleotides, and since there are four different types of nucleotides, there are $4^3 = 64$ possible codons. Sixty-one of these code for specific amino acids and are called **sense codons**, while the remaining three are **nonsense** or **stop codons** that act as signals for the termination of the translation process. Since there are 61 sense codons and only 20 primary amino acids in proteins, most amino acids are encoded by more than one codon. Such a code is referred to as a **degenerate code**. Different codons specifying the same amino acid are called **synonymous codons**. Synonymous codons that differ from each other at the third position only are referred to as a **codon family**. For example, the four codons for valine form a four-codon family. In contrast, the six codons for serine are divided into a four-codon and a two-codon family.

The first amino acid in most eukaryotic proteins is a methionine encoded by the **initiation codon** AUG. This amino acid is usually removed in the mature protein. Most prokaryotic genes also use the AUG codon for initiation, but the amino acid initiating the translation process is a methionine derivative called formylmethionine.

The universal genetic code is also used in the independent process of translation employed by the genomes of plastids, such as the chloroplasts of vascular plants. In contrast, most mitochondrial genomes and a few nuclear ones (e.g., *Mycoplasma* and *Tetrahymena*) use codes that are different from the universal genetic code. Usually, however, there are only minor differences between these codes and the universal genetic code. One such example, the mammalian mitochondrial code, is shown in Table 3. Note that two of the codons that specify serine in the universal genetic code are used as termination codons and that tryptophan and methionine are each encoded by two codons rather than one.

Table 2. The universal genetic code.

Codon	Amino acid	Codon	Amino acid	Codon	Amino acid	Codon	Amino acid
UUU	Phe	UCU	Ser	UAU	Tyr	UGU	Cys
UUC	Phe	UCC	Ser	UAC	Tyr	UGC	Cys
UUA	Leu	UCA	Ser	UAA	Stop	UGA	Stop
UUG	Leu	UCG	Ser	UAG	Stop	UGG	Trp
CUU	Leu	CCU	Pro	CAU	His	CGU	Arg
CUC	Leu	CCC	Pro	CAC	His	CGC	Arg
CUA	Leu	CCA	Pro	CAA	Gln	CGA	Arg
CUG	Leu	CCG	Pro	CAG	Gln	CGG	Arg
AUU	Ile	ACU	Thr	AAU	Asn	AGU	Ser
AUC	Ile	ACC	Thr	AAC	Asn	AGC	Ser
AUA	Ile	ACA	Thr	AAA	Lys	AGA	Arg
AUG	Met	ACG	Thr	AAG	Lys	AGG	Arg
GUU	Val	GCU	Ala	GAU	Asp	GGU	Gly
GUC	Val	GCC	Ala	GAC	Asp	GGC	Gly
GUA	Val	GCA	Ala	GAA	Glu	GGA	Gly
GUG	Val	GCG	Ala	GAG	Glu	GGG	Gly

Table 3. The mammalian mitochondrial genetic code.[a]

Codon	Amino acid	Codon	Amino acid	Codon	Amino acid	Codon	Amino acid
UUU	Phe	UCU	Ser	UAU	Tyr	UGU	Cys
UUC	Phe	UCC	Ser	UAC	Tyr	UGC	Cys
UUA	Leu	UCA	Ser	UAA	Stop	**UGA**	**Trp**
UUG	Leu	UCG	Ser	UAG	Stop	UGG	Trp
CUU	Leu	CCU	Pro	CAU	His	CGU	Arg
CUC	Leu	CCC	Pro	CAC	His	CGC	Arg
CUA	Leu	CCA	Pro	CAA	Gln	CGA	Arg
CUG	Leu	CCG	Pro	CAG	Gln	CGG	Arg
AUU	Ile	ACU	Thr	AAU	Asn	AGU	Ser
AUC	Ile	ACC	Thr	AAC	Asn	AGC	Ser
AUA	**Met**	ACA	Thr	AAA	Lys	**AGA**	**Stop**
AUG	Met	ACG	Thr	AAG	Lys	**AGG**	**Stop**
GUU	Val	GCU	Ala	GAU	Asp	GGU	Gly
GUC	Val	GCC	Ala	GAC	Asp	GGC	Gly
GUA	Val	GCA	Ala	GAA	Glu	GGA	Gly
GUG	Val	GCG	Ala	GAG	Glu	GGG	Gly

[a]Differences from the universal genetic code are shown in boldface.

MUTATION

DNA sequences are normally copied exactly during the process of chromosome replication. Rarely, however, errors occur that give rise to new sequences. These errors are called **mutations**. Mutations can occur in either somatic or germ-line cells. Since somatic mutations are not inherited, we can disregard them in an evolutionary context, and throughout this book the term "mutation" will denote mutations in germ-line cells.

Mutations may be classified by the length of the DNA sequence affected by the mutational event. For instance, mutations may affect a single nucleotide (**point mutations)** or several adjacent nucleotides. We may also classify mutations by the type of change caused by the mutational event into (1) **substitutions**, the replacement of one nucleotide by another, (2) **deletions**, the removal of one or more nucleotides from the DNA, (3) **insertions**, the addition of one or more nucleotides to the sequence, and (4) **inversions**, the reversal of polarity of a sequence involving two or more nucleotides (Figure 5).

Nucleotide substitutions

Nucleotide substitutions are divided into **transitions** and **transversions**. Transitions are substitutions between A and G (purines) or between C and T (pyrimidines). Transversions are substitutions between a purine and a pyrimidine.

Nucleotide substitutions occurring in protein-coding regions can also be characterized by their effect on the product of translation, the protein. A substitution is said to be **synonymous** or **silent** if it causes no amino acid change (Figure 6a). Otherwise, it is **nonsynonymous**. Nonsynonymous or amino-acid-altering mutations are further classified into **missense** and **nonsense** mutations. A missense mutation changes the affected codon into a codon that specifies a different amino acid from the one previously encoded (Figure 6b). A nonsense mutation changes a codon into one of the termination codons, thus prematurely ending the translation process and ultimately resulting in the production of a truncated protein (Figure 6c).

Each of the sense codons can mutate to nine other codons by means of a single nucleotide substitution. For example, CCU (Pro) can experience six nonsynonymous substitutions, to UCU (Ser), ACU (Thr), GCU (Ala), CUU (Leu), CAU (His), or CGU (Arg), and three synonymous substitutions, to CCC, CCA, or CCG. Since the universal genetic code consists of 61 sense codons, there are $61 \times 9 = 549$ possible nucleotide substitutions. If we

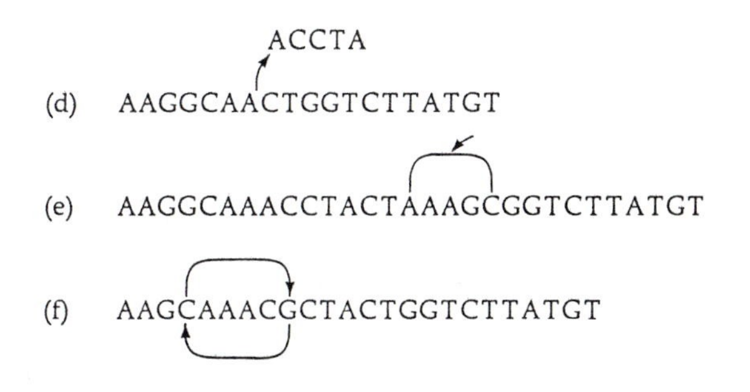

Figure 5. Types of mutations. (a) Original sequence; (b) transition from C to T; (c) transversion from G to C; (d) deletion of the sequence ACCTA; (e) insertion of the sequence AAAGC; (f) inversion of 5′—GCAAAC—3′ to 5′—CAAACG—3′.

(a)

Ile	Cys	Ile	Lys	Ala	Leu	Val	Leu	Leu	Thr
ATA	TGT	ATA	AAG	GCA	CTG	GTC	CTG	TTA	ACA
						↓			
ATA	TGT	ATA	AAG	GCA	CTG	GTA	CTG	TTA	ACA
Ile	Cys	Ile	Lys	Ala	Leu	Val	Leu	Leu	Thr

(b)

Ile	Cys	Ile	Lys	Ala	Asn	Val	Leu	Leu	Thr
ATA	TGT	ATA	AAG	GCA	AAC	GTC	CTG	TTA	ACA
						↓			
ATA	TGT	ATA	AAG	GCA	AAC	TTC	CTG	TTA	ACA
Ile	Cys	Ile	Lys	Ala	Asn	Phe	Leu	Leu	Thr

(c)

Ile	Cys	Ile	Lys	Ala	Asn	Val	Leu	Leu	Thr
ATA	TGT	ATA	AAG	GCA	AAC	GTC	CTG	TTA	ACA
			↓						
ATA	TGT	ATA	TAG	GCAAACGTCCTGTTAACA					
Ile	Cys	Ile	Ter						

Figure 6. Types of substitutions in a coding region: (a) synonymous, (b) missense, and (c) nonsense.

assume that nucleotide substitutions occur at random and that all codons are equally frequent in coding regions, we can compute the expected proportion of the different types of nucleotide substitutions from the genetic code. These are shown in Table 4. Because of the structure of the genetic code, synonymous substitutions occur mainly at the third position of codons. Indeed, almost 70% of all the possible nucleotide changes at the third position are synonymous. In contrast, all the substitutions at the second position of codons are nonsynonymous, and so are the vast majority of nucleotide changes at the first position (96%).

Deletions and insertions

Deletions and insertions can occur by several mechanisms. One mechanism is **unequal crossing over**. Figure 7 shows a simple model in which an unequal crossing over between two chromosomes results in the deletion of a DNA segment in one chromosome and a reciprocal addition in the other. Another mechanism is **replication slippage** or **slipped-strand mispairing**. This type of event occurs in DNA regions containing contiguous short repeats. Figure 8a shows that, during DNA replication, slippage can occur because of mispairing between neighboring repeats and that slippage can result in either

Table 4. Relative frequencies of different types of mutational substitutions in a random protein-coding sequence.

Substitution	Number	Percent
Total in all codons	549	100
Synonymous	134	25
Nonsynonymous	415	75
Missense	392	71
Nonsense	23	4
Total in first position	183	100
Synonymous	8	4
Nonsynonymous	175	96
Missense	166	91
Nonsense	9	5
Total in second position	183	100
Synonymous	0	0
Nonsynonymous	183	100
Missense	176	96
Nonsense	7	4
Total in third position	183	100
Synonymous	126	69
Nonsynonymous	57	31
Missense	50	27
Nonsense	7	4

Figure 7. Unequal crossing-over resulting in the deletion of a DNA sequence in one of the daughter strands and in the duplication of the same sequence in the other strand. A box denotes a particular stretch of DNA.

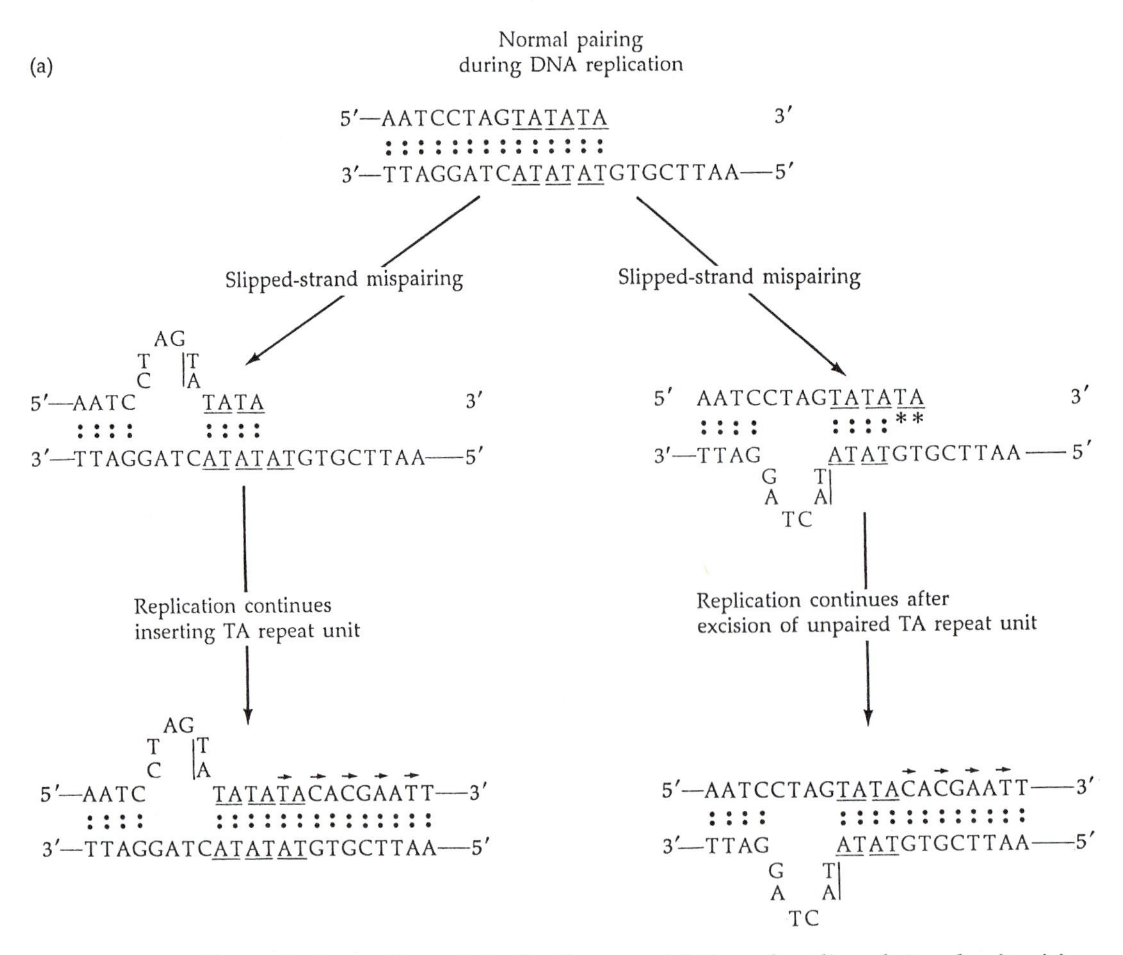

Figure 8. Generation of duplications or deletions by slipped-strand mispairing between contiguous repeats (underlined). Small arrows indicate direction of DNA synthesis. Dots indicate base pairing. (a) A two-base slippage in a TA repeat during DNA replication. Slippage in the 3′ → 5′ direction results in the insertion of one TA unit (left panel). Slippage in the other direction results in the deletion of one repeat unit (right panel). The deletion shown in the right panel results from excision

deletion or duplication of a DNA segment, depending on whether the slippage occurs in the 5′ → 3′ direction or in the opposite direction. Figure 8b shows that slipped-strand mispairing can also occur in nonreplicating DNA. A third mechanism responsible for the insertion or deletion of DNA sequences is DNA transposition, which will be dealt with in Chapter 7.

Deletions and insertions are collectively referred to as **gaps**, because when a sequence involving either a deletion or an insertion is compared with the original sequence a "gap" will appear in one of the two sequences (Chapter

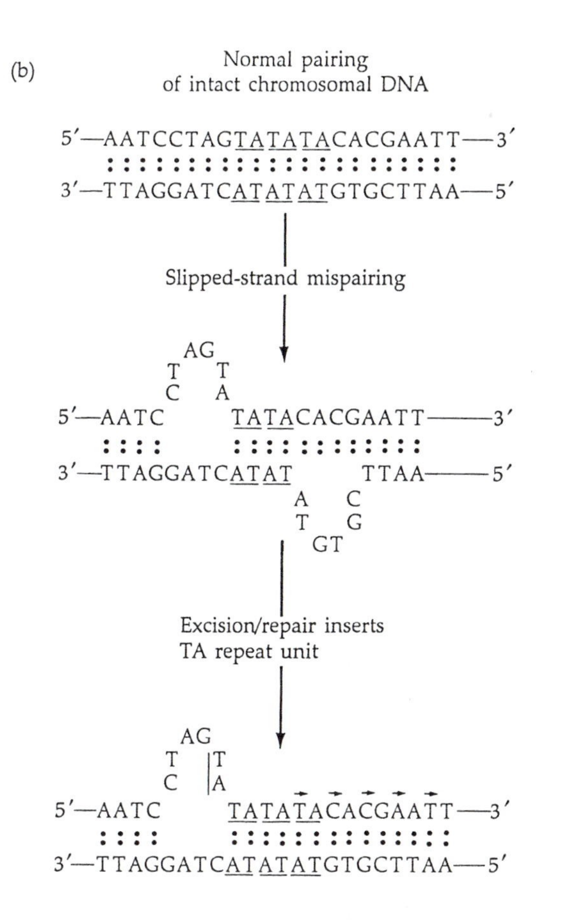

of the unpaired repeat unit (asterisks) at the 3′ end of the growing strand, presumably by the 3′ → 5′ exonuclease activity of DNA polymerase. (b) A two-base slippage in a TA repeat in nonreplicating DNA. Mismatched regions form single-stranded loops, which may be targets of excision and mismatch repair. The outcome (a deletion or an insertion) will depend on which strand is excised and repaired and which strand is used as template in the DNA repair process. Modified from Levinson and Gutman (1987).

3). The number of nucleotides involved in a gap event ranges from one or a few nucleotides to contiguous stretches involving thousands of nucleotides. The lengths of the gaps essentially exhibit a bimodal type of frequency distribution, with short gaps (up to 20–30 nucleotides) most often being caused by errors in the process of DNA replication, such as the slipped-strand mispairing discussed above, and with long insertions and deletions occurring mainly because of unequal crossing-over or DNA transposition.

In a coding region, a gap event involving a number of nucleotides that is

(a) Lys Ala Leu Val Leu Leu Thr Ile Cys Ile Ter
AAG GCA CTG GTC CTG TTA ACA ATA TGT ATA TAA TACCATCGCAATAGGG
↓
G

AAG GCA CTG TCC TGT TAA CAATATGTATATAATACCATCGCAATAGGG
Lys Ala Leu Phe Cys Ter

(b) Lys Ala Asn Val Leu Leu Thr Ile Cys Ile Ter
AAG GCA AAC GTC CTG TTA ACA ATA TGT ATA TAA TACCATCGCAATAGGG
↑
G

AAG GCA AAC GGT CCT GTT AAC AAT ATG TAT ATA ATA CCA TCG CAA TAG GG
Lys Ala Asn Gly Pro Val Asn Asn Met Tyr Ile Ile Pro Ser Gln Ter

Figure 9. Examples of frameshifts in reading frames caused by deletion or insertion. (a) A deletion of a G causes premature termination. (b) An insertion of a G obliterates a stop codon. Termination codons are underlined.

not a multiple of three will cause a shift in the reading frame so that the coding sequence downstream of the gap will be read in the wrong phase. Such a mutation is known as a **frameshift** mutation. Consequently, a gap not only introduces numerous amino acid changes, but may also obliterate the termination codon or bring into phase a new stop codon, thus resulting in a protein of abnormal length (Figure 9).

Spatial distribution of mutations

Mutations do not occur randomly throughout the genome. Some regions are more prone to mutate than others, and they are called **hotspots** of mutation. One such hotspot of mutation is the dinucleotide 5′—CG—3′ in which the cytosine is frequently methylated and replicated with error, changing it to 5′—TG—3′. The dinucleotide 5′—TT—3′ is a hotspot of mutation in prokaryotes but usually not in eukaryotes. In bacteria, regions within the DNA containing short **palindromes** (i.e., sequences that read the same on the complementary strand, such as 5′—GCCGGC—3′, 5′—GGCGCC—3′, and 5′—GGGCCC—3′) were found to be more prone to mutate than other regions. In eukaryotic genomes, short tandem repeats are often hotspots for deletions and insertions, probably as a result of slipped-strand mispairing.

PROBLEMS

1. Find two completely sequenced protein-coding genes from a mammalian species. (You can use either a data bank or a journal.) Determine the length ratio between introns and coding exons in each of them. Repeat the procedure with two protein-coding genes from *Drosophila*. Which genes have larger introns relative to the size of the exons?

2. Find a complete mRNA sequence from a mammalian species. Delete the first CT you encounter in the coding region. Determine whether a premature termination codon comes into the reading frame, or whether the translation will extend beyond the original stop codon.

3. Using the sequence in Problem 2, insert an A after the sixth C in the coding region. How does the insertion affect translation?

4. Using the sequence in Problem 2, determine how many codons can be mutated into a termination codon by one nucleotide substitution.

5. Using the sequence in Problem 2, determine how many transitions in the third position of codons will be synonymous and how many transversions will be synonymous.

FURTHER READINGS

Darnell, J. E., H. F. Lodish and D. Baltimore. 1990. *Molecular Cell Biology*, 2nd Ed. Scientific American Books, New York.

Lewin, B. 1990. *Genes IV*. Oxford University Press, New York.

Stryer, L. 1988. *Biochemistry*, 3rd Ed. Freeman, New York.

Suzuki, D. T., A. J. F. Griffiths, J. H. Miller and R. C. Lewontin. 1989. *An Introduction to Genetic Analysis*, 4th Ed. Freeman, New York.

Watson, J. D., N. H. Hopkins, J. W. Roberts, J. A. Steitz and A. M. Weiner. 1987. *Molecular Biology of the Gene*, 4th Ed. Benjamin/Cummings, Menlo Park, CA.

2

DYNAMICS OF GENES IN POPULATIONS

Population genetics deals with genetic changes that occur within populations. In this chapter we review some basic principles of population genetics that are essential for understanding molecular evolution. A basic problem in population genetics is to determine how the frequency of a mutant gene will change with time under the effect of various evolutionary forces. In addition, from the long-term point of view, it is important to determine the probability that a new mutant variant will completely replace an old one in the population and to estimate how fast the replacement process will be. Unlike morphological changes, many molecular changes are likely to have only a small effect on the phenotype of the organism, and so the frequencies of molecular variants are subject to strong chance effects. Therefore, the element of chance should be taken into account when dealing with molecular evolution.

CHANGES IN ALLELE FREQUENCIES

The chromosomal or genomic location of a gene is called a **locus**, and alternative forms of the gene at a given locus are called **alleles**. In a population, more than one allele may be present at a locus, and their relative proportions are referred to as the **allele frequencies** or **gene frequencies**. For example, let us assume that there are two alleles with n_1 and n_2 copies at a certain locus in a haploid population of size N. Then, their allele frequencies are equal to n_1/N and n_2/N, respectively. Note that $n_1 + n_2 = N$, and $n_1/N + n_2/N = 1$.

Evolution is a process of change in the genetic makeup of populations. Therefore, the most basic component of the evolutionary process is the change in allele frequencies with time. In fact, for a new mutation to become significant from an evolutionary point of view, it must increase its frequency and ultimately become **fixed** in the population (i.e., all the individuals in a subsequent generation will share the same mutant allele). Without increasing its frequency, a mutation will have little effect on the evolutionary history of the species. For a mutant allele to increase in frequency, factors other than mutation must come into play. These factors include natural selection, random genetic drift, recombination, and migration.

To understand the process of evolution, we must study how the above factors govern the changes of allele frequencies. In this book, we discuss only natural selection and random genetic drift. In classical evolutionary studies involving morphological traits, natural selection has been considered as the major driving force of evolution. In contrast, random genetic drift is thought to have played an important role in evolution at the molecular level.

There are two mathematical approaches to studying genetic changes in populations: deterministic and stochastic. The **deterministic model** is simpler. It assumes that changes in the frequencies of alleles in a population from generation to generation occur in a unique manner and can be unambiguously predicted from knowledge of initial conditions. Strictly speaking, this approach applies only when two conditions are met: (1) the population is infinite in size and (2) the environment either remains constant with time or changes according to deterministic rules. These conditions are obviously never met in nature, and therefore a purely deterministic approach may not be sufficient to describe the temporal changes in allele frequencies in populations. Random or unpredictable fluctuations in allele frequencies must also be taken into account.

Dealing with random fluctuations requires a different mathematical approach. **Stochastic models** assume that changes in allele frequencies occur in a probabilistic manner, that is, from knowledge of the conditions in one generation we cannot predict unambiguously the allele frequencies in the next generation, but can only determine the probabilities with which certain allele frequencies will be attained. Obviously, stochastic models are preferable to deterministic ones, since they are based on more realistic assumptions. However, deterministic models are much easier to treat mathematically and, under certain circumstances, they yield sufficiently accurate approximations.

In the following, we shall deal with natural selection in a deterministic fashion.

NATURAL SELECTION

Natural selection is defined as the differential reproduction of genetically distinct individuals or genotypes within a population. Differential reproduction is caused by differences among individuals in such factors as mortality, fertility, fecundity, mating success, and the viability of offspring. Natural selection is predicated on the availability of genetic variation among individuals in characters related to reproduction. When a population consists of individuals that do not differ from one another in such traits, it is not subject to natural selection. Selection leads to changes in allele frequencies over time. However, a mere change in allele frequencies from generation to generation does not necessarily indicate that natural selection is at work. Other processes, such as random genetic drift, can bring about temporal changes in allele frequencies as well (see below).

The **fitness** of a genotype, commonly denoted as w, is a measure of the individual's ability to survive and reproduce. However, since the size of a population is usually constrained by the carrying capacity of the environment in which the population resides, the evolutionary success of an individual is determined not by its **absolute fitness**, but by its **relative fitness** in comparison to the other genotypes in the population. In nature, the fitness of a genotype is not expected to remain constant for all generations and under all environmental circumstances. However, by assigning a constant value of fitness to each genotype, we are able to formulate simple theories, which are useful for understanding the dynamics of change in the genetic structure of populations brought about by natural selection. In the simplest class of models, we assume that the fitness of an organism is determined solely by its genetic makeup. We also assume that all loci contribute independently to the fitness of the individual, so that each locus can be treated separately.

Most new mutants arising in a population reduce the fitness of their carriers. Such mutations will be selected against and eventually removed from the population. This type of selection is called **negative** or **purifying selection**. Occasionally, a new mutation may be as fit as the best allele in the population. Such a mutation is selectively **neutral** and its fate is not determined by selection. Rarely, a mutant that confers a selective advantage on its carriers may arise. Such a mutation will be subjected to **positive** or **advantageous selection**. If the new mutant is advantageous only in heterozygotes but not in homozygotes, the resulting selective regime will be **overdominant selection**.

In the following, we shall consider the case of one locus with two alleles, A_1 and A_2. Each allele can be assigned an intrinsic fitness value; it can be advantageous, deleterious, or neutral. However, this assignment is only applicable to haploid organisms. In diploid organisms the fitness is ultimately determined by the interaction between the two alleles at the locus. With two alleles, there are three possible diploid genotypes: A_1A_1, A_1A_2 and A_2A_2, and their fitnesses can be denoted by w_{11}, w_{12}, and w_{22}, respectively.

Given that the frequency of allele A_1 in a population is p, and the frequency of the complementary allele, A_2, is $q = 1 - p$, we can show that, under random mating, the frequencies of A_1A_1, A_1A_2, and A_2A_2 are p^2, $2pq$, and q^2, respectively. A population in which such genotypic ratios are maintained is said to be at **Hardy-Weinberg equilibrium**. Note that $p = p^2 + ½ (2pq)$ and $q = ½ (2pq) + q^2$.

In the general case, the three genotypes are assigned the following fitness values and initial frequencies:

Genotype	A_1A_1	A_1A_2	A_2A_2
Fitness	w_{11}	w_{12}	w_{22}
Frequency	p^2	$2pq$	q^2

Let us now consider the dynamics of allele-frequency changes following selection. Given the frequencies of the three genotypes and their fitnesses as above, the relative contributions of the three genotypes to the next generation will be p^2w_{11}, $2pqw_{12}$ and q^2w_{22}, for A_1A_1, A_1A_2, and A_2A_2, respectively. Therefore, the frequency of allele A_2 in the next generation will become

$$q' = \frac{pqw_{12} + q^2w_{22}}{p^2w_{11} + 2pqw_{12} + q^2w_{22}} \quad \textbf{(2.1)}$$

The extent of change in the frequency of allele A_2 per generation is denoted as $\Delta q = q' - q$. We can show that

$$\Delta q = \frac{pq[p(w_{12} - w_{11}) + q(w_{22} - w_{12})]}{p^2w_{11} + 2pqw_{12} + q^2w_{22}} \quad \textbf{(2.2)}$$

In the following, we shall assume that A_1 is the original or "old" allele in the population. We shall then consider the dynamics of change in allele frequencies following the appearance of a new mutant allele, A_2. For mathematical convenience, we shall assign a fitness value of 1 to the A_1A_1 genotype. The fitness of the newly created genotypes, A_1A_2 and A_2A_2, will depend on the mode of interaction between A_1 and A_2. For example, if A_2 is completely dominant to A_1, then w_{11}, w_{12} and w_{22} can be written as 1, $1+ s$, and $1 + s$, respectively. If A_2 is completely recessive, the fitnesses become 1, 1, and $1 + s$, respectively, where s is the difference between the fitness of an A_2-carrying genotype and the fitness of A_1A_1. A positive value of s denotes an increase in the fitness in comparison with A_1A_1, while a negative value denotes a decrease in fitness.

Two common modes of interaction will be considered: (1) codominance, or genic selection, and (2) overdominance. Codominance represents a case of directional selection and is mathematically the simplest mode of interaction, while overdominance represents a type of balancing selection.

Codominance

In the **codominant mode of selection**, or **genic selection**, the two homozygotes have different fitness values, while the fitness of the heterozygote is the mean of the fitnesses of the two homozygous genotypes. The relative fitness values for the three genotypes can be written as:

Genotype	A_1A_1	A_1A_2	A_2A_2
Fitness	1	$1 + s$	$1 + 2s$

From Equation 2.2, we obtain the following change in the frequency of allele A_2 per generation under codominance:

$$\Delta q = \frac{spq}{1 + 2spq + 2sq^2} \tag{2.3}$$

Figure 1 illustrates the increase in the frequency of allele A_2 for $s = 0.01$. We see that codominant selection always increases the frequency of one allele at the expense of the other, regardless of the relative allele frequencies in the populations. Therefore, genic selection is a type of **directional selection**.

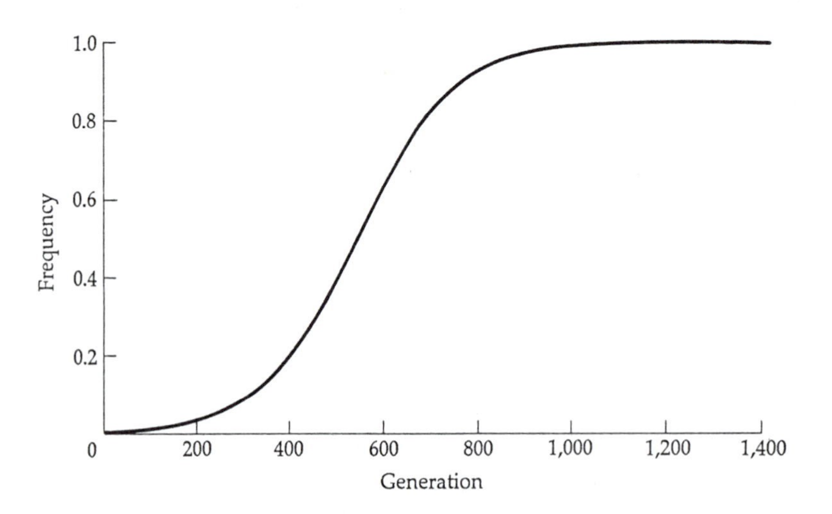

Figure 1. Frequency of a codominant advantageous allele with $s = 0.01$ following its appearance as a result of mutation in generation 0.

Note that, at low frequencies, selection for a codominant allele is not very efficient (i.e., the change in allele frequencies is slow). The reason is that, at low frequencies of A_2, the proportion of A_2 alleles residing in heterozygotes is large. For example, when the frequency of A_2 is 0.5, 50% of A_2 alleles are carried by heterozygotes, whereas when the frequency of A_2 is 0.01, 99% of all such alleles reside in heterozygotes. Because heterozygotes, which contain both alleles, are subject to weaker selective pressure than are A_2A_2 homozygotes (i.e., s versus $2s$) the overall change in allele frequencies at low values of q will be small.

Overdominance

In the **overdominant mode of selection**, the heterozygote has the highest fitness. Thus:

Genotype	A_1A_1	A_1A_2	A_2A_2
Fitness	1	$1 + s$	$1 + t$

In this case, $s > 0$ and $s > t$. Depending on whether the fitness of A_2A_2 is higher, equal, or lower than that of A_1A_1, t can be positive, zero, or negative, respectively. The change in allele frequencies is expressed as

$$\Delta q = -\frac{pq(2sq - tq - s)}{1 + 2spq + tq^2} \quad \textbf{(2.4)}$$

Figure 2 illustrates the changes in the frequency of an allele subject to overdominant selection. In contrast to the codominant selection regime, in which one of the alleles is eventually eliminated from the population, under overdominant selection the population sooner or later will reach an equilibrium in which the two alleles coexist. After equilibrium is reached no further change in allele frequencies will be observed (i.e., $\Delta q = 0$). Thus, overdominant selection belongs to a class of selection regimes called **balancing** or **stabilizing selection**.

The frequency of allele A_2 at equilibrium is obtained by solving Equation 2.4 for $\Delta q = 0$:

$$\hat{q} = s/(2s - t) \quad \textbf{(2.5)}$$

When $t = 0$ (i.e., both homozygotes have identical fitness values), the equilibrium frequencies of both alleles will be 50%.

RANDOM GENETIC DRIFT

As noted above, natural selection is not the only factor that can cause changes in allele frequency. Allele frequency changes can also occur by chance,

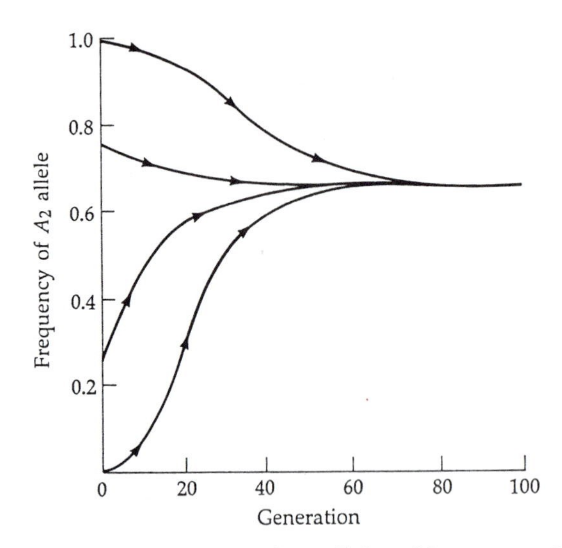

Figure 2. Changes in the frequency of an allele subject to overdominant selection. Initial frequencies from top to bottom: 0.99, 0.75, 0.25, and 0.01; $s = 0.250$ and $t = 0.125$. Since the s and t values are exceptionally large, the change in allele frequency is rapid. Note that there is a stable equilibrium at $q = 0.667$. Modified from Hartl and Clark (1989).

though in this case the changes are not directional but random. An important factor in producing random fluctuations in allele frequencies is the random sampling of gametes in the process of reproduction (Figure 3). Sampling occurs because, in the vast majority of cases in nature, the number of gametes available in any generation is much larger than the number of adult individuals produced in the next generation. In other words, only a minute fraction of gametes succeed in developing into adults. In a diploid population subject to Mendelian segregation, sampling can still occur even if there is no excess of gametes, i.e., even if each individual contributes exactly two gametes to the next generation. The reason is that heterozygotes can produce two types of gametes, but the two gametes passing on to the next generation may by chance be of the same type.

To see the effect of sampling, let us consider an idealized situation in which all the individuals in the population have the same fitness and selection does not operate. We further simplify the problem by considering a population with nonoverlapping generations (i.e., a group of individuals that reproduce simultaneously), such that any given generation can be unambiguously distinguished from both previous and subsequent generations. The population under consideration is diploid and consists of N individuals, so

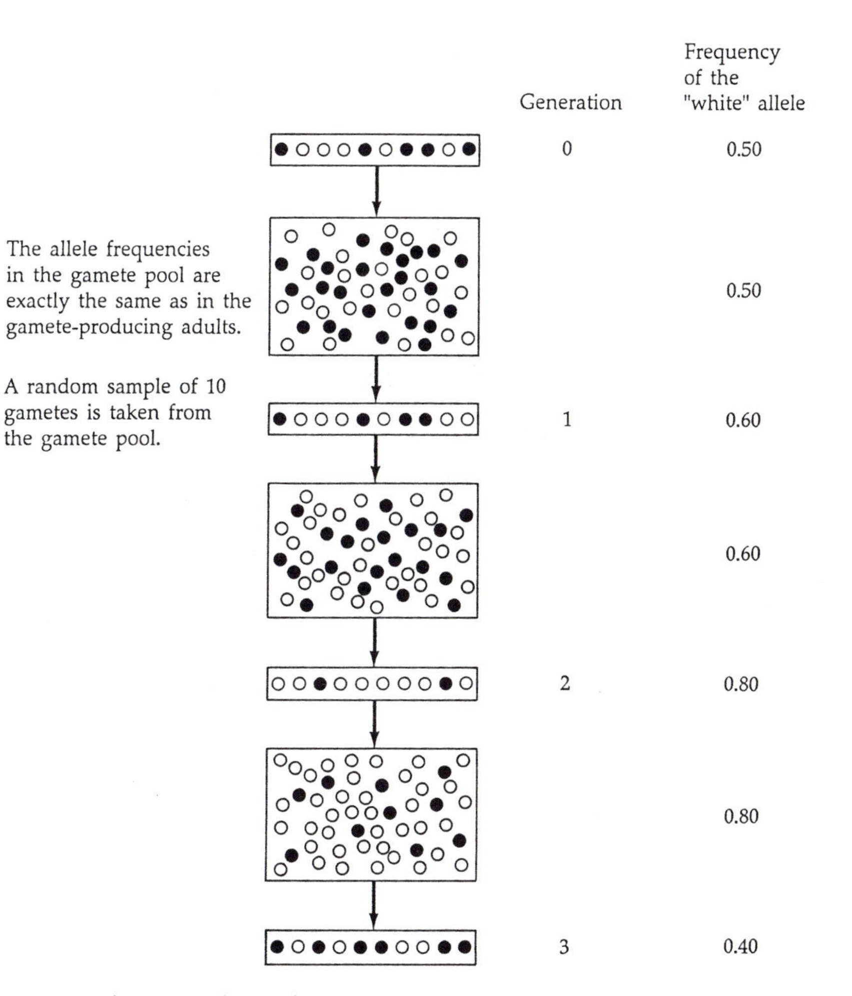

Figure 3. Random sampling of gametes. Allele frequencies in the gamete pools (large boxes) in each generation are assumed to reflect exactly the allele frequencies in the adults of the parental generation (small boxes). Since the population size is finite, allele frequencies fluctuate up and down. Modified from Bodmer and Cavalli-Sforza (1976).

that, at any given locus, this population contains $2N$ genes. Let us again consider the simple case of one locus with two alleles, A_1 and A_2, with frequencies p and $q = 1 - p$, respectively. When $2N$ gametes are sampled from the infinite gamete pool, the probability, P_i, that the sample contains exactly i genes of type A_1 is given by the binomial probability function:

$$P_i = \frac{(2N)!}{i!(2N - i)!} p^i q^{2N-i} \quad \textbf{(2.6)}$$

Since P_i is always greater than 0 for populations in which the two alleles coexist (i.e., $0 < p < 1$), the allele frequencies may change from generation to generation without the aid of selection.

The process of change in allele frequency due solely to chance effects is called **random genetic drift**. One should note, however, that random genetic drift can also be caused by processes other than the sampling of gametes. For example, stochastic changes in selection intensity can also bring about random changes in allele frequencies.

In Figure 4, we illustrate the effects of random sampling on the frequencies of alleles in populations of different sizes. The allele frequencies change from generation to generation, but the direction of the change is random at any point in time. The most obvious feature of random drift is that the fluctuations in allele frequencies are much more pronounced in small populations than in larger ones.

Let us follow the dynamics of change in allele frequencies due to the process of random genetic drift in succeeding generations. The frequencies of allele A_1 are written as $p_0, p_1, p_2, \ldots, p_t$, where the subscripts denote the generation number. The initial frequency of allele A_1 is p_0. Similarly, the frequencies of allele A_2 are $q_0, q_1, q_2, \ldots, q_t$. In the absence of selection, we

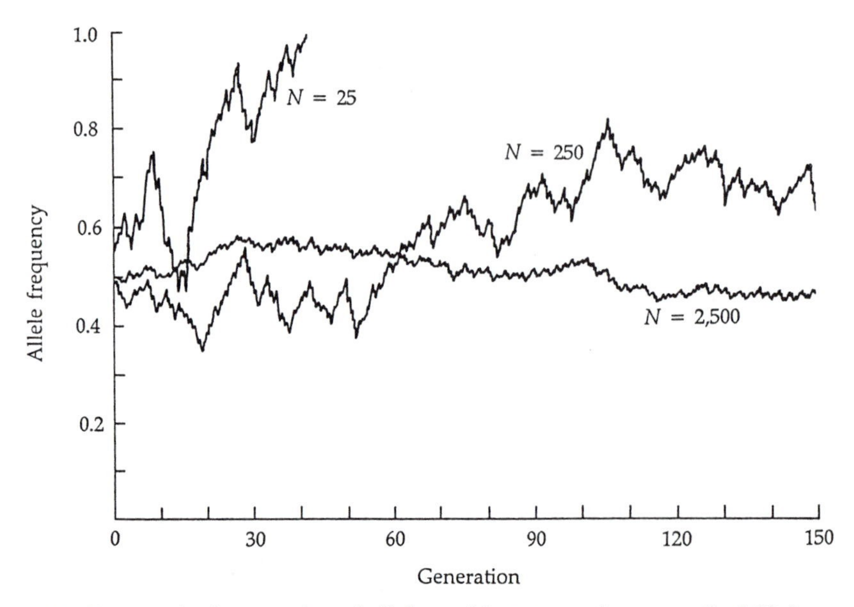

Figure 4. Changes in frequencies of alleles subject to random genetic drift in populations of different sizes. The smallest population reached fixation after 42 generations. The other two populations were still polymorphic after 150 generations but will ultimately reach fixation (allele frequency = 0% or 100%) if the experiment is continued long enough. From Bodmer and Cavalli-Sforza (1976).

expect p_1 to be equal to p_0, q_1 to be equal to q_0, and so on for all subsequent generations. The fact that the population is finite, however, means that p_1 will be equal to p_0 only on the average (i.e., when repeating the sampling process an infinite number of times). In reality, sampling occurs only once in each generation, and p_1 is usually different from p_0. In the second generation, the frequency of A_1 (p_2) will no longer depend on p_0 but on p_1. Similarly, in the third generation, the frequency of A_1 (p_3) will depend on neither p_1 nor p_0 but only on p_2. Thus, the most important property of random genetic drift is its **cumulative behavior**; i.e., from generation to generation, the frequency of an allele will tend to deviate more and more from its initial frequency.

To see the cumulative effect of random genetic drift, let us consider the following numerical example. A certain population is composed of five diploid individuals in which the frequencies of two alleles at a locus, A_1 and A_2, are 50%. We now ask, "What is the probability of obtaining the same allele frequencies in the next generation?" By using Equation 2.6, we obtain a probability of 0.25. In other words, in 75% of the cases the allele frequencies in the second generation will be different from the initial frequencies. Moreover, the probability of retaining the initial allele frequencies in subsequent generations will no longer be 0.25 but will become progressively smaller. For example, the probability of having 50% A_1's in the population is about 18% in the third generation and only about 5% after 10 generations (Figure 5). Concomitantly, the probability of either A_1 or A_2 being lost increases with time, because in every generation there is a finite probability that all the chosen gametes happen to carry the same allele. Once the frequency of an allele reaches either 0 or 1, its frequency will not change in subsequent generations. The first case is referred to as **loss** or **extinction**, and the second, **fixation**. If the process of sampling continues for a long period of time, the probability of such an eventuality reaches certainty.

The ultimate result of random genetic drift is the fixation of one allele and the loss of all the others. This will happen unless there is a constant input of alleles into the population by such processes as mutation or migration, or unless polymorphism is actively maintained by a balancing type of selection.

EFFECTIVE POPULATION SIZE

A basic parameter in population biology is the **population size**, N, defined as the total number of individuals in a population. From the point of view of population genetics and evolution, however, the relevant number to be considered consists of only those individuals that actively participate in the reproductive process. Since not all individuals take part in reproduction, the

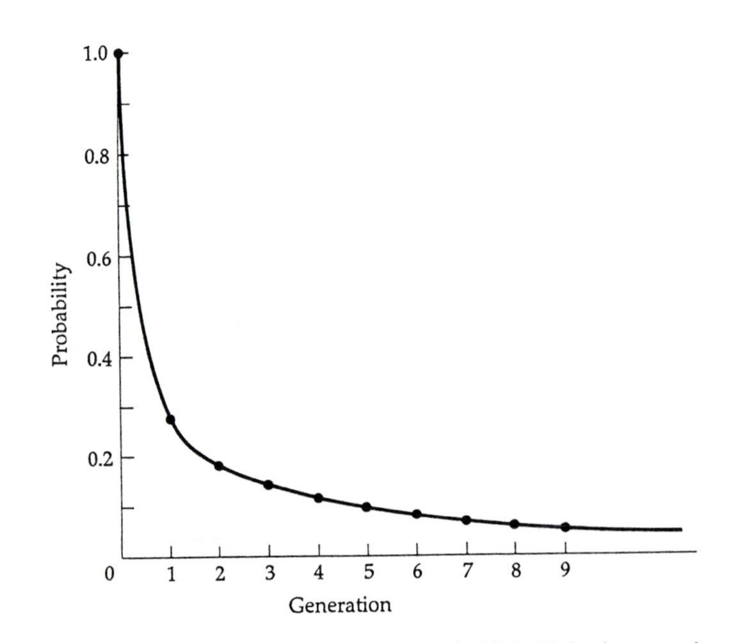

Figure 5. Probability of maintaining the same initial allele frequencies over time for two selectively neutral alleles with N = 5 and p = 0.5.

population size that matters in the evolutionary process is different from the census size. This part is called the **effective population size** and is denoted by N_e.

In general, N_e is smaller, sometimes much smaller, than N. Various factors can contribute to this difference. For example, in a population with overlapping generations, at any given time, part of the population will consist of individuals in either their pre- or postreproductive stage. Due to this developmental stratification, the effective size can be considerably smaller than the census size. For example, according to Nei and Imaizumi (1966), in humans N_e is only slightly larger than $N/3$.

Reduction in the effective population size in comparison to the census size can also occur if the number of males involved in reproduction is different from the number of females. This is especially pronounced in polygamous species, such as social mammals and territorial birds, or in species in which a nonreproducing caste exists (e.g., bees, ants, and termites). If a population consists of N_m males and N_f females, N_e is given by

$$N_e = \frac{4N_mN_f}{N_m + N_f} \qquad \textbf{(2.7)}$$

Note that, unless the number of females equals the number of males, N_e will always be smaller than N. For an extreme example, let us assume that in a population of size N, all females ($N/2$) and only one male take part in the reproductive process. By using Equation 2.7, we see that $N_e = 2N/(1 + N/2)$. If N is considerably larger than 1, N_e becomes 4, regardless of the census population size.

The effective population size can also be much reduced due to long-term variations in the population size, which in turn are caused by such factors as environmental catastrophes, cyclical modes of reproduction, and local extinction and recolonization events. For example, the **long-term effective population size** in a species for a period of n generations is given by

$$N_e = n/(1/N_1 + 1/N_2 + \ldots + 1/N_n) \quad \textbf{(2.8)}$$

where N_i is the population size of the ith generation. In other words, N_e equals the harmonic mean of the N_i values, and consequently it is closer to the smallest value of N_i than to the largest one. Similarly, if a population goes through a bottleneck, the effective population size is greatly reduced.

GENE SUBSTITUTION

Gene substitution is defined as the process whereby a mutant allele completely replaces the predominant or **"wild type"** allele in a population. In this process, a mutant allele arises in a population, usually as a single copy, and becomes **fixed** after a certain number of generations. The time it takes for a new allele to become fixed is called the **fixation time**. Not all new mutants, however, reach fixation. In fact, the majority of them are lost after a few generations. Thus, we also need to address the issue of **fixation probability** and discuss the factors affecting the chance that a new mutant allele will reach fixation in a population.

New mutations arise continuously within populations. Consequently, gene substitutions occur in succession, with one allele replacing another and being itself replaced in time by a new allele. Thus, we can speak of the **rate of gene substitution**, i.e., the number of substitutions or fixations per unit time.

Fixation probability

The probability that a particular allele will become fixed in a population depends on (1) its initial frequency, (2) its selective advantage or disadvantage, s, and (3) the effective population size, N_e. In the following, we shall consider the case of genic selection and assume that the relative fitness of the three genotypes A_1A_1, A_1A_2, and A_2A_2 are 1, $1 + s$, and $1 + 2s$, respectively.

Kimura (1962) showed that the probability of fixation of A_2 is given by

$$P = \frac{1 - e^{-4N_esq}}{1 - e^{-4N_es}} \qquad \textbf{(2.9)}$$

where q is the initial frequency of allele A_2. Since $e^{-x} \approx 1 - x$ when x is small, Equation 2.9 reduces to $P \approx q$ as s approaches 0. Thus, for a neutral allele, the fixation probability equals its frequency in the population. For example, a neutral allele with a frequency of 40% will become fixed in 40% of the cases and become lost in 60% of the cases. This is intuitively understandable because in the case of neutral alleles, fixation occurs by random genetic drift, which favors neither allele.

We note that a new mutant arising as a single copy in a diploid population of size N has an initial frequency of $1/(2N)$. The probability of fixation of an individual mutant allele, P, is thus obtained by replacing q with $1/(2N)$ in Equation 2.9. When $s \neq 0$,

$$P = \frac{1 - e^{-(2N_es)/N}}{1 - e^{-4N_es}} \qquad \textbf{(2.10)}$$

For a neutral mutation, Equation 2.10 becomes

$$P = \frac{1}{2N} \qquad \textbf{(2.11)}$$

If the population size is equal to the effective population size, Equation 2.10 reduces to

$$P = \frac{1 - e^{-2s}}{1 - e^{-4Ns}} \qquad \textbf{(2.12)}$$

If the absolute value of s is small, we obtain

$$P \approx \frac{2s}{1 - e^{-4Ns}} \qquad \textbf{(2.13)}$$

For positive values of s and large values of N, Equation 2.13 reduces to

$$P \approx 2s \qquad \textbf{(2.14)}$$

Thus, if an advantageous mutation arises in a large population and its selective advantage over the rest of the alleles is small, say up to 5%, the probability of its fixation is approximately twice its selective advantage. For example, if a new mutation with $s = 0.01$ arises in a population, the probability of its eventual fixation is 2%.

Let us now consider a numerical example. A new mutant arises in a population of 1,000 individuals. What is the probability that this allele will become fixed in the population if (1) it is neutral, (2) it confers a selective advantage of 0.01, or (3) it has a selective disadvantage of 0.001? For simplicity, we assume that $N = N_e$. For the neutral case, the probability of fixation is $1/(2N) = 0.05\%$. From Equation 2.12, we obtain probabilities of 2% and 0.004% for the advantageous and deleterious cases, respectively. These results are quite noteworthy, since they essentially mean that an advantageous mutation does not always become fixed in the population. In fact, 98% of all the mutations with a selective advantage of $s = 0.01$ will be lost by chance. This theoretical result is of great importance, since it shows that the perception of adaptive evolution as a process in which advantageous mutations arise in populations and invariably take over subsequent generations is a naive concept. Moreover, even slightly deleterious mutations have a finite probability of becoming fixed in a population, albeit a small one. The mere fact that a deleterious allele may become fixed in a population illustrates in a powerful way the importance of chance effects in determining the fate of mutations during evolution.

Fixation time

The time required for the fixation or loss of an allele depends on the frequency of the allele and the size of the population. The mean time to fixation or loss becomes shorter as the frequency of the allele approaches 1 or 0, respectively.

When dealing with new mutations, it is more convenient to treat fixation and loss separately. In the following, we shall deal with the mean fixation time of those mutants that will eventually become fixed in the population. This variable is called the **conditional fixation time**. In the case of a new mutation whose initial frequency in a diploid population is by definition $q = 1/(2N)$, the mean conditional fixation time, $\bar{t}$, was calculated by Kimura and Ohta (1969). For a neutral mutation, it is approximated by

$$\bar{t} = 4N \text{ generations} \qquad \textbf{(2.15)}$$

and for a mutation with a selective advantage of s, it is approximated by

$$\bar{t} = (2/s)\ell n(2N) \text{ generations} \qquad \textbf{(2.16)}$$

where ℓn denotes the natural logarithmic function.

To illustrate the difference between different types of mutation, let us assume that a mammalian species has an effective population size of about 10^6 and a mean generation time of 2 years. Under these conditions, it will take a neutral mutation, on average, 8 million years to become fixed in the population. In comparison, a mutation with a selective advantage of 1% will become fixed in the same population in only about 5,800 years. Interestingly, the conditional fixation time for a deleterious allele with a selective disadvantage s is exactly the same as that for an advantageous allele with a selective advantage s (Maruyama and Kimura 1974). This is intuitively understandable

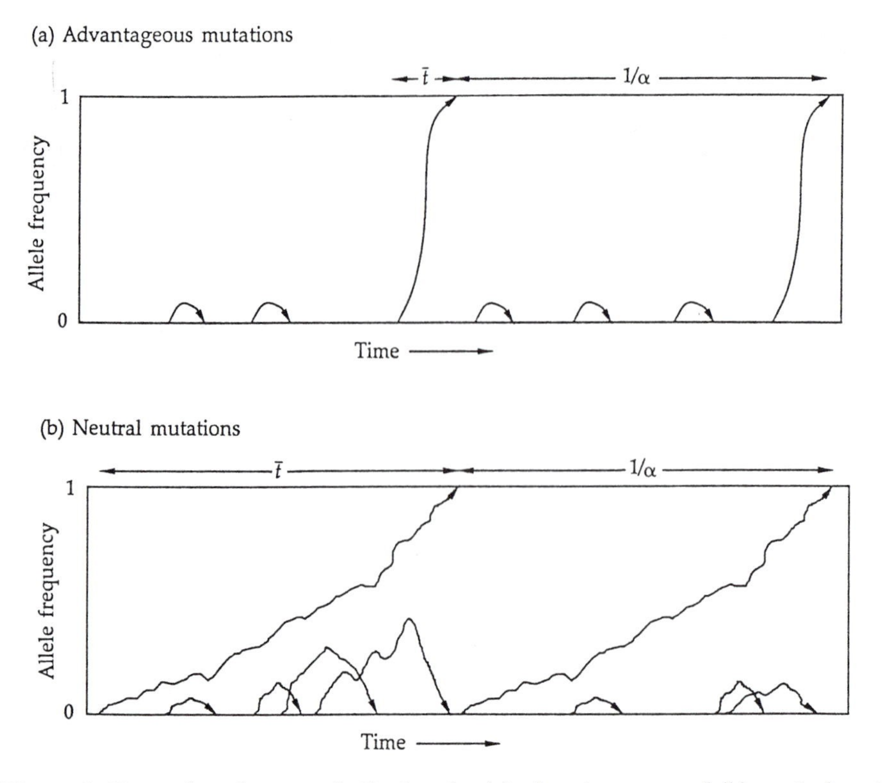

Figure 6. Dynamics of gene substitution for (a) advantageous and (b) neutral mutations. Advantageous mutations are either quickly lost from the population or quickly fixed, so that their contribution to genetic polymorphism is small. The frequency of neutral alleles, on the other hand, changes very slowly by comparison, so a large amount of transient polymorphism is generated. $\bar{t}$ is the conditional fixation time and $1/\alpha$ is the mean time between consecutive fixation events. From Nei (1987).

given the high probability of loss for a deleterious allele. That is, for a deleterious allele to become fixed in a population, fixation must occur very quickly.

In Figure 6, we present the dynamics of gene substitution for advantageous and neutral mutations. We note that advantageous mutations are either rapidly lost or rapidly fixed in the population. In contrast, the frequency changes for neutral alleles are slow, and the fixation time is much longer than for advantageous mutants.

Rate of gene substitution

Let us now consider the **rate of substitution**, defined as the number of mutants reaching fixation per unit time. We shall first consider neutral mutations. If neutral mutations occur at a rate of u per gene per generation, then the number of mutants arising in a diploid population of size N is $2Nu$ per generation. Since the probability of fixation for each of these mutations is $1/(2N)$, we obtain the rate of substitution of neutral alleles by multiplying the total number of mutations by the probability of their fixation:

$$K = u \tag{2.17}$$

Thus, for neutral mutations, the rate of substitution is equal to the rate of mutation, a remarkably simple result (Kimura 1968a). This result can be intuitively understood by noting that, in a large population, the number of mutations arising every generation is high but the fixation probability of each mutation is low. In comparison, in a small population, the number of mutations arising every generation is low, but the fixation probability of each mutation is high. As a consequence, the rate of substitution for neutral mutations is independent of population size.

For advantageous mutations, the rate of substitution can also be obtained by multiplying the rate of mutations by the probability of fixation for such alleles as given in Equation 2.14. For genic selection with $s > 0$, we obtain

$$K = 4Nsu \tag{2.18}$$

In other words, the rate of substitution for the case of genic selection depends on the population size (N) and the selective advantage (s), as well as on the rate of mutation (u).

GENETIC POLYMORPHISM

A locus is said to be **polymorphic** if two or more alleles coexist in the population. However, if one of the alleles has a very high frequency, say 99% or more, then none of the other alleles is likely to be observed in a sample unless the sample size is very large. Thus, for practical purposes, a locus is commonly defined as polymorphic if the frequency of the most common allele is less than 99%. This definition is obviously arbitrary, and in the literature one may find other criteria used.

One of the simplest ways to measure the extent of polymorphism in a population is to compute the average proportion of polymorphic loci by dividing the number of polymorphic loci by the total number of loci sampled. This measure, however, is dependent on the number of individuals studied. A more appropriate measure of genetic variability within populations is the mean **expected heterozygosity** or **gene diversity**. This measure does not depend on an arbitrary delineation of polymorphism, can be computed directly from knowledge of the gene frequencies, and is less affected by sampling effects. Gene diversity at a locus is defined as

$$h = 1 - \sum_{i=1}^{m} x_i^2 \qquad \textbf{(2.19)}$$

where x_i is the frequency of allele i and m is the total number of alleles at the locus. For any given locus, h is the probability that two alleles chosen at random from the population are different from each other. The average of the h values over all the loci studied, H, can be used as an estimate of the extent of genetic variability within the population.

The gene diversity measures h and H have been used extensively for electrophoretic data and restriction enzyme data. However, they may not be suitable for DNA sequence data, since the extent of genetic variation at the DNA level in nature is quite extensive. In particular, when long sequences are considered, each sequence in the sample is likely to be different by one or more nucleotides from the other sequences, and both h and H will be close to 1 in most cases. Thus, these gene-diversity measures will not discriminate among different loci or populations and will no longer be informative measures of polymorphism.

For DNA sequence data, a more appropriate measure of polymorphism in a population is the average number of nucleotide differences per site between any two randomly chosen sequences. This measure is called **nucleotide diversity** (Nei and Li 1979) and is denoted by π:

$$\pi = \sum_{ij} x_i x_j \pi_{ij} \qquad \textbf{(2.20)}$$

where x_i and x_j are the frequencies of the ith and jth type of DNA sequences, respectively, and π_{ij} is the proportion of different nucleotides between the ith and jth type of DNA sequences.

At the present time, there are only a few studies on nucleotide diversity at the DNA sequence level. One such study concerns the alcohol dehydrogenase (*Adh*) locus in *Drosophila melanogaster*. Eleven sequences spanning the *Adh* region were sequenced by Kreitman (1983). The aligned sequences were 2,379 nucleotides long. Disregarding deletions and insertions, there were nine different alleles, one of which was represented by three sequences, while the rest were represented by one each (Figure 7). Thus, the frequencies x_1–x_8 were each 1/11, while the frequency x_9 was 3/11. Forty-three nucleotide sites were polymorphic. We first calculate the proportion of different nucleotides for each pair of alleles. For example, alleles 1-*S* and 2-*S* differ from each other by 3 nucleotides out of 2,379, or $\pi_{12} = 0.13\%$. The π_{ij} values for all the pairs in the sample are listed in Table 1. By using Equation 2.20, the nucleotide diversity is estimated to be $\pi = 0.007$. Six of the alleles studied were slow migrating electrophoretic variants (*S*), and five were fast (*F*). *S* and *F* are distinguished from each other by one amino acid replacement that

	5′ flanking sequence / Exon 1	Intron I			Larval leader	Exon 2	Intron II	Exon 3	Exon 3 / Intron III / Exon 4			Exon 4	3′ untranslated region		3′ flanking sequence		
Consensus	CCG	CAATATGGG	C	G	C	T	AC	C	CCC	GGAATCTCCACTA		G	A	C	AGC	C	T
1-S	...	......AT..	.	.	.	.	..	T	T.A	CA.TAAC......		.	.	.	...	.	.
2-S	..C	....	.	.	.	.	..	T	T.A	CA.TAAC......		.	.	.	...	.	.
3-S	...	....	.	.	.	.	..	.	...	....		A	.	.	..T	.	A
4-S	...	....	.	.	.	.	GT	.	...	....		A	.	.	TA.	.	.
5-S	...	AG...A.TC	.	.	A	G	GT	.	...	....		.	C	.	...	.	.
6-S	..C	....	.	.	.	G	..	.	...	........T.T.C		A	C	.	...	T	.
7-F	..C	....	.	.	.	G	..	.	...	.......GTCTCC		.	C	.	...	.	.
8-F	TGC	AG...A.TC	G	.	.	G	..	.	...	.......GTCTCC		.	C	G	...	.	.
9-F	TGC	AG...A.TC	G	.	.	G	..	.	...	.......GTCTCC		.	C	G	...	.	.
10-F	TGC	AG...A.TC	G	.	.	G	..	.	...	.......GTCTCC		.	C	G	...	.	.
11-F	TGC	AGGGGA...	.	T	.	G	..	.	.A.	..G....GTCTCC		.	C	.	...	.	.
										*							

Figure 7. Polymorphic nucleotide sites among 11 sequences of the alcohol dehydrogenase gene in *Drosophila melanogaster*. Only differences from the consensus sequence are shown. Dots indicate identity with the consensus sequence. The asterisk in exon 4 indicates the site of the lysine-for-threonine replacement that is responsible for the fast/slow mobility differences between the two electrophoretic alleles. Modified from Hartl and Clark (1989).

Table 1. Pairwise percent nucleotide differences among 11 alleles of the alcohol dehydrogenase locus in *Drosophila melanogaster*.[a]

Allele	*1-S*	*2-S*	*3-S*	*4-S*	*5-S*	*6-S*	*7-F*	*8-F*	*9-F*	*10-F*
1-S										
2-S	0.13									
3-S	0.59	0.55								
4-S	0.67	0.63	0.25							
5-S	0.80	0.84	0.55	0.46						
6-S	0.80	0.67	0.38	0.46	0.59					
7-F	0.84	0.71	0.50	0.59	0.63	0.21				
8-F	1.13	1.10	0.88	0.97	0.59	0.59	0.38			
9-F	1.13	1.10	0.88	0.97	0.59	0.59	0.38	0.00		
10-F	1.13	1.10	0.88	0.97	0.59	0.59	0.38	0.00	0.00	
11-F	1.22	1.18	0.97	1.05	0.84	0.67	0.46	0.42	0.42	0.42

From Nei (1987). Data from Kreitman (1983).
[a]Total number of compared sites is 2,379. *S* and *F* denote the slow and fast migrating electrophoretic alleles, respectively.

confers a different electrophoretic mobility to the proteins. The nucleotide diversity for each of these electrophoretic classes was calculated separately. We obtain $\pi = 0.006$ for *S* and $\pi = 0.003$ for *F*. These results indicate that the *S* alleles are twice as polymorphic as the *F* alleles.

THE NEO-DARWINIAN THEORY AND THE NEUTRAL MUTATION HYPOTHESIS

Darwin proposed his theory of evolution by natural selection without knowledge of the sources of variation in populations. After Mendel's laws were rediscovered and genetic variation was shown to be generated by mutation, Darwinism and Mendelism were used as the framework of what came to be called the **synthetic theory** of evolution, or **neo-Darwinism**. According to this theory, although mutation is recognized as the ultimate source of genetic variation, natural selection is given the dominant role in shaping the genetic makeup of populations and in the process of gene substitution.

In time, neo-Darwinism became a dogma in evolutionary biology, and selection came to be considered the only force capable of driving the evolutionary process. Other factors, such as mutation and random drift, were thought of as minor contributors at best. This particular brand of neo-Darwinism was called **selectionism**.

According to the selectionist or neo-Darwinian perception of the evolutionary process, gene substitutions occur as a consequence of selection for advantageous mutations. Polymorphism, on the other hand, is maintained by balancing selection. Thus, neo-Darwinists regard substitution and polymorphism as two separate phenomena driven by different evolutionary forces. Gene substitution is the end result of a positive adaptive process whereby a new allele takes over future generations of the population if and only if it improves the fitness of the organism, while polymorphism is maintained when the coexistence of two or more alleles at a locus is advantageous for the organism or the population. Neo-Darwinian theories maintain that most genetic polymorphism in nature is stable.

The late 1960s witnessed a revolution in population genetics. The availability of protein sequence data removed the species boundary in population-genetics studies and for the first time provided adequate empirical data for examining theories pertaining to the process of gene substitution. In 1968, Kimura postulated that the majority of molecular changes in evolution are due to the random fixation of neutral or nearly neutral mutations (Kimura 1968a; see also King and Jukes 1969). This hypothesis, now known as the **neutral theory of molecular evolution**, contends that at the molecular level the majority of evolutionary changes and much of the variability within species are caused neither by positive selection of advantageous alleles nor by balancing selection, but by random genetic drift of mutant alleles that are selectively neutral or nearly so. Neutrality, in the sense of the theory, does not imply strict equality in fitness for all alleles. It only means that the fate of alleles is determined largely by random genetic drift. In other words, selection may operate, but its intensity is too weak to offset the influences of chance effects. For this to be true, the absolute value of the selective advantage or disadvantage of an allele must be smaller than $1/(2N_e)$, where N_e is the effective population size.

According to the neutral theory, the frequency of alleles is determined by purely stochastic rules, and the picture that we obtain at any given time is merely a transient state representing a temporary frame from an ongoing dynamic process. Consequently, polymorphic loci consist of alleles that are either on their way to fixation or are about to become extinct. Viewed from this perspective, all molecular manifestations that are relevant to the evolutionary process should be regarded as the result of a continuous process of mutational input and a concomitant random extinction or fixation of alleles. Thus, the neutral theory regards substitution and polymorphism as two facets of the same phenomenon. Substitution is a long and gradual process whereby the frequencies of mutant alleles increase or decrease randomly, until the alleles are ultimately fixed or lost by chance. At any given time, some loci will possess alleles at frequencies that are neither 0% nor 100%. These are the polymorphic loci. According to the neutral theory, most genetic polymorphism in populations is transient in nature.

The essence of the dispute between neutralists and selectionists essentially concerns the distribution of fitness values of mutant alleles. Both theories agree that most new mutations are deleterious and that these mutations are quickly removed from the population so that they contribute neither to the rate of substitution nor to the amount of polymorphism within populations. The difference concerns the relative proportion of neutral mutations among nondeleterious mutations. While selectionists maintain that very few mutations are selectively neutral, neutralists maintain that most nondeleterious mutations are effectively neutral.

The heated controversy over the neutral mutation hypothesis during the last two decades has had a strong impact on molecular evolution. First, it has led to the general recognition that the effect of random drift cannot be neglected when considering the evolutionary dynamics of molecular changes. Second, the synthesis between molecular biology and population genetics has been greatly strengthened by the introduction of the concept that molecular evolution and genetic polymorphism are but two facets of the same phenomenon (Kimura and Ohta 1971). Although the controversy still continues, it is now recognized that any adequate theory of evolution must be consistent with both of these aspects of the evolutionary process at the molecular level.

In a series of studies, Nei et al. (1978) have examined the neutral mutation hypothesis from this point of view. More recently, Hudson et al. (1987) have proposed a method for testing whether DNA regions that evolve at high rates, as revealed by interspecific DNA sequence comparisons, also exhibit high levels of polymorphism within species, as predicted by the neutral mutation hypothesis.

PROBLEMS

1. Derive Equation 2.3 from Equation 2.2.

2. If A_2 is completely dominant to A_1, what will be the change in the frequency of allele A_2 per generation?

3. Derive the equilibrium frequency in Equation 2.5 from Equation 2.4.

4. Given a population of five diploid individuals in which the frequency of A_1 is 0.5, and A_1 and A_2 have the same fitness, what is the probability that, in the next generation, the frequency of A_1 will be (a) 0.0, (b) 0.5, or (c) 1.0?

5. What is the ratio of the effective population size to census population size in a population in which females outnumber males by 2:1?

6. A population runs through a bottleneck such that in six consecutive generations its population size is: 10^4, 10^4, 10^4, 10, 10^4, and 10^4. What is its long-term effective population size?

7. What is the fixation probability of a new mutation with a selective disadvantage of 0.01 in a population in which the effective population size is 100 and $N_e = N$?

8. Using the sequences in Figure 7, calculate the nucleotide diversity (a) among alleles 1-*S*, 2-*S*, and 3-*S*, and (b) between 1-*S* and 7-*F*, in the coding region only (black boxes). The coding region is 771 nucleotides long.

FURTHER READINGS

Christiansen, F. B. and M. W. Feldman. 1986. *Population Genetics*. Blackwell Scientific Publications, Cambridge, MA.

Crow, J. F. and M. Kimura. 1970. *An Introduction to Population Genetics Theory*. Harper & Row, New York.

Hartl, D. L. and A. G. Clark. 1989. *Principles of Population Genetics*, 2nd Ed. Sinauer Associates, Sunderland, MA.

Hedrick, P. W. 1983. *Genetics of Populations*. Science Books International, Portola Valley, CA.

Kimura, M. 1983. *The Neutral Theory of Molecular Evolution*. Cambridge University Press, Cambridge.

Nei, M. and D. Graur. 1984. Extent of protein polymorphism and the neutral mutation theory. Evol. Biol. 17: 73–118.

3

EVOLUTIONARY CHANGE IN NUCLEOTIDE SEQUENCES

A basic process in the evolution of DNA sequences is the change in nucleotides with time. This process deserves a detailed consideration since changes in nucleotide sequences are used in molecular evolutionary studies both for estimating the rate of evolution and for reconstructing the evolutionary history of organisms. However, as the process of nucleotide substitution is usually extremely slow, it cannot be observed within a researcher's life. Therefore, to detect evolutionary changes in a DNA sequence, we resort to comparative methods whereby a given sequence is compared with another sequence with which it shared a common ancestry in the evolutionary past. Such comparisons require statistical methods, several of which are discussed in this chapter.

NUCLEOTIDE SUBSTITUTION IN A DNA SEQUENCE

In the previous chapter, we described the evolutionary process as a series of gene substitutions in which new alleles arising as single mutations progressively increase their frequency and ultimately become fixed in the population. We now look at the process from a different point of view. We note that the alleles that become fixed are different in their sequences from the alleles that they replace. If we use a time scale in which one time unit is larger than the time of fixation, the DNA sequence at any given locus will appear to change continuously. For this reason, it is interesting to study how the nucleotides of a DNA sequence change with time. As explained later, the results of this study can be used to develop methods for estimating the number of substitutions between two sequences.

To study the dynamics of nucleotide substitution, we must make several assumptions regarding the probability of substitution of one nucleotide by another. Numerous such mathematical schemes have been proposed in the literature. We shall restrict our discussion to only the simplest and most frequently used ones: Jukes and Cantor's (1969) **one-parameter model** and Kimura's (1980) **two-parameter model**. For a review of more general models, readers may consult Li et al. (1985a).

Jukes and Cantor's one-parameter model

The substitution scheme of Jukes and Cantor's (1969) model is shown in Figure 1. This model assumes that substitutions occur randomly among the four types of nucleotides. In other words, there is no bias in the direction of change. For example, if the nucleotide under consideration is A, it will change to T, C, or G with equal probability. In this model, the rate of substitution for each nucleotide is 3α per unit time, and the rate of substitution in each of the three possible directions of change is α. Since the model involves only one parameter, α, it is also called the one-parameter model.

Let us assume that the nucleotide residing at a certain site in a DNA

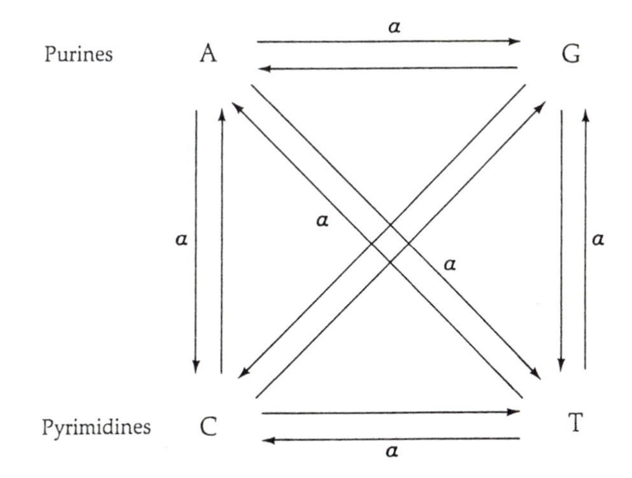

Figure 1. One-parameter model of nucleotide substitution. In this model, the rate of substitution in each direction is α.

sequence is A at time 0. First, we ask, "What is the probability that this site will be occupied by A at time t?" This probability is denoted by $P_{A(t)}$.

Since we start with A, the probability that this site is occupied by A at time 0 is $P_{A(0)} = 1$. At time 1, the probability of still having A at this site is given by

$$P_{A(1)} = 1 - 3\alpha \quad \textbf{(3.1)}$$

which reflects the probability, $1 - 3\alpha$, that the nucleotide has remained unchanged.

The probability of having A at time 2 is

$$P_{A(2)} = (1 - 3\alpha)P_{A(1)} + \alpha[1 - P_{A(1)}] \quad \textbf{(3.2)}$$

To derive this equation, we consider two possible scenarios: (1) the nucleotide has remained unchanged, and (2) the nucleotide has changed to T, C, or G but has subsequently reverted to A (Figure 2). The probability of the nucleotide being A at time 1 is $P_{A(1)}$, and the probability that it has remained A at time 2 is $1 - 3\alpha$. The product of these two independent variables gives us the probability for the first scenario, which constitutes the first term in Equation 3.2. The probability of the nucleotide not being A at time 1 is $1 - P_{A(1)}$ and its probability of changing to A at time 2 is α. The product of these two probabilities gives us the probability for the second scenario and constitutes the second term in Equation 3.2.

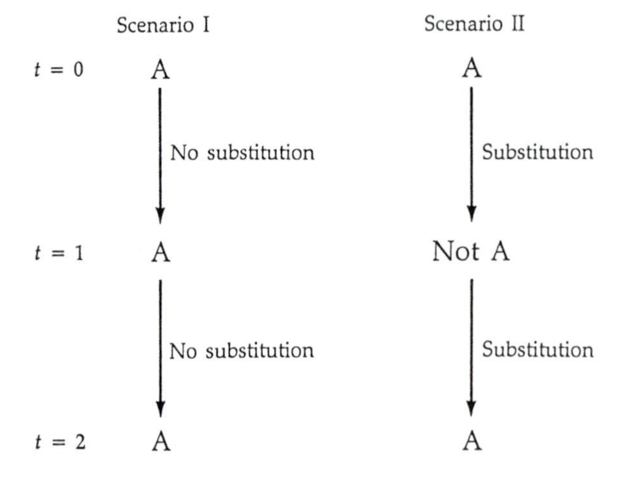

Figure 2. Two possible scenarios for having A at a site at time $t = 2$, given that the site had A at time 0.

Using the above formulation, we can show that the following recurrence equation applies to any t:

$$P_{A(t+1)} = (1 - 3\alpha)P_{A(t)} + \alpha[1 - P_{A(t)}] \qquad \textbf{(3.3)}$$

We can rewrite Equation 3.3 in terms of the amount of change in $P_{A(t)}$ per unit time as

$$P_{A(t+1)} - P_{A(t)} = -3\alpha P_{A(t)} + \alpha[1 - P_{A(t)}] \qquad \textbf{(3.4a)}$$

or

$$\Delta P_{A(t)} = -3\alpha P_{A(t)} + \alpha[1 - P_{A(t)}] = -4\alpha P_{A(t)} + \alpha \qquad \textbf{(3.4b)}$$

So far we have considered a discrete-time process. We can, however, approximate this process by a continuous-time model, by regarding $\Delta P_{A(t)}$ as the rate of change at time t. With this approximation, Equation 3.4b is rewritten as

$$\frac{\mathrm{d}P_{A(t)}}{\mathrm{d}t} = -4\alpha P_{A(t)} + \alpha \qquad \textbf{(3.5)}$$

This is a first-order linear differential equation, and the solution is given by

$$P_{A(t)} = \tfrac{1}{4} + (P_{A(0)} - \tfrac{1}{4})e^{-4\alpha t} \qquad \textbf{(3.6)}$$

Since we started with A, $P_{A(0)} = 1$. Therefore,

$$P_{A(t)} = \tfrac{1}{4} + (\tfrac{3}{4})e^{-4\alpha t} \qquad \textbf{(3.7)}$$

Actually, Equation 3.6 holds regardless of the initial conditions. For example, if the initial nucleotide is not A, then $P_{A(0)} = 0$, and the probability of having A at this position at time t is

$$P_{A(t)} = \tfrac{1}{4} - (\tfrac{1}{4})e^{-4\alpha t} \qquad \textbf{(3.8)}$$

Equations 3.7 and 3.8 are sufficient for describing the substitution process. From Equation 3.7, we can see that, if the initial nucleotide is A, then $P_{A(t)}$ decreases exponentially from 1 to ¼ (Figure 3). On the other hand, from Equation 3.8 we see that if the initial nucleotide is not A, then $P_{A(t)}$ will increase monotonically from 0 to ¼. Thus, regardless of the initial condition, $P_{A(t)}$ will eventually reach ¼ (Figure 3). This also holds true for T, C, and G. Therefore, under the Jukes-Cantor model the equilibrium frequency of each of the four nucleotides is ¼. After reaching equilibrium, there will be no further change in probabilities, i.e., $P_{A(t)} = P_{T(t)} = P_{C(t)} = P_{G(t)} = \tfrac{1}{4}$ for all t's. The frequencies of the nucleotides, however, will remain unchanged only in DNA sequences of infinite length. In practice, the lengths of DNA sequences are finite, and so fluctuations in nucleotide frequencies are likely to occur.

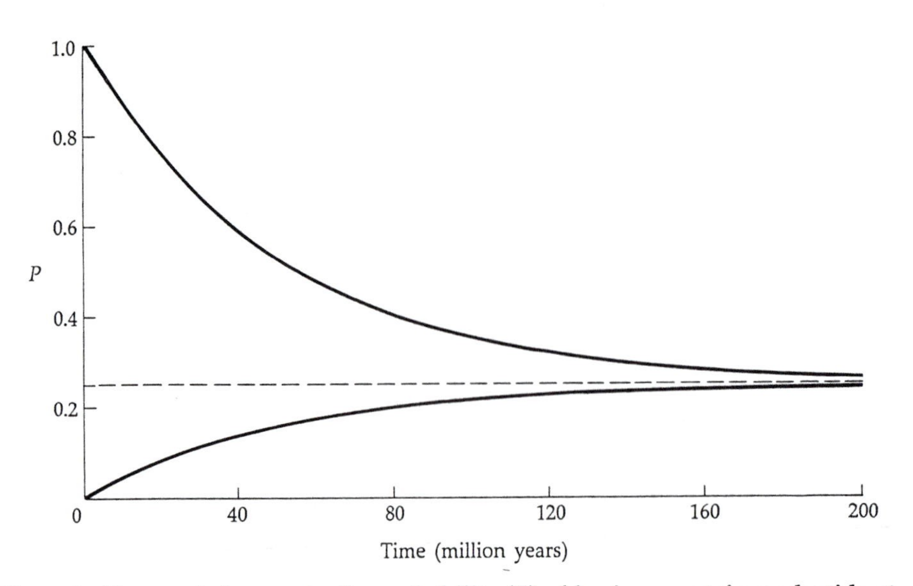

Figure 3. Temporal changes in the probability (P) of having a certain nucleotide at a position starting with the same nucleotide (upper line) or with a different nucleotide (lower line). The dashed line denotes the equilibrium frequency (0.25). α = 5×10^{-9} substitutions/site/year.

In the above, we focused on a particular nucleotide site and treated $P_{A(t)}$ as a probability. However, $P_{A(t)}$ can also be interpreted as the frequency of A in a DNA sequence. For example, if we start with a sequence made of adenines only, then $P_{A(0)} = 1$, and $P_{A(t)}$ is the expected frequency of A in the sequence at time t.

We can rewrite Equation 3.7 in a more explicit form to take into account the facts that the initial nucleotide is A and the nucleotide at time t is also A:

$$P_{AA(t)} = \tfrac{1}{4} + (\tfrac{3}{4})e^{-4\alpha t} \qquad \textbf{(3.9)}$$

If the initial nucleotide is G instead of A, then from Equation 3.8 we obtain

$$P_{GA(t)} = \tfrac{1}{4} - (\tfrac{1}{4})e^{-4\alpha t} \qquad \textbf{(3.10)}$$

Since all the nucleotides are equivalent under the Jukes-Cantor model, $P_{GA(t)} = P_{CA(t)} = P_{TA(t)}$. In fact, we can consider a general probability, $P_{ij(t)}$, which is the probability that a nucleotide will become j at time t, given that the initial nucleotide is i. By using this generalized notation and Equation 3.9, we obtain

$$P_{ii(t)} = \tfrac{1}{4} + (\tfrac{3}{4})e^{-4\alpha t} \qquad \textbf{(3.11)}$$

and from Equation 3.10:

$$P_{ij(t)} = \tfrac{1}{4} - (\tfrac{1}{4})e^{-4\alpha t} \qquad \textbf{(3.12)}$$

where $i \neq j$.

Kimura's two-parameter model

The assumption that all nucleotide substitutions occur randomly, as in Jukes and Cantor's model, is unrealistic in most cases. For example, transitions (i.e., changes between A and G or between C and T) are generally more frequent than transversions (i.e., all the other types of changes) (Chapter 4). To take this fact into account, Kimura (1980) has proposed a two-parameter model, which is shown in Figure 4. In this scheme, the rate of transitional substitution at each nucleotide site is α per unit time, whereas the rate of each type of transversional substitution is β per unit time.

This model is more complicated than the Jukes-Cantor model, and we shall only present the final result. We note from Equation 3.11 that in the Jukes-Cantor model the probability that the nucleotide at a site at time t is identical to that at time 0 is the same for all four nucleotides. In other words, $P_{AA}(t) = P_{GG}(t) = P_{CC}(t) = P_{TT}(t)$. Because of the symmetry of the substitution scheme, this equality also holds for Kimura's two-parameter model. We shall denote this probability by $X_{(t)}$. It can be shown that

$$X_{(t)} = 1/4 + (1/4)e^{-4\beta t} + (1/2)\, e^{-2(\alpha+\beta)t} \qquad \textbf{(3.13)}$$

Under the Jukes-Cantor model, Equation 3.12 holds, regardless of whether the change from nucleotide i to nucleotide j is a transition or a transversion. In contrast, in Kimura's two-parameter model, we need to distinguish between transitional and transversional changes. We denote by $Y_{(t)}$ the probability that the initial nucleotide and the nucleotide at time t differ from each other by a transition. We note that, because of the symmetry of the substi-

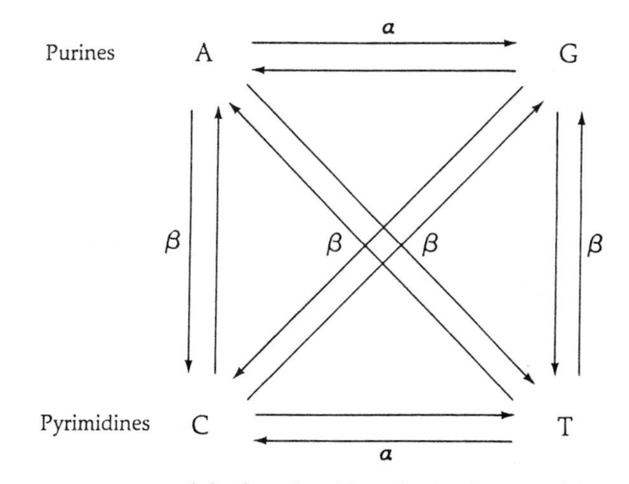

Figure 4. Two-parameter model of nucleotide substitution. In this model, the rate of transitions (α) may not be equal to the rate of a transversion (β).

tution scheme, $Y_{(t)} = P_{AG(t)} = P_{GA(t)} = P_{TC(t)} = P_{CT(t)}$. It can be shown that

$$Y_{(t)} = 1/4 + (1/4)e^{-4\beta t} - (1/2)e^{-2(\alpha+\beta)t} \quad \textbf{(3.14)}$$

The probability $Z_{(t)}$ that the nucleotide at time t and the initial nucleotide differ by a specific type of transversion is given by

$$Z_{(t)} = 1/4 - (1/4)e^{-4\beta t} \quad \textbf{(3.15)}$$

Note that each nucleotide is subject to two types of transversion as opposed to only one type of transition. For example, if the initial nucleotide is A, then the two possible transversional changes are A → C and A → T. Therefore, the probability that the initial nucleotide and the nucleotide at time t differ by one of the two types of transversion is twice the probability given by Equation 3.15. Note also that $X_{(t)} + Y_{(t)} + 2Z_{(t)} = 1$.

NUMBER OF NUCLEOTIDE SUBSTITUTIONS BETWEEN TWO DNA SEQUENCES

The substitution of alleles in a population generally takes thousands or even millions of years to complete (Chapter 2). For this reason, we cannot deal with the process of nucleotide substitution by direct observation, and nucleotide substitutions are always inferred from pairwise comparisons of DNA molecules that share a common origin.

After two nucleotide sequences diverge from each other, each of them will start accumulating nucleotide substitutions. Thus, the number of nucleotide substitutions that have occurred since two sequences diverged from each other is the most commonly used variable in molecular evolution.

When the degree of divergence between two nucleotide sequences is small, the chance for more than one substitution to have occurred at any site is negligible, and the number of observed differences between the two sequences should be close to the actual number of substitutions. On the other hand, if the degree of divergence is substantial, then the observed number of differences is likely to be smaller than the actual number of substitutions due to **multiple substitutions** or **multiple "hits"** at the same site. For example, if the nucleotide at a certain site changed from A to C and then to T in one sequence and from A to T in the other sequence, then the two sequences are identical at this site, despite the fact that there were three substitutions (Figure 5). Several methods have been proposed in the literature to correct for such distortions.

The number of substitutions is usually expressed in terms of the number of substitutions per nucleotide site rather than as the total number of sub-

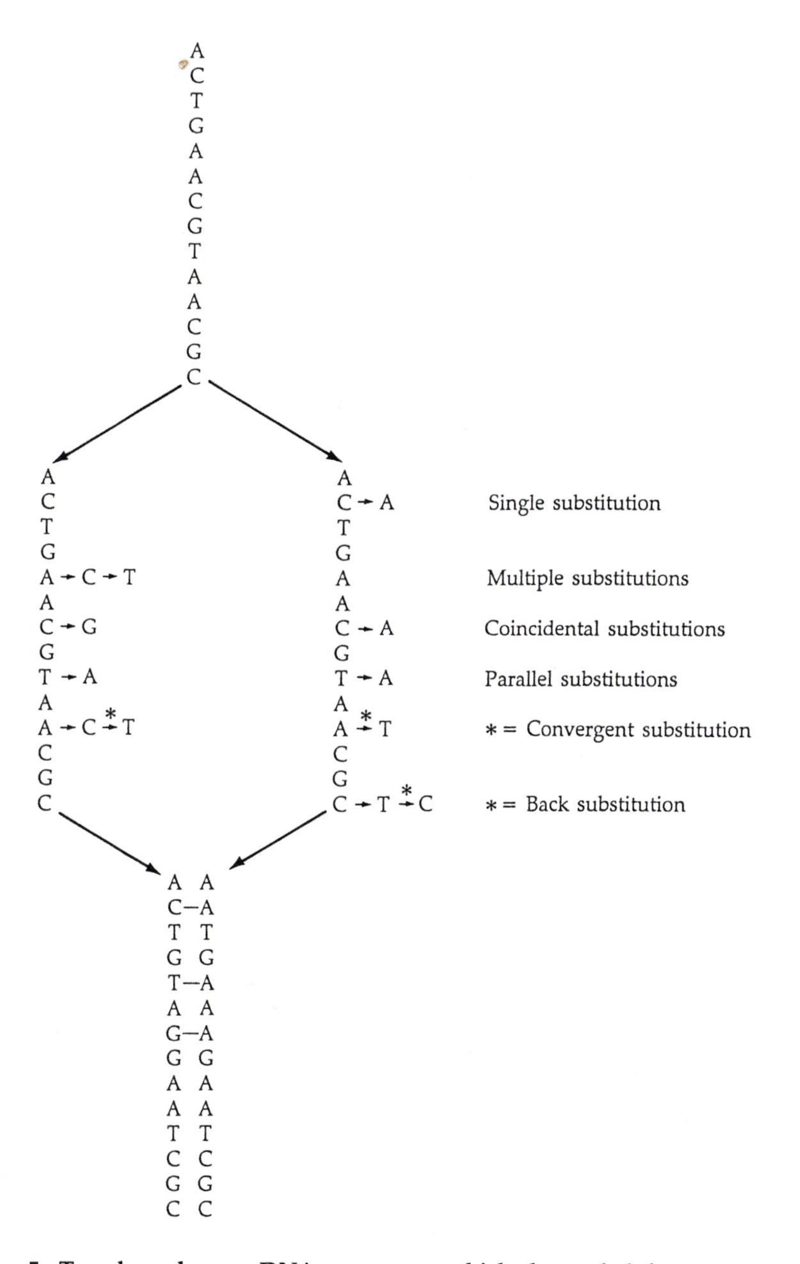

Figure 5. Two homologous DNA sequences which descended from an ancestral sequence and accumulated mutations since their divergence from each other. Note that although 12 mutations have accumulated, differences can be detected at only three nucleotide sites. Note further that "coincidental substitutions," "parallel substitutions," "convergent substitutions," and "back substitutions" all involve multiple substitutions at the same site, though maybe in different lineages.

stitutions between the two sequences. This facilitates comparison of the degrees of divergence among sequence pairs that differ in length from each other.

Protein-coding and noncoding sequences should be treated separately, because they usually evolve at different rates. In the former case, it is advisable to distinguish between synonymous and nonsynonymous substitutions, since they are known to evolve at markedly different rates (Chapter 4). In noncoding regions, on the other hand, one may assume that all sites evolve at the same rate.

Number of substitutions between two noncoding sequences

The results that we obtained earlier in this chapter for a single DNA sequence can be applied to study the nucleotide divergence between two sequences that share a common origin. Let us start with the one-parameter model. In this model, it is sufficient to consider only $I_{(t)}$, which is the probability that the nucleotide at a given site at time t is the same in both sequences. Suppose that the nucleotide at a given site was A at time 0. At time t, the probability that a descendant sequence will have A at this site is $P_{AA(t)}$, and consequently the probability that both descendant sequences have A at this site is $P_{AA(t)}^2$. Similarly, the probabilities that both sequences have T, C or G at that site are $P_{AT(t)}^2$, $P_{AC(t)}^2$, and $P_{AG(t)}^2$, respectively. Therefore,

$$I_{(t)} = P_{AA(t)}^2 + P_{AT(t)}^2 + P_{AC(t)}^2 + P_{AG(t)}^2 \quad \textbf{(3.16)}$$

From Equations 3.11 and 3.12, we obtain

$$I_{(t)} = 1/4 + (3/4)e^{-8\alpha t} \quad \textbf{(3.17)}$$

Equation 3.17 also holds for T, C, or G. Therefore, regardless of the initial nucleotide at a site, $I_{(t)}$ represents the proportion of identical nucleotides between two sequences that diverged t time units ago. Note that the probability that the two sequences are different at a site at time t is $p = 1 - I_{(t)}$. Thus,

$$p = 3/4(1 - e^{-8\alpha t}) \quad \textbf{(3.18a)}$$

or

$$8\alpha t = -\ell n(1 - 4/3 p) \quad \textbf{(3.18b)}$$

The time of divergence between two sequences is usually not known, and thus we cannot estimate α. Instead, we compute K, which is the number of substitutions per site since the time of divergence between the two sequences. In the case of the one-parameter model, $K = 2(3\alpha t)$, where $3\alpha t$ is the number of substitutions per site in each of the two lineages. By using Equation 3.18b we can calculate K as

$$K = -\tfrac{3}{4}\ell n(1 - \tfrac{4}{3}p) \tag{3.19}$$

where p is the proportion of different nucleotides between the two sequences (Jukes and Cantor 1969). For sequences of length L, the sampling variance is approximately given by

$$V(K) = \frac{p(1 - p)}{L(1 - \tfrac{4}{3}p)^2} \tag{3.20}$$

(Kimura and Ohta 1972).

In the case of the two-parameter model, the differences between two sequences are classified into transitions and transversions. Let P and Q be the proportions of transitional and transversional differences between the two sequences, respectively. Then, the number of nucleotide substitutions between the two sequences, K, is estimated by

$$K = \tfrac{1}{2}\,\ell n\,(a) + \tfrac{1}{4}\,\ell n\,(b) \tag{3.21}$$

where $a = 1/(1 - 2P - Q)$, and $b = 1/(1 - 2Q)$. The sampling variance is approximately given by

$$V(K) = [a^2P + c^2Q - (aP + cQ)^2]/L \tag{3.22}$$

where $c = (a + b)/2$ and L is the length of the sequences (Kimura 1980).

Let us now consider a hypothetical numerical example of two sequences of length 200 that differ from each other by 20 transitions and 4 transversions. Thus, $L = 200$, $P = 20/200 = 0.1$, and $Q = 4/200 = 0.02$. In this case, according to the two-parameter model we have $a = 1/(1 - 0.2 - 0.02) = 1.28$, $b = 1/(1 - 0.04) = 1.04$, and $K = \tfrac{1}{2}\,\ell n(1.28) + \tfrac{1}{4}\,\ell n(1.04) \approx 0.13$. The total number of substitutions can be obtained by multiplying the number of substitutions per site, K, by the number of sites, L. In this case we obtain an estimate of about 26 substitutions resulting in 24 differences between the two sequences. According to the one-parameter model, $p = 24/200 = 0.12$, and $K \approx 0.13$. Thus, by using the one-parameter model, we arrive at the same result as in the case of the two-parameter model.

In the above example, the two models give essentially the same estimate because the degree of divergence is small such that the corrected degree of divergence ($K = 0.13$) is only slightly larger than the uncorrected value ($p = 24/200 = 0.12$). In such cases, one may use Jukes and Cantor's model, which is simpler.

When the degree of divergence between the two sequences is large, the estimates by the two models may differ considerably . For example, consider two sequences with $L = 200$ that differ from each other by 50 transitions and 16 transversions. Thus, $P = 50/200 = 0.25$, and $Q = 16/200 = 0.08$. According to the two-parameter model we have $a = 2.38$, $b = 1.19$, and $K \approx 0.48$. According to the one-parameter model, $p = 66/200 = 0.33$, and $K \approx 0.43$. Thus, the estimate of K according to the one-parameter model is smaller than that obtained by using the two-parameter model. When the degree of divergence between two sequences is large, and especially in cases where there are a priori reasons to believe that the rate of transition greatly differs from the rate of transversion, the two-parameter model tends to be more accurate than the one-parameter model.

Number of substitutions between two protein-coding sequences

In studying protein-coding sequences we usually exclude the initiation and the termination codons from analysis because these two codons almost never change with time.

In order to treat synonymous and nonsynonymous substitutions separately, we first classify the remaining nucleotide sites as follows. Consider a particular position in a codon. Let i be the number of possible synonymous changes at this site. Then this site is counted as $i/3$ synonymous and $(3 - i)/3$ nonsynonymous. For example, in the codon TTT (Phe), the first two positions are counted as nonsynonymous because no synonymous change can occur at these two positions and the third position is counted as one-third synonymous and two-thirds nonsynonymous because one of the three possible changes at this position is synonymous. As another example, the codon ACT (Thr) has two nonsynonymous sites (the first two positions) and one synonymous site (the third position) because all possible changes at the first two positions are nonsynonymous while all possible changes at the third position are synonymous. When comparing two sequences, one first counts the number of synonymous sites and the number of nonsynonymous sites in each sequence and then computes the averages between the two sequences. Let us denote the average number of synonymous sites by N_S and that of nonsynonymous sites by N_A.

Second, we classify nucleotide differences into synonymous and nonsynonymous differences. For two codons that differ by only one nucleotide, the difference is easily inferred. For example, the difference between the two codons GTC (Val) and GTT (Val) is synonymous, while the difference between the two codons GTC (Val) and GCC (Ala) is nonsynonymous. For two codons that differ by more than one nucleotide, we need to consider all possible evolutionary pathways that can lead to the observed changes. For example, for the two codons AAT (Asn) and ACG (Thr), there are two possible pathways:

Pathway I: AAT (Asn) ↔ ACT (Thr) ↔ ACG (Thr)
Pathway II: AAT (Asn) ↔ AAG (Lys) ↔ ACG (Thr)

Pathway I requires one synonymous and one nonsynonymous change, whereas pathway II requires two nonsynonymous changes. It is known that synonymous substitutions occur considerably more often than nonsynonymous substitutions (Chapter 4), and so we may assume that pathway I is more likely than pathway II. For example, if we assume a weight of 0.7 for pathway I and a weight of 0.3 for pathway II, then the number of synonymous differences between the two codons is estimated to be $0.7 \times 1 + 0.3 \times 0 = 0.7$, and the number of nonsynonymous differences is $0.7 \times 1 + 0.3 \times 2 = 1.3$. Here, the weights used are hypothetical. The weights for all possible codon pairs have been estimated empirically by Miyata and Yasunaga (1980) from protein sequence data and by Li et al. (1985b) from DNA sequence data. If we assume that both pathways are equally likely, then for the above example the number of nonsynonymous differences is $(1 + 2)/2 = 1.5$, and the number of synonymous differences is $(1 + 0)/2 = 0.5$. Thus, the weighted and unweighted approaches may give somewhat different results. In practice, the differences in estimates between the two approaches are usually small (Nei and Gojobori 1986), but they can be important for genes coding for highly conserved proteins such as histones and actins (Li et al. 1985b). Using either approach, we can estimate the number (M_S) of synonymous differences and the number (M_A) of nonsynonymous differences between two coding sequences.

From the above results, we can compute the number of synonymous differences per synonymous site by $p_S = M_S/N_S$ and the number of nonsynonymous differences per nonsynonymous site by $p_A = M_A/N_A$. These formulas obviously do not take into account the effect of multiple hits at the same site. We can make such corrections by using Jukes and Cantor's formula:

$$K_S = -\tfrac{3}{4} \ln [1 - (4M_S/3N_S)] \qquad \textbf{(3.23)}$$

and

$$K_A = -\tfrac{3}{4} \ln [1 - (4M_A/3N_A)] \qquad \textbf{(3.24)}$$

An alternative way of treating coding regions is to classify nucleotide sites into **nondegenerate**, **twofold degenerate**, and **fourfold degenerate** sites (Li et al. 1985b). A site is nondegenerate if all possible changes at this site are nonsynonymous, twofold degenerate if one of the three possible changes is synonymous, and fourfold degenerate if all possible changes at the site are synonymous. For example, the first two positions of the codon TTT (Phe) are nondegenerate while the third position is twofold degenerate (Table 2 in

Chapter 1). In comparison the third position of the codon GTT (Val) is fourfold degenerate. The third position in the three isoleucine codons is treated for simplicity as a twofold degenerate site although in reality the degeneracy at this position is threefold. In mammalian mitochondrial genes, there are only two codons for isoleucine, and the third position is indeed a twofold degenerate site (Table 3 in Chapter 1).

From the above classification of nucleotide sites into **degeneracy classes**, we can calculate the number of substitutions between two coding sequences for the three types of sites separately. Note that by definition all the substitutions at nondegenerate sites are nonsynonymous. Similarly, all the substitutions at fourfold degenerate sites are synonymous. At twofold degenerate sites, transitional changes (C ↔ T and A ↔ G) are synonymous, whereas all the other changes, which are transversions, are nonsynonymous. There are no exceptions to this rule in the mammalian mitochondrial genetic code. In the universal nuclear genetic code, on the other hand, there are two exceptions: the first position of the arginine codons (CGA, CGG, AGA, and AGG), in which one type of transversion is synonymous, while the other type and the transition are nonsynonymous, and the last position in the three isoleucine codons (AUU, AUC, and AUA).

A computer program for computing the rates of substitution according to both methods is available from the authors upon request; please send a formatted IBM PC-compatible floppy disk.

ALIGNMENT OF NUCLEOTIDE AND AMINO ACID SEQUENCES

Comparison of two homologous sequences involves the identification of the locations of deletions and insertions that might have occurred in either of the two lineages since their divergence from a common ancestor. This process is referred to as **sequence alignment**. Comparisons of two DNA sequences usually cannot tell us whether a deletion had occurred in one sequence or an insertion had occurred in the other. Therefore, the outcomes of both types of events are collectively referred to as **gaps**.

Although we illustrate the process of alignment by using DNA sequences, the same principle and procedure can be used to align amino acid sequences. As a matter of fact, one usually obtains more reliable alignments by using amino acid sequences than by using DNA sequences.

An alignment consists of a series of paired bases, one base from each sequence. There are three types of aligned pairs: (1) pairs of matched bases, (2) pairs of mismatched bases, and (3) pairs consisting of a base from one sequence and a **null base** from the other. Null bases are denoted by —. A matched pair implies a site that has not changed since the divergence between the two sequences, a mismatched pair denotes a substitution, and a null pair indicates that a deletion or an insertion has occurred at this position in one of the two sequences.

Consider the case of two DNA sequences, A and B, of lengths m and n, respectively. If we denote the number of matched pairs by x, the number of mismatched pairs by y, and the number of pairs containing a null base by z, we obtain

$$n + m = 2(x + y) + z \tag{3.25}$$

The dot-matrix method

When there are few gaps and the two sequences are not too different from each other in any other respects, a reasonable alignment can be obtained by either visual inspection or by an approach called the **dot matrix method**. In this method, the two sequences to be aligned are written out as column and row headings of a matrix (Figure 6). Dots are put in the matrix when the nucleotides in the two sequences are identical. If the two sequences are identical, there will be dots in all the diagonal elements of the matrix (Figure 6a). If the two sequences are different but can be aligned without gaps, there will be dots in most of the diagonal elements (Figure 6b). If a gap occurred in one of the two sequences, the alignment diagonal would be shifted vertically or horizontally (Figure 6c). If the two sequences differ from each other by both gaps and substitutions (Figure 6d), it may be difficult to identify the location of gaps and to choose between several alternative alignments. In such cases, visual inspection and the dot matrix method are not reliable, and several algorithms have been developed for objective alignment.

The sequence-distance method

The best possible alignment between two sequences is the one in which the number of mismatches and gaps is minimized according to certain criteria. Unfortunately, reducing the number of mismatches usually results in an increase in the number of gaps, and vice versa.

For example, consider the following two sequences:

A: TCAGACGATTG ($m = 11$)
B: TCGGAGCTG ($n = 9$)

We can reduce the number of mismatches to zero as follows:

(I)
TCAG-ACG-ATTG
TC-GGA-GC-T-G

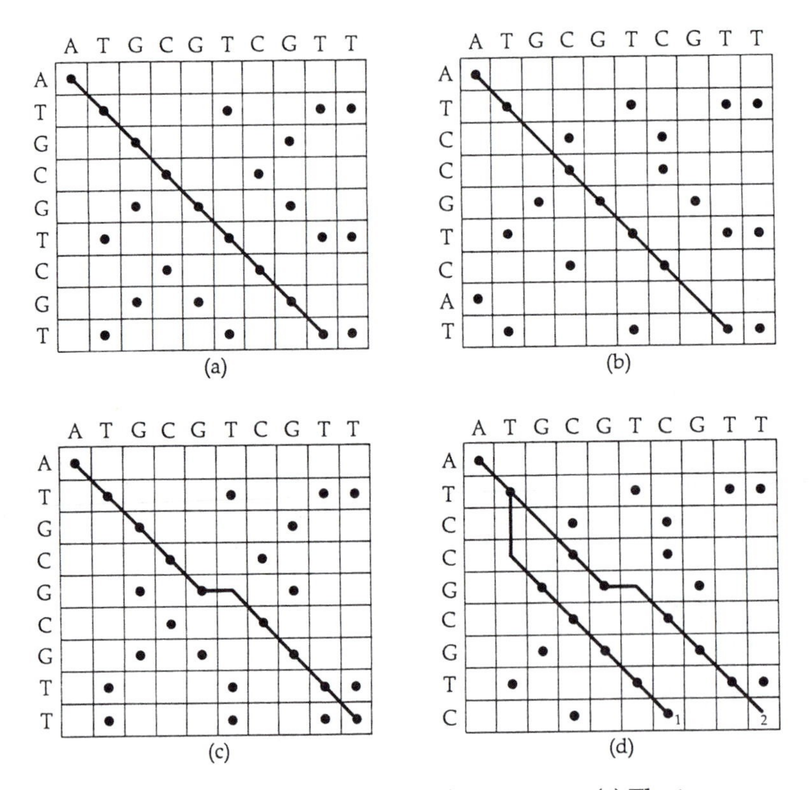

Figure 6. Dot matrices for aligning nucleotide sequences. (a) The two sequences are identical; (b) the two sequences differ from each other but contain no gaps; (c) the two sequences contain a gap, but are otherwise identical to each other; (d) the two sequences contain both substitutions and gaps. In (d), path 1 consists of six diagonal steps, none of which is empty, and 2 vertical steps. Path 2 contains eight diagonal steps, two of which are empty, and one horizontal step. The decision between paths 1 and 2 depends on the gap penalty, i.e., on which evolutionary sequence is more probable: a two-nucleotide deletion (path 1) or a one-nucleotide deletion and two substitutions (path 2).

The number of gaps in this case is 6. Conversely, the number of gaps can be reduced to a single gap of $|m - n|$ nucleotides, with a consequent increase in the number of mismatches:

```
      TCAGACGATTG
(II)    *  ****
      TCGGAGCTG- -
```

In this example, we have only one terminal and, hence, unavoidable gap, but the number of mismatches (indicated by asterisks) is 5.

Alternatively, we can choose an alignment that minimizes neither the number of gaps nor the number of substitutions. For example,

```
(III) TCAG-ACGATTG
           * *
      TC-GGA-GCTG-
```

In this case the number of mismatches is 2, and the number of gaps is 4.

So, which of the three alignments is preferable? It is obvious that comparing point substitutions with gaps is like comparing apples with oranges. As a consequence, we must find a common denominator with which to compare gaps and substitutions. This common denominator is called the **gap penalty**.

There are several systems of assigning gap penalties. All are based on certain a priori notions of how frequently deletions and insertions occur relative to point substitutions. In the first system, the total length of gaps (z) is multiplied by a constant gap penalty (w). The assumption behind this system is that the probability of having a gap is inversely proportional to the size of the gap. For instance, the probability of having a gap of two nucleotides equals the probability of having two gaps each of which is one nucleotide long. Thus, for any alignment, we can calculate a measure of distance (D) between the two sequences as

$$D = y + wz \tag{3.26}$$

In the second penalty system, we assume that long deletions and insertions have a different likelihood of occurrence in evolution than short ones. In this case, the penalties for different gap lengths may or may not be proportional to the gap length. The measure of distance associated with a particular alignment according to this system is

$$D = y + \Sigma w_k z_k \tag{3.27}$$

where z_k is the number of gaps of length k and w_k is the penalty for gaps of length k.

Let us now compare alignments I, II, and III using the first system with $w = 2$. The distances (D) obtained are $0 + (2 \times 6) = 12$, $5 + (2 \times 2) = 9$, and $2 + (2 \times 4) = 10$, for alignments I, II, and III, respectively. We choose alignment II. If we use the second system with $w_1 = 2$ and $w_2 = 6$. The values of D turn out to be 12, 11, and 10 for I, II, and III, respectively. In this case, we choose alignment III.

The purpose of any alignment algorithm is to choose the alignment associated with the smallest D from among all possible alignments. Among the most frequently used methods are those of Needleman and Wunsch (1970) and Sellers (1974). In the former method, the **similarity** between two sequences is measured by a **similarity index**, and the alignment with the maximum similarity is chosen from among all the alternatives. In Sellers' method, the **dissimilarity** between the two sequences is measured by a

distance index, and the alignment with the minimum distance is chosen. The two methods have been shown to be equivalent under certain conditions (Smith et al. 1981).

When there are many alternative alignments, the task of finding the best one cannot usually be accomplished without the aid of a computer. Many of the commonly used computer programs for aligning sequences are based on the algorithm by Needleman and Wunsch (1970) or its modifications.

The most important thing to remember is that the resulting alignment frequently depends on the choices of gap penalties, which in turn depend on crucial assumptions about how frequently gap events occur in the evolution of DNA and proteins relative to the frequency of point substitutions.

INDIRECT ESTIMATION OF THE NUMBER OF NUCLEOTIDE SUBSTITUTIONS

In estimating the number of nucleotide substitutions between two sequences, the highest resolution is obtained by comparing the nucleotide sequences themselves. However, the number of substitutions can also be inferred indirectly from other types of molecular data, such as those obtained by restriction enzyme mapping or DNA–DNA hybridization.

Restriction endonuclease fragment patterns and site maps

Restriction endonucleases or **restriction enzymes** recognize specific double-stranded DNA sequences called **recognition sequences** and cleave the DNA at or near the recognition sequence. Recognition sequences are usually four or six base pairs in length, and many of them are palindromes (i.e., they are rotationally symmetrical). Recognition sequences can be either unique (e.g., *Eco*RI), or equivocal (e.g., *Hind*II) (Table 1). The point of cleavage is called the **splicing site** or the **restriction site**. Many restriction endonucleases cut the double-stranded DNA in a staggered manner and therefore produce "**sticky ends**" that can be subsequently **ligated** to one another by the enzyme **ligase**. This is the reason why restriction enzymes constitute such a powerful tool in genetic engineering. Table 1 lists the recognition sequences and splicing sites of several restriction enzymes.

When a double-stranded piece of DNA is digested, fragments of different lengths are produced. These can be separated on an electrophoretic gel by length because the shorter fragments run faster and migrate further on the gel than the longer ones. By using DNA fragments of known lengths for calibration, the lengths of the restriction fragments can be estimated. Different sequences of DNA will be cut differently by restriction enzymes depend-

Table 1. Recognition sequences and cleavage sites of several restriction endonucleases.

Enzyme (Source organism)	Restriction site[a]	Recognition sequence (RS)				Cleavage	
		Size	Ambi-guity	Palin-drome	Contig-uous	In RS	Stag-gered
*Eco*RI (*Escherichia coli*)	5′—G↓A—A—T—T—C—3′ 3′—C—T—T—A—A—G—5′	6	−	+	+	+	+
Hind II (*Haemophilus influenzae*)	5′—G—T—Py↓Pu—A—C—3′ 3′—C—A—Pu↑Py—T—G—5′	6	+	+	+	+	−
*Hae*III (*Haemophilus aegyptus*)	5′—G—G↓C—C—3′ 3′—C—C↑G—G—5′	4	−	+	+	+	−
*Bbv*I (*Bacillus brevis*)	5′—G—C—A—G—C—(N_8)↓3′ 3′—C—G—T—C—G—(N_{12})↑5′	5	−	−	+	−	+
*Nci*I (*Neisseria cinerea*)	5′—C—C↓C/G—G—G—3′ 3′—G—G—G/C↑C—C—5′	5	+	+	+	+	+
*Not*I (*Nocardia otitidis-caviarum*)	5′—G—C↓G—G—C—C—G—C—3′ 3′—C—G—C—C—G—G↑C—G—5′	8	−	+	+	+	+
*Hinf*I (*Haemophilus influenzae*)	5′—G↓A—N—T—C—3′ 3′—C—T—N—A↑G—5′	4	−	+	−	+	+

[a]Recognition sequences are in boldface letters. Cleavage sites are marked by arrows. Ambiguities are marked as Pu, purine; Py, pyrimidine; C/G, C or G; and N, any nucleotide. N_n means a sequence of n arbitrary nucleotides.

ing on the number and location of the recognition sites. The number and sizes of the fragments resulting from the digestion of a DNA sequence is called the **restriction-fragment pattern**. By sequentially and reciprocally using several restriction enzymes that will digest the DNA into overlapping fragments, it is often possible to deduce the approximate location of the restriction sites on the DNA (Figure 7). The schematic representation showing the location of restriction sites on a DNA sequence is called a **restriction map**.

The rationale behind deducing the number of substitutions between two sequences by using restriction enzymes is that the greater the similarity of two DNA sequences, the more similar their restriction-fragment patterns will

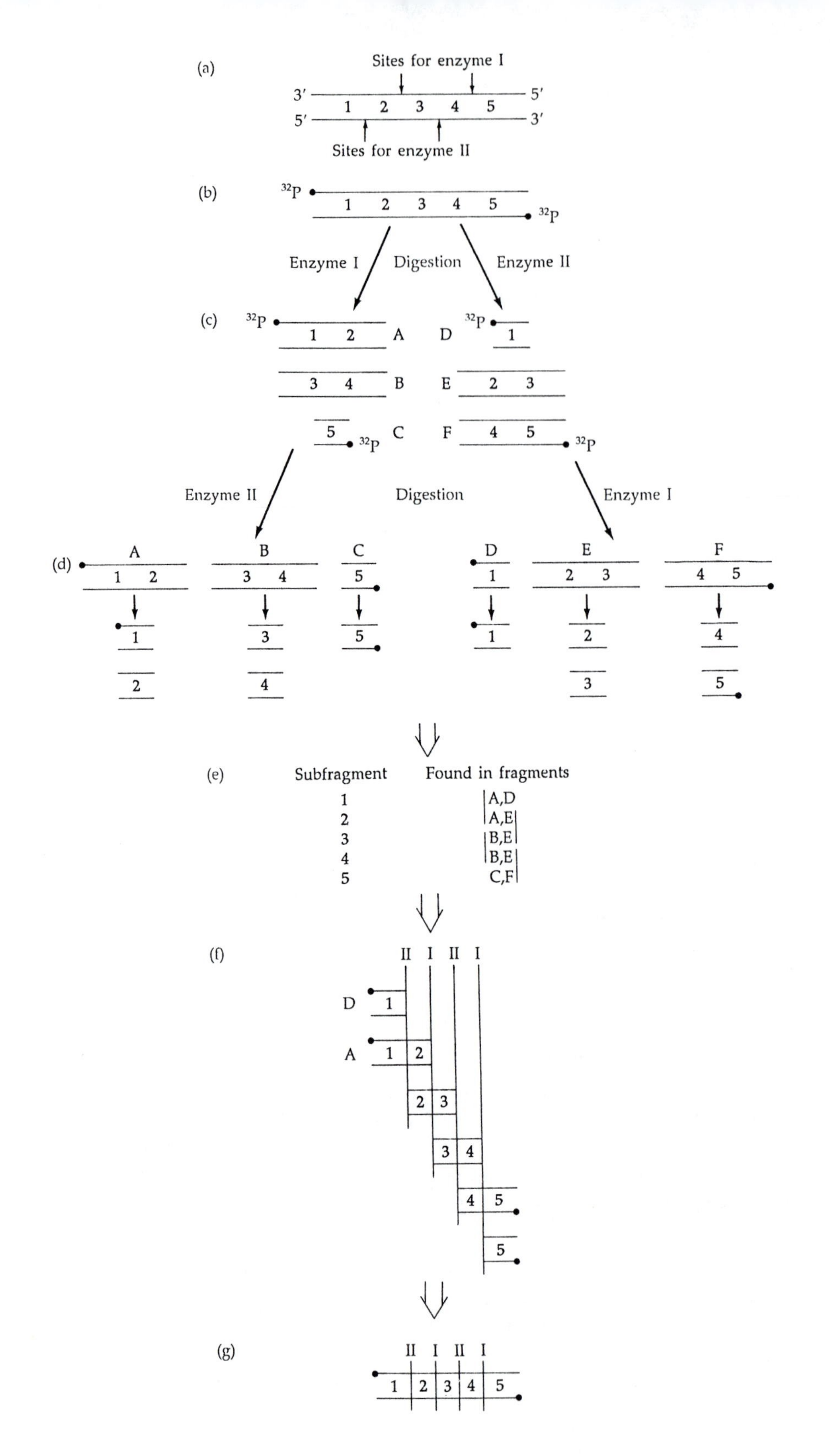
(a)
Sites for enzyme I
3′
5′
1 2 3 4 5
5′
3′
Sites for enzyme II
(b)
^{32}P
1 2 3 4 5
^{32}P
Enzyme I
Digestion
Enzyme II
(c)
A
B
C
D
E
F
Enzyme II
Digestion
Enzyme I
(d)
(e)
Subfragment
Found in fragments
1 A,D
2 A,E
3 B,E
4 B,E
5 C,F
(f)
II I II I
(g)
II I II I

be. By making certain assumptions about the distribution of restriction sites within DNA sequences, such as, for instance, that the four nucleotides are equally frequent and that their spatial distribution within the sequence is random, it is possible to study the evolutionary changes in the restriction pattern and, consequently, to estimate the number of nucleotide substitutions per site (K) between DNA sequences from restriction-site data.

First, we consider the estimation of K from restriction-fragment patterns. Estimating K from the number of shared fragments requires that we directly compare the electrophoretic patterns of the DNAs digested by the restriction endonucleases. The method presented here was developed by Nei and Li (1979). Two other methods have been proposed in the literature (Upholt 1977, Engels 1981a), and a study by Kaplan (1983) has shown that all three methods give similar results.

The expected proportion of shared DNA fragments between two sequences of DNA (F) can be estimated by

$$\hat{F} = \frac{2m_{XY}}{m_X + m_Y} \tag{3.28}$$

where m_X and m_Y are the numbers of restriction fragments resulting from the digestion of sequences X and Y, respectively, and m_{XY} is the number of fragments shared by the two sequences.

Nei and Li (1979) have shown that the expected proportion of shared fragments (F) can be expressed in terms of the probability (G) that a restriction site has remained unaltered during time t by the approximate formula

$$F \approx G^4/(3 - 2G) \tag{3.29}$$

where $G = e^{-r\lambda t}$, in which r is the number of nucleotides in the recognition site, λ is the rate of nucleotide substitution, and t is the time of divergence between the two sequences. The number of substitutions per site between

◄Figure 7. Mapping of restriction sites on a DNA sequence. (a) A hypothetical DNA with recognition sites for two different restriction enzymes. The restriction map of the sequence is unknown. (b) Radioactive labeling of 3′ ends. (c) Digestion of the DNA by enzyme I produces fragments A, B, and C, whereas digestion with enzyme II produces fragments D, E, and F. (d) Each of the fragments obtained with one restriction enzyme is digested with the other restriction enzyme to produce subfragments 1–5. (e) The subfragments are used to identify overlapping fragments. (f) The subfragments are arranged in the order indicated by the pattern of overlapping between them. (g) Inferrred restriction map of the DNA sequence. The fragments and subfragments can be identified individually according to their lengths as deduced from their position on an electrophoretic gel. Terminal fragments and subfragments are identified by their radioactive labeling. Modified from Suzuki et al. (1989).

the sequences is $K = 2\lambda t$. To estimate G, we rearrange Equation 3.29 and obtain

$$G = [F(3 - 2G)]^{1/4} \qquad \textbf{(3.30)}$$

This equation can be solved by a process of iteration. As the first trial value, Nei (1987) recommends $G = \hat{F}^{1/4}$. Usually, very few iteration cycles are required. The estimate of G allows us to obtain an estimate of K, as follows:

$$K = -(2/r)\,\ell n\,(G) \qquad \textbf{(3.31)}$$

Let us consider the following example in which corresponding mitochondrial DNA segments from two species of wild wheat (*Aegilops sharonensis* and *Ae. bicornis*) were digested with three restriction enzymes, *Bam*I, *Hind*III, and *Eco*RI, the recognition sequences of which are six base pairs long (data from Graur et al. 1989a). The digestion of *Ae. sharonensis* yielded 4 fragments, while that of *Ae. bicornis* yielded 5 fragments. Two fragments were shared by both species. By using Equation 3.29, we estimate F to be $2/9 = 0.222$. We can now start the iteration process as given in Equation 3.30. The first value of G that we use is $0.222^{1/4} = 0.687$. In the first cycle of the iteration we obtain $G = 0.775$, and in the next cycle $G = 0.753$. The oscillations get smaller, and in both the fifth and the sixth cycles of the iteration we obtain $G = 0.758$. Therefore, we end the iteration process. To obtain an estimate of the number of substitutions between the two sequences, we use Equation 3.31. The final result is that the two mitochondrial sequences are estimated to differ from each other by $K = 0.092$ substitutions per nucleotide site.

Let us now consider the estimation of the number of nucleotide substitutions between two sequences from restriction-site maps. In the previous case, the location of the restriction sites was unknown. If the restriction sites are mapped onto the DNA sequences, we can look for shared and unshared sites and estimate the number of substitutions directly from the map. Let m_X and m_Y be the numbers of restriction sites in DNA sequences X and Y, respectively, and let m_{XY} be the number of shared restriction recognition sites between the two sequences. The probability that X and Y share the same recognition sequence at a given site is denoted by S, which can be estimated by

$$\hat{S} = \frac{2m_{XY}}{m_X + m_Y} \qquad \textbf{(3.32)}$$

(Nei and Li 1979). The proportion of nucleotide differences, p, can be estimated by

$$\hat{p} = 1 - \hat{S}^{1/r} \qquad \textbf{(3.33)}$$

where r is the number of nucleotides in the recognition sequence. The number of substitutions per site between the two sequences is estimated from knowledge of $\hat{p}$ by using Equation 3.19.

The restriction-site-map method is more tedious than the restriction-fragment-pattern method, but it is considerably more reliable. The former can be used for K values of up to 0.25; the latter may not be accurate for $K > 0.05$.

DNA–DNA hybridization

The technique of **DNA–DNA hybridization** is based on the fact that the thermal stability of double-stranded DNA molecules depends on the proportion of nucleotide matches between the two strands. As the proportion of matches decreases, the duplex becomes less thermally stable. The proportion of matches in a double-stranded DNA in which the two strands are from the same sequence (i.e., **homoduplex** molecules) is by definition 100%. On the other hand, the proportion of matches in a double-stranded DNA in which the two strands come from different sources (i.e., **heteroduplex** molecules) is less than 1 by an amount that depends on how many nucleotide differences these two sequences have accumulated since their separation from a common ancestor. Consequently, heteroduplex DNA will denature or melt into single strands at lower temperatures than homoduplex DNA.

The basic experimental procedure for DNA–DNA hybridization tests is shown in Figure 8. In essence, the procedure involves the creation of artificial hybrid DNA molecules by slowly cooling a mixture of denatured DNA from two different species, after repetitive sequences have been removed. Subsequently, the mixture is heated gradually, and the percentage of single-stranded DNA in the solution is determined at each temperature. For details on one such method (the TEACL method), see, e.g., Hunt et al. (1981).

The thermal stability of the hybrid DNA is measured by the temperature at which 50% of the hybrid DNA is dissociated into single strands. This median melting temperature is then compared with the temperature at which 50% of double-stranded homoduplex DNA becomes single-stranded. Note that for each interspecific comparison we have two types of homoduplexes, one from each species, so it is customary to use the average value of their median melting temperatures. The median melting temperatures are denoted as T_m. The difference between the median melting temperatures of homoduplexes and heteroduplexes, ΔT_m, has been empirically shown to be approximately linearly related to the proportion of base-pair mismatches (Britten et al. 1974). We express this relationship as

$$p = c\,\Delta T_m \tag{3.34}$$

where p is the proportion of mismatches and c is a constant. The value of c has been obtained empirically by conducting DNA–DNA hybridization tests

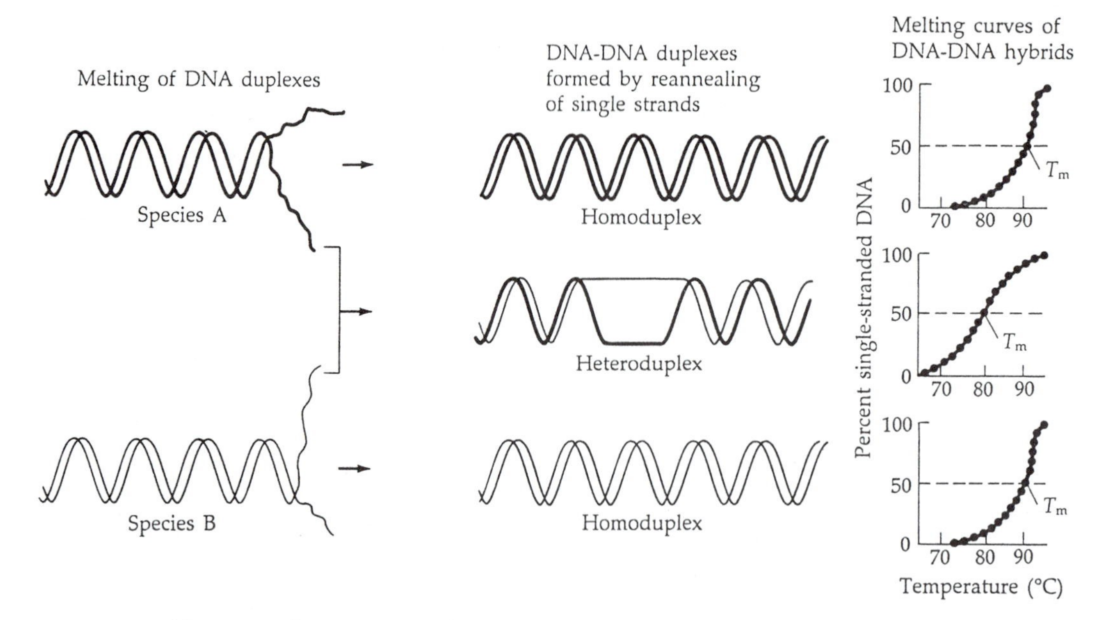

Figure 8. Sequence similarity deduced from DNA–DNA hybridization studies. Duplexes are melted to single-stranded DNA. Homoduplexes and heteroduplexes are formed by reannealing of single strands. The temperature at which 50% of the DNA melts into single strands, T_m, is determined for both homoduplexes and the heteroduplex. T_m values may differ between the two homoduplexes, as well as between the two reciprocal heteroduplex types. Modified from Avers (1989).

with heteroduplexes for which the degree of base-pair mismatch is known. The c values were found to vary with experimental conditions between $c = 0.01$ and $c = 0.015$. The experimental error of ΔT_m is known to be very large, and hence, many replicate observations for the same pair of species should be made.

Let us now consider the following numerical example (data from Caccone and Powell 1989). The average T_m values for homoduplex DNA were 59.50°C and 59.12°C for human and pygmy chimpanzee (*Pan paniscus*) males, respectively. Thus, the mean T_m for homoduplex molecules was 59.31°C. The mean T_m for the two reciprocal heteroduplex DNAs was 57.59°C. ΔT_m is therefore 1.72°C. From Equation 3.34, we obtain a difference of about 0.017 – 0.026 substitutions per nucleotide site.

PROBLEMS

1. Show that Equation 3.3 holds for $t = 0$, i.e., that it reduces to Equation 3.1 if $t = 0$.

2. Derive Equation 3.10 and show that $P_{GA(t)} = P_{CA(t)} = P_{TA(t)}$ under the Jukes-Cantor model.

3. Kimura's two-parameter model should become identical to Jukes and Cantor's one-parameter model when $\alpha = \beta$. To verify this, show that when this condition is met, Equation 3.13 becomes identical to Equation 3.11, and Equations 3.14 and 3.15 both become identical to Equation 3.12.

4. Use Equations 3.13, 3.14, and 3.15 to show that in Kimura's two- parameter model the equilibrium frequencies of the four nucleotides in a sequence are equal (i.e., 1/4) as in the one-parameter model.

5. Derive Equation 3.17 from Equation 3.16.

6. Calculate (a) the number of synonymous substitutions per synonymous site and (b) the number of nonsynonymous substitutions per nonsynonymous site for the following two sequences.

Ser	Thr	Glu	Met	Cys	Leu	Met	Gly	Gly
TCA	ACT	GAG	ATG	TGT	TTA	ATG	GGG	GGA
TCG	ACA	GGG	ATA	TAT	CTA	ATG	GGT	ATA
Ser	Thr	Gly	Ile	Tyr	Leu	Met	Gly	Ile

7. By noting that the number of differences between two sequences according to Kimura's two-parameter model is $P + Q$, show that Equation 3.21 reduces to Equation 3.19 when transitions and transversions are considered together.

8. Use the dot-matrix method to align the following two sequences:

   AATGCTTGCATGGGGCTAGTT
   ATTGCTGCATGAGGCGCGCTAGT

   Choose two possible alignments and decide which one is better by using a constant gap penalty of 2 per nucleotide. Will the choice be affected by using a much bigger gap penalty, say 10?

9. From the restriction-site maps of the two sequences in Figure 9, estimate the number of nucleotide substitutions between the two sequences by using (a) the proportion of shared fragments and (b) the proportion of shared restriction sites. The recognition sequence of the restriction endonuclease was four nucleotides. What would have been the result with a restriction endonuclease with a six-nucleotide recognition site? What is the reason for the difference?

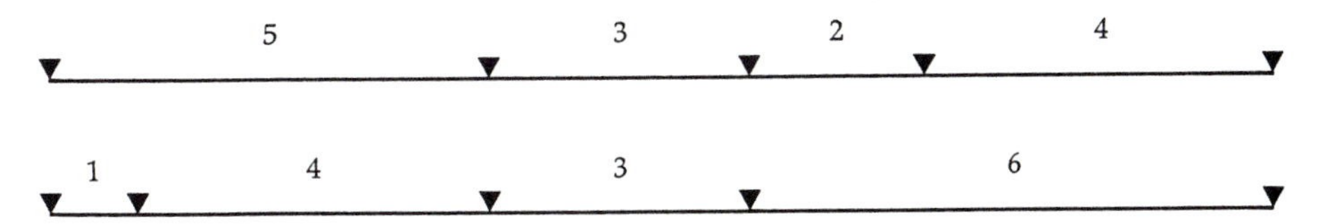

Figure 9. Hypothetical example of two restriction endonuclease maps. The numbers on the sequence represent the length of the fragments (in kb).

FURTHER READINGS

Doolittle, R. F. 1990. *Molecular Evolution: Computer Analysis of Protein and Nucleic Acid Sequences*. Academic Press, San Diego, CA.

Li, W.-H., C.-C. Luo and C.-I. Wu. 1985. Evolution of DNA sequences. pp. 1–94. *In* R. J. MacIntyre (ed.), *Molecular Evolutionary Genetics*, Plenum, New York.

Nei, M. 1987. *Molecular Evolutionary Genetics*. Columbia University Press, New York.

4

RATES AND PATTERNS OF NUCLEOTIDE SUBSTITUTION

The mathematical theory developed in the preceding chapter can be used to study the rate of nucleotide substitution, which is a basic quantity in the study of molecular evolution. Indeed, in order to characterize the evolution of a DNA sequence, we need to know how fast it evolves and what are the rates of nucleotide substitution of its constituent parts. It is also interesting to compare the substitution rates among genes and different DNA regions, because this can help us understand the mechanism of nucleotide substitution in evolution. Knowing the rate of nucleotide substitution may also enable us to date evolutionary events such as the divergence between species. However, to do this we need to know whether the rate estimated from one group of species is applicable to another group. This raises the issue of how variable the rate is among different evolutionary lineages.

RATES OF NUCLEOTIDE SUBSTITUTION

The **rate of nucleotide substitution** is defined as the number of substitutions per site per year and can be calculated by dividing the number of substitutions between two homologous sequences, K, by $2T$, where T is the time of divergence between the two sequences (Figure 1). That is,

$$r = K/(2T) \quad \textbf{(4.1)}$$

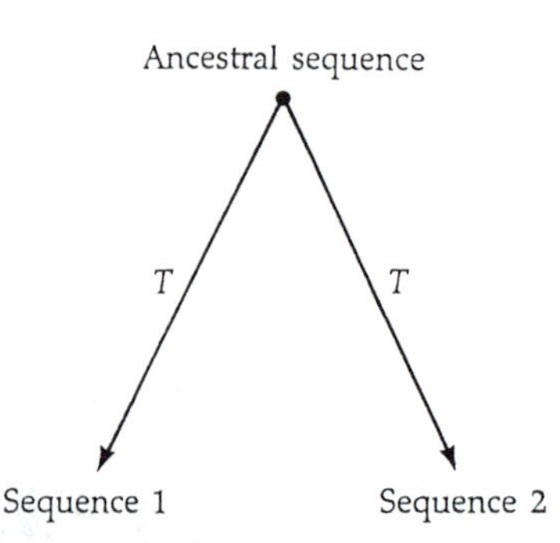

Figure 1. Divergence of two homologous sequences from a common ancestral sequence *T* years ago.

The divergence time, *T*, is assumed to be the same as the time of divergence between the two species from which the two sequences were taken and is usually inferred from paleontological data.

In this section we will deal with the issue of rate variation among genes and among different regions in a gene. For this purpose, it is advisable to use the same species pair for all the genes under consideration. The reason is twofold. First, there are usually considerable uncertainties about paleontological estimates of divergence times. By using the same pair of species we can compare rates of substitution among genes without knowledge of the divergence time. Second, the rate of substitution may vary considerably among lineages (see page 79). In this case, differences in rates between two genes may be due to differences between lineages rather than to differences that are attributable to the genes themselves.

At the present time the best data for studying the rate of nucleotide substitution are from mammals because DNA sequence data from mammals are most abundant and because the fossil record for mammals is relatively well-characterized and fairly reliable times of divergence between mammalian species are available.

Coding regions

In Table 1 we list the rates of synonymous and nonsynonymous substitutions for 36 protein-coding genes. The rates were calculated from comparisons between human and rodent homologous genes. Based on paleontological evidence pertaining to the radiation of eutherian mammals, the time for the human–rodent divergence event has been set at 80 million years ago.

We note that the rate of nonsynonymous substitution is extremely variable among genes. It ranges from effectively zero in histones 3 and 4 to 2.79×10^{-9} substitutions per nonsynonymous site per year in interferon γ.

Some hormones (e.g., somatostatin-28 and insulin) are extremely conservative, but others evolve at either intermediate rates (e.g., erythropoietin) or at high rates (e.g., interleukin I and relaxin). Hemoglobins and myoglobin evolve at intermediate rates, while apolipoproteins and immunoglobulins evolve very rapidly.

The rate of synonymous substitution also varies considerably, though much less than the nonsynonymous rate. It can be shown that the variation

Table 1. Rates of synonymous and nonsynonymous substitutions in various mammalian protein-coding genes.[a]

Gene	L^b	Nonsynonymous rate ($\times 10^9$)	Synonymous rate ($\times 10^9$)
HISTONES			
Histone 3	135	0.00 ± 0.00	6.38 ± 1.19
Histone 4	101	0.00 ± 0.00	6.12 ± 1.32
CONTRACTILE SYSTEM PROTEINS			
Actin α	376	0.01 ± 0.01	3.68 ± 0.43
Actin β	349	0.03 ± 0.02	3.13 ± 0.39
HORMONES, NEUROPEPTIDES, AND OTHER ACTIVE PEPTIDES			
Somatostatin-28	28	0.00 ± 0.00	3.97 ± 2.66
Insulin	51	0.13 ± 0.13	4.02 ± 2.29
Thyrotropin	118	0.33 ± 0.08	4.66 ± 1.12
Insulin-like growth factor II	179	0.52 ± 0.09	2.32 ± 0.40
Erythropoietin	191	0.72 ± 0.11	4.34 ± 0.65
Insulin C-peptide	35	0.91 ± 0.30	6.77 ± 3.49
Parathyroid hormone	90	0.94 ± 0.18	4.18 ± 0.98
Luteinizing hormone	141	1.02 ± 0.16	3.29 ± 0.60
Growth hormone	189	1.23 ± 0.15	4.95 ± 0.77
Urokinase-plasminogen activator	435	1.28 ± 0.10	3.92 ± 0.44
Interleukin I	265	1.42 ± 0.14	4.60 ± 0.65
Relaxin	54	2.51 ± 0.37	7.49 ± 6.10
HEMOGLOBINS AND MYOGLOBIN			
α-globin	141	0.55 ± 0.11	5.14 ± 0.90
Myoglobin	153	0.56 ± 0.10	4.44 ± 0.82
β-globin	144	0.80 ± 0.13	3.05 ± 0.56

(Continued on next page)

Table 1. (*Continued*)

Gene	L^b	Nonsynonymous rate (× 10^9)	Synonymous rate (× 10^9)
APOLIPOPROTEINS			
E	283	0.98 ± 0.10	4.04 ± 0.53
A-I	243	1.57 ± 0.16	4.47 ± 0.66
A-IV	371	1.58 ± 0.12	4.15 ± 0.47
IMMUNOGLOBULINS			
Ig V_H	100	1.07 ± 0.19	5.66 ± 1.36
Ig γ1	321	1.46 ± 0.13	5.11 ± 0.64
Ig k	106	1.87 ± 0.26	5.90 ± 1.27
INTERFERONS			
α1	166	1.41 ± 0.13	3.53 ± 0.61
β1	159	2.21 ± 0.24	5.88 ± 1.08
γ	136	2.79 ± 0.31	8.59 ± 2.56
OTHER PROTEINS			
Aldolase A	363	0.07 ± 0.03	3.59 ± 0.51
Hydroxanthine phosphoribosyltransferase	217	0.13 ± 0.04	2.13 ± 0.35
Creatine kinase M	380	0.15 ± 0.03	3.08 ± 0.37
Glyceradehyde-3-phosphate dehydrogenase	331	0.20 ± 0.05	2.84 ± 0.37
Lactate dehydrogenase A	331	0.20 ± 0.04	5.03 ± 0.61
Acetylcholine receptor γ subunit	540	0.29 ± 0.04	3.23 ± 0.31
Fibrinogen γ	411	0.55 ± 0.06	5.82 ± 0.67
Albumin	590	0.91 ± 0.07	6.63 ± 0.61
Average[c]		0.85 (0.73)	4.61 (1.44)

[a] All rates are based on comparisons between human and rodent genes and the time of divergence was set at 80 million years ago. Rates are in units of substitutions per site per 10^9 years.
[b] L = number of codons compared.
[c] Average is the arithmetic mean, and values in parentheses are the standard deviations, computed over all genes.

in the synonymous substitution rate among genes is significantly greater than expected on the basis of statistical fluctuations alone.

For the vast majority of the genes in Table 1, the synonymous substitution rate greatly exceeds the nonsynonymous rate. In the most extreme case, for example, the synonymous rate in the histone 3 gene is very high, though in terms of its amino acid sequence, this protein is one of the most evolutionarily conservative proteins. The mean rate of nonsynonymous substitution for the genes in Table 1 is 0.85×10^{-9} substitutions per nonsynonymous site per year. The mean rate of synonymous substitution is 4.6×10^{-9} substitutions per synonymous site per year, i.e., five times higher than the mean rate of nonsynonymous substitution.

Noncoding regions

Data from noncoding regions are much less abundant than data from coding regions, and so only a limited comparative analysis can be done at the present time. (Note that in order to estimate the rate of substitution in a sequence we must have data from at least two species.) Since most published sequences are mRNAs, which do not include introns and flanking regions, the 5′ and 3′ untranslated regions are the only noncoding regions that can be studied in detail. Table 2 shows the substitution rates in these two regions for 16

Table 2. Rates of nucleotide substitution in 5′ and 3′ untranslated regions and at fourfold degenerate sites of protein-coding genes, based on comparisons between human and mouse or rat genes.[a]

	5′ untranslated		3′ untranslated		Fourfold degenerate	
Gene	L[b]	Rate	L	Rate	L	Rate
ACTH	99	1.87 ± 0.41	97	2.32 ± 0.49	275	2.78 ± 0.34
Aldolase A	124	1.08 ± 0.26	154	1.73 ± 0.32	195	3.16 ± 0.48
Apolipoprotein A-IV	83	3.06 ± 0.68	134	1.73 ± 0.33	160	3.38 ± 0.50
Apolipoprotein E	23	1.27 ± 0.69	84	1.70 ± 0.42	153	4.00 ± 0.60
Na,K-ATPase β	118	2.45 ± 0.45	1,117	0.57 ± 0.06	118	2.87 ± 0.54
Creatine kinase M	70	1.71 ± 0.46	168	1.79 ± 0.30	178	2.81 ± 0.41
α-fetoprotein	47	3.64 ± 1.13	144	2.79 ± 0.49	225	4.14 ± 0.54
α-globin	34	1.56 ± 0.65	90	2.21 ± 0.50	81	4.47 ± 0.98
β-globin	50	1.30 ± 0.46	126	2.85 ± 0.49	78	2.42 ± 0.56
Glyceraldehyde-3 phosphate dehydrogenease	70	1.34 ± 0.38	121	1.74 ± 0.36	170	2.43 ± 0.39
Growth hormone	21	1.79 ± 0.85	91	1.83 ± 0.41	83	3.82 ± 0.78
Insulin	56	2.92 ± 0.80	53	3.09 ± 0.81	62	4.19 ± 1.00
Interleukin I	59	1.09 ± 0.38	1,046	2.02 ± 0.14	105	2.97 ± 0.60
Lactate dehydrogenase A	95	2.79 ± 0.55	470	2.48 ± 0.23	152	3.64 ± 0.60
Metallothionein II	61	1.88 ± 0.52	111	2.57 ± 0.48	23	2.37 ± 1.00
Parathyroid hormone	84	1.79 ± 0.43	228	2.21 ± 0.30	38	3.85 ± 1.21
Average[c]		1.96 (0.78)		2.10 (0.61)		3.33 (0.69)

[a] Rates are in units of substitutions per site per 10^9 years.
[b] L = number of sites.
[c] Average is the arithmetic mean, and values in parentheses are the standard deviations, computed over all genes.

genes based on comparisons between humans and rodents. In both regions the rates vary greatly among genes, but this variation may largely represent sampling effects due to the fact that both of these regions are usually very short. In almost all genes, the rates in the 5′ and 3′ untranslated regions are lower than the rate of substitution at fourfold degenerate sites (i.e., sites at which all possible nucleotide substitutions are synonymous). The average rates for the 5′ and 3′ untranslated regions are 1.96×10^{-9} and 2.10×10^{-9} substitutions per year, respectively, which are both about 60% of the average rate of 3.55×10^{-9} substitutions per year at fourfold degenerate sites.

Pseudogenes are DNA sequences that were derived from functional genes but have been rendered nonfunctional by mutations that prevent their proper expression (Chapters 6 and 7). Since they are subject to no functional constraints, they are expected to evolve at a high rate. Table 3 shows a comparison between the rate of substitution in cow and goat $\psi\beta^X$ and $\psi\beta^Z$ pseudogenes and the rates in the noncoding regions and fourfold degenerate sites in the β- and γ-globin genes. The rate in these pseudogenes is indeed slightly higher than that in the other regions. This seems to be generally true for pseudogenes, though currently, pseudogene data are limited.

In Figure 2 we present a comparison of the rates of substitution in different regions of the gene, as well as in pseudogenes. The rates for the 5′ and 3′ untranslated regions, nondegenerate sites, twofold degenerate sites, and fourfold degenerate sites are the average rates for the genes in Table 2. The rate for the 5′ flanking region was computed by assuming that the ratio of this rate to that at fourfold degenerate sites is 5.3/8.6 (as suggested by the values in Table 3) and that the average rate at fourfold degenerate sites is 3.33×10^{-9} substitutions per year (Table 2). The rates for introns, the 3′ flanking region, and pseudogenes were computed in the same manner. Since these are rough estimates based on limited data and since the rate in a region varies from gene to gene, the rates shown in Figure 2 may not be applicable to any particular gene but are meant to provide a rough general comparison

Table 3. Divergence between cow and goat β- and γ-globin genes and between cow and goat β-globin pseudogenes.

Statistic	β- and γ-globin genes[a]						
	5′FL	5′UT	Fourfold	Introns	3′UT	3′FL	Pseudogenes
Percent divergence	5.3	4.0	8.6	8.1	8.8	8.0	9.1
Standard error	1.2	2.0	2.5	0.7	2.2	1.5	0.9

[a] FL = flanking region; UT = untranslated region; fourfold = fourfold degenerate sites.

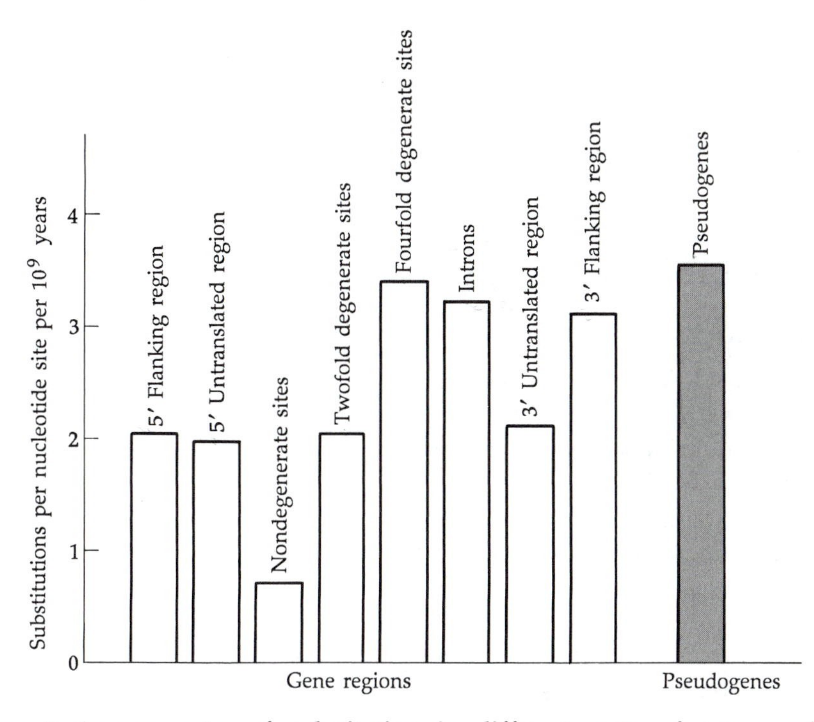

Figure 2. Average rates of substitution in different parts of genes and in pseudogenes.

of the substitution rates in different DNA regions. With this precaution, we note that the substitution rate in a gene is highest at fourfold degenerate sites, slightly lower in introns and the 3′ flanking region, intermediate for the 3′ untranslated region, the 5′ flanking and untranslated regions, and twofold degenerate sites, and lowest at nondegenerate sites. Pseudogenes have on average the highest rate of substitution, although only slightly higher than that at the fourfold degenerate sites of a functional gene.

CAUSES OF VARIATION IN SUBSTITUTION RATES

To infer the causes of variation in substitution rates among DNA regions, we note that the rate of substitution is determined by two factors: (1) the rate of mutation and (2) the probability of fixation of a mutation (Chapter 2). The latter depends on whether the mutation is advantageous, neutral, or deleterious. Since the rate of mutation is unlikely to vary much within a gene but may vary among genes, we shall discuss the rate variation among different regions of a gene and the variation among genes separately.

Variation among different gene regions

Let us first consider the large difference between the synonymous and nonsynonymous rates in a gene. Since the rate of mutation at synonymous and nonsynonymous sites within a gene should be the same, or at least very similar, the difference in substitution rates can be attributed to differences in the intensity of purifying selection between the two types of sites. This is understandable in light of the neutral theory of molecular evolution (Chapter 2). Mutations that result in an amino acid substitution have a higher chance of causing deleterious effects on the function of the protein than do synonymous changes. Consequently, the majority of nonsynonymous mutations will be eliminated from the population by purifying selection. The result will be a reduction in the rate of substitution at nonsynonymous sites. In contrast, synonymous changes have a better chance of being neutral, and more of them will be fixed in a population.

Of course, nonsynonymous substitutions may have a better chance of improving the function of a protein. However, if advantageous selection plays a major role in the evolution of proteins, the rate of nonsynonymous substitution should exceed that of synonymous substitution. Indeed, in some immunoglobulin genes, the nonsynonymous rate in the complementarity-determining regions (CDRs; also known as hypervariable regions) is higher than the synonymous rate. The higher rate has been attributed to overdominant selection for antibody diversity (Tanaka and Nei 1989). However, when the entire immunoglobulin gene is considered, the nonsynonymous rate is still considerably lower than the synonymous rate (Table 1). This result indicates that, even in immunoglobulins, most nonsynonymous mutations are disadvantageous and are eliminated from the population. Hughes and Nei (1989) reported a similar situation in certain regions of the major histocompatibility complex genes, i.e., the rate of nonsynonymous substitution exceeds the rate of synonymous substitution. They attributed the higher rates of nonsynonymous substitution to overdominant selection.

The contrast between synonymous and nonsynonymous rates in a gene demonstrates a well-known principle in molecular evolution, namely that the stronger the functional constraints on a macromolecule, the slower the rate of evolution. Kimura (1983) has formulated this principle by a simple model. Suppose that a certain fraction, f_0, of all mutations in a certain molecule are selectively neutral or nearly neutral and that the rest are deleterious. (Advantageous mutations are assumed to occur only rarely, such that their relative frequency is effectively zero, and they do not contribute significantly to the overall rate of molecular evolution.) If we denote by v_T the total mutation rate per site per unit time, then the rate of neutral mutation is $v_0 = v_T f_0$. According to the neutral theory of molecular evolution, the rate of substitution is $k = v_0$ (Chapter 2). Hence,

$$k = v_T f_0 \tag{4.2}$$

Within any given gene, the v_T value can be assumed to be the same for both synonymous sites and nonsynonymous sites. However, the f_0 value is higher for synonymous sites than for nonsynonymous sites and so the former should evolve faster than the latter. Thus, although the model is oversimplified, it is helpful for explaining the rate differences among different DNA regions.

According to the above model, the highest rate is expected to occur in a sequence that does not have any function, so that all mutations in it are neutral (i.e., $f_0 = 1$). Indeed, pseudogenes do seem to have the highest rate of nucleotide substitution (Table 3 and Figure 2). The observation that 5′ and 3′ untranslated regions have lower substitution rates than the rate of synonymous substitutions in coding regions further supports the neutral line of reasoning, because these regions contain important signals for transcription initiation and termination.

Within a protein, the different structural or functional domains are likely to be subject to differential functional constraints and to evolve at different rates. A good example is provided by proinsulin, which consists of three segments: A, B, and C (Figure 3). Segment C resides in the middle of the

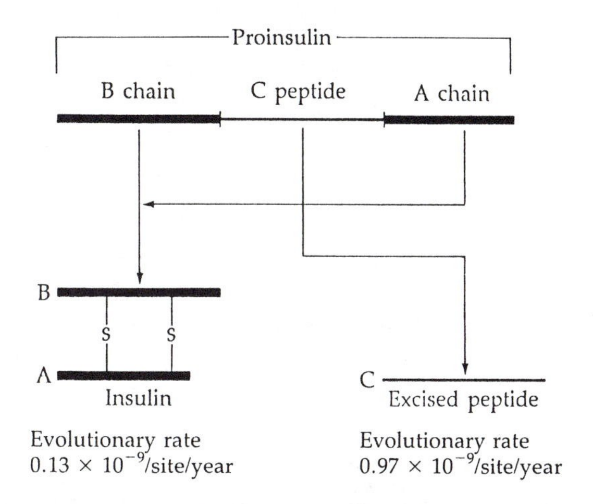

Figure 3. Comparison between the rates of nucleotide substitution for DNA regions coding for functional insulin (A and B chains) and the C peptide. A mature insulin molecule consists of one A and one B chain, linked by disulfide (S) bonds. Modifed from Kimura (1983).

proinsulin molecule and is removed during the formation of the active hormone (insulin), which consists of the two remaining segments, A and B. Segment C does not take part in the hormonal activity of insulin and is thought only to facilitate the creation of the proper tertiary structure of the hormone. Consequently, the nonsynonymous substitution rate for the region coding for the C segment is seven times higher than the average nonsynonymous rate for the regions coding for the A and B chains (Figure 3). Nevertheless, considerable constraints must still operate on the C segment, because the nonsynonymous substitution rate in this region is rather low, comparable to that in β-globin (Table 1).

Variation among genes

To explain the large variation in the rates of nonsynonymous substitution among genes, we must again consider the two possible culprits: the rate of mutation and the intensity of selection. The assumption of equal rates of mutation for different genes may not hold in this case, since different regions of the genome may have different propensities to mutate. Wolfe et al. (1989a) suggested that different regions of the mammalian nuclear genome may differ from each other by a factor of two in their rates of mutation. However, a twofold difference in mutation rates among different genomic regions cannot even partially account for the close to 1,000-fold range in nonsynonymous substitution rates. Thus, the most important factor in determining the rates of nonsynonymous substitution seems to be the selection intensity, which in turn is determined by functional constraints.

To illustrate the effects of functional constraints, let us consider apolipoproteins and histone 3, which exhibit markedly different rates of nonsynonymous substitution. Apolipoproteins are the major carriers of various lipids in the blood of vertebrates, and their lipid binding domains consist mostly of hydrophobic residues. Comparative analysis of apolipoprotein sequences from various mammalian orders suggests that in these domains the substitution of a hydrophobic amino acid (e.g., valine, leucine) for another hydrophobic amino acid is acceptable at many sites (Luo et al. 1989). This lax structural requirement may explain why the nonsynonymous rates in these genes are fairly high (Table 1).

At the other extreme, we have histone 3. Since most amino acids in histone 3 interact directly with either the DNA or other core histones in the formation of the nucleosome (Figure 4), it is reasonable to assume that there are very few possible substitutions that can occur without impeding the function of this protein. In addition, histone 3 must retain its strict compactness and its high alkalinity, which are necessary for interaction with the acidic DNA molecule. As a consequence, histone 3 is very intolerant of most molecular

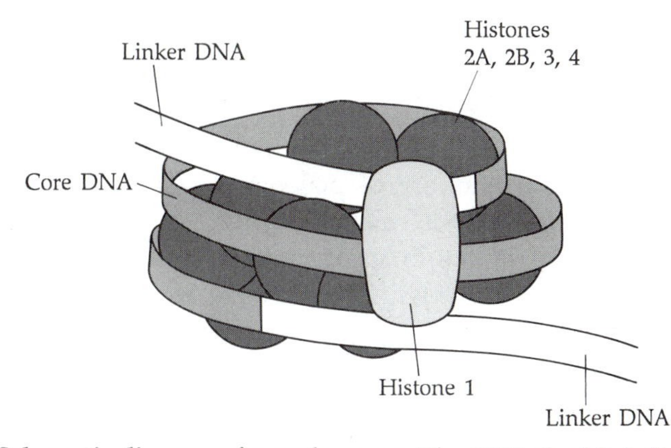

Figure 4. Schematic diagram of a nucleosome. The DNA double helix (dark ribbon) is wound around the core histones (two each of histones 2A, 2B, 3, and 4). Histone 1 (light shading) binds to the outside of this core particle and to the linker DNA (white ribbon). Modified from Stryer (1988).

changes. Indeed, this protein is one of the slowest evolving proteins known, more than 1,000 times slower than the apolipoproteins.

Why the rate of synonymous substitution also varies from gene to gene is less clear. There may be two reasons for this variation. First, the rate of mutation may differ among different regions of the genome, and the variation in rates of synonymous substitution may simply reflect the chromosomal position of the gene (Wolfe et al. 1989a). This possibility is further supported by the fact that the genome of eukaryotes is made up of segments of distinct GC content called isochores which may be replicated independently and consequently may exhibit different rates of mutation (Chapter 8). The second reason may be that, in some genes, not all synonymous codons are equivalent in fitness. As a result, some synonymous substitutions may be selected against. Such purifying selection will create variation in the rate of synonymous substitution among genes. However, although purifying selection has been shown to affect the synonymous substitution rate as well as the pattern of usage of synonymous codons in the genomes of bacteria, yeast, and *Drosophila*, it is not clear whether this type of selection operates in mammals (see page 91).

It has also been noticed that there is a positive correlation between the synonymous and nonsynonymous substitution rates in a gene (Graur 1985; Li et al. 1985b). This may be explained by assuming either that the rate of mutation varies among genes (and hence some genes will have both high synonymous and nonsynonymous rates of substitution) or that the extent of selection at synonymous positions is affected by the nucleotide composition at adjacent nonsynonymous positions (Ticher and Graur 1989).

A CASE OF POSITIVE SELECTION: LYSOZYME IN COWS AND LANGURS

As discussed in the preceding sections, the rates and patterns of nucleotide substitutions in the vast majority of genes and nongenic regions of the genome can be explained by a combination of (1) mutational input, (2) random genetic drift of neutral or nearly neutral alleles, and (3) purifying selection against deleterious alleles. In the case of lysozyme, however, positive selection for advantageous mutations has been shown to have played a role in several mammalian lineages.

Foregut fermentation has independently arisen twice in the evolution of placental mammals, once in the ruminants (e.g., cows) and once in the colobine monkeys (e.g., langurs). In both cases, lysozyme, which in other mammals is not normally secreted in the stomach, has been recruited to degrade the walls of bacteria, which carry on the fermentation in the foregut. Stewart and Wilson (1987) compared the amino acid sequences of lysozyme from cows, langurs, baboons, humans, rats, horses, and chickens (Table 4). They noted that there are four uniquely shared amino acids between cows and langurs. There are two possible explanations for this observation. First, it may be that cows are evolutionarily more closely related to langurs than to horses, so that the uniquely shared amino acids merely represent the unchanged amino acid sequence in their common ancestor. This assumption of a close phylogenetic relationship between cows and langurs is known to be wrong. Alternatively, the uniquely shared amino acids in these two species could be the result of a series of parallel substitutions that occurred independently in both lineages. Indeed, when the sequence of amino acid substitu-

Table 4. Pairwise comparison of lysozyme sequences among different species.[a]

	Species					
Species	Langur	Baboon	Human	Rat	Cow	Horse
Langur		14	18	38	32	65
Baboon	0		14	33	39	65
Human	0	1		37	41	64
Rat	0	1	0		55	64
Cow	4	0	0	0		71
Horse	0	0	0	0	1	

From Stewart and Wilson (1987).

[a] The numbers above the diagonal are the numbers of amino acid differences between species and those below the diagonal are the numbers of uniquely shared residues between species.

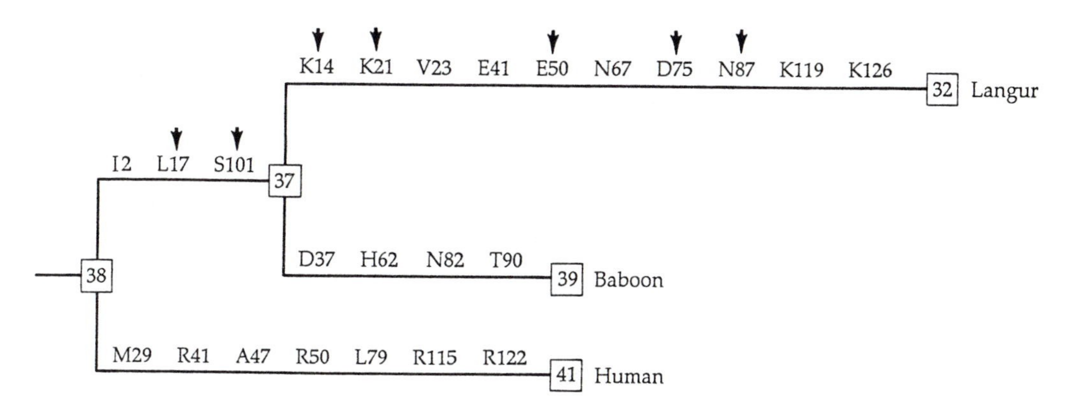

Figure 5. Parallel or convergent amino acid replacements in cow and langur lysozymes. The lengths of the lineages are proportional to the numbers of amino acid replacements along them. Each replacement is denoted by a one-letter abbreviation of the resultant amino acid (Table 1 in Chapter 1) followed by the position number at which the replacement occurred. Arrows point to seven replacements in langurs that occurred in parallel or in convergence to those in the cow lineage. The numbers of amino acid differences from cow stomach lysozyme (not shown) are in boxes. From Stewart and Wilson (1987).

tions was reconstructed, Stewart and Wilson (1987) found seven parallel or convergent substitutions in the cow and langur lineages (Figure 5).

Moreover, it has been determined that some of these substitutions contribute to a better performance of lysozyme at low pH values, such as those found in the ruminant digestive system. Conversely, both langur and cow lysozyme performed less well than human lysozyme at higher pH values. In conclusion, it seems safe to deduce that we are dealing here with a case of parallel occurrence of advantageous substitutions in different evolutionary lines showing parallel adaptations to similar selective agents.

MOLECULAR CLOCKS

In their comparative studies of hemoglobin and cytochrome *c* protein sequences from different species, Zuckerkandl and Pauling (1962, 1965) and Margoliash (1963) first noticed that the rates of amino acid substitution in these proteins were approximately the same among various mammalian lineages. Zuckerkandl and Pauling (1965) therefore proposed that for any given protein the rate of molecular evolution is approximately constant over time in all lineages or, in other words, that there exists a **molecular clock**. This proposal immediately stimulated a great deal of interest in the use of macromolecules in evolutionary studies. Indeed, if proteins evolve at constant rates, then they can be used to determine dates of species divergence and to reconstruct phylogenetic relationships among organisms. This would be similar to the dating of geological times by measuring the decay of radioactive elements.

The molecular clock hypothesis also stimulated a great deal of controversy. Classical evolutionists, for instance, argued against it because the suggestion of rate constancy does not sit well with the erratic tempo of evolution at the morphological and physiological levels. The hypothesis met with particularly heated opposition when the rate-constancy assumption was used to obtain an estimate of 5 million years for the divergence time between humans and the African apes (Sarich and Wilson 1967), in sharp contrast to the prevailing view among paleontologists at that time that humans and apes diverged at least 15 million years ago. Many molecular evolutionists have also challenged the validity of the molecular clock hypothesis. In particular, Goodman (1981) and his associates (Czelusniak et al. 1982) contended that the rate of evolution often accelerates following gene duplication and that protein sequences evolve much more rapidly at times of adaptive radiation. For example, they claimed that extremely high rates of amino acid substitution occurred following the gene duplication separating α and β hemoglobins and that the high rates were due to advantageous mutations that improved the function of hemoglobin.

Although the rate-constancy assumption has always been controversial, it has been widely used in the estimation of divergence times and in the reconstruction of phylogenetic trees (Nei 1975; Wilson et al. 1977). Thus, the validity of the molecular clock hypothesis is a vital issue in molecular evolution. The rapid accumulation of DNA sequence data in recent years affords an unprecedented opportunity for testing the hypothesis. Such data allow a closer examination of the hypothesis than do protein sequences and can be interpreted more directly than DNA–DNA hybridization and immunological distance data.

Relative-rate tests

The controversy over the molecular clock hypothesis often involves disagreements on dates of species divergence. To avoid this problem, Sarich and Wilson (1973) proposed a test that does not require knowledge of divergence times. The test, called the **relative-rate test**, is illustrated in Figure 6. Suppose that we want to compare the rates in lineages A and B. Then, we use a third species, C, as a reference. We should be certain that the reference species branched off earlier than the divergence between species A and B. For example, to compare the rates in the human and orangutan lineages we can use a monkey species as a reference.

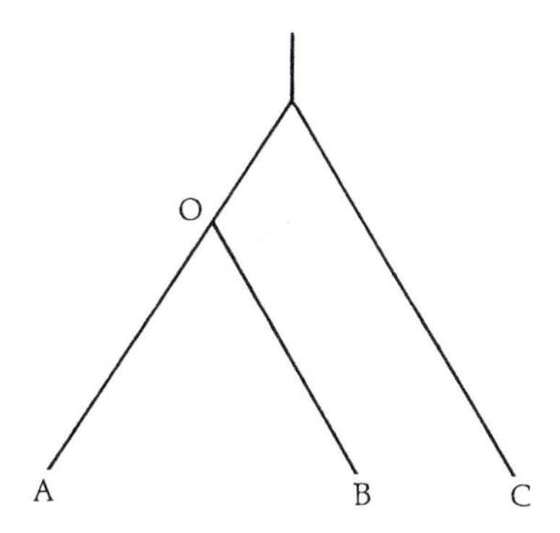

Figure 6. Phylogenetic tree used in the relative rate test. O denotes the common ancestor of species A and B.

From Figure 6, it is easy to see that the number of substitutions between species A and C, K_{AC}, is equal to the sum of substitutions that have occurred from point O to point A (K_{OA}) and from point O to point C (K_{OC}). That is,

$$K_{AC} = K_{OA} + K_{OC} \qquad \textbf{(4.3a)}$$

Similarly,

$$K_{BC} = K_{OB} + K_{OC} \qquad \textbf{(4.3b)}$$

and

$$K_{AB} = K_{OA} + K_{OB} \qquad \textbf{(4.3c)}$$

Since K_{AC}, K_{BC}, and K_{AB} can be directly estimated from the nucleotide sequences (Chapter 3), we can easily solve the three equations to find the values of K_{OA}, K_{OB}, and K_{OC}:

$$K_{OA} = (K_{AC} + K_{AB} - K_{BC})/2 \qquad \textbf{(4.4a)}$$

$$K_{OB} = (K_{AB} + K_{BC} - K_{AC})/2 \qquad \textbf{(4.4b)}$$

$$K_{OC} = (K_{AC} + K_{BC} - K_{AB})/2 \qquad \textbf{(4.4c)}$$

We can now decide whether the rates of substitution are equal in lineages A and B by comparing the value of K_{OA} with that of K_{OB}. The time that has passed since species A and B last shared a common ancestor is by definition equal for both lineages. Thus, according to the molecular clock hypothesis, K_{OA} and K_{OB} should be equal, i.e., $K_{OA} - K_{OB} = 0$. From Equations 4.3a and 4.3b, we note that $K_{OA} - K_{OB} = K_{AC} - K_{BC}$. Therefore, we can compare the rates of substitution in A and B directly from K_{AC} and K_{BC}.

Table 5. Differences in the number of synonymous substitutions per 100 sites ($K_{AC} - K_{BC}$) between mice (species A) and rats (species B).[a]

Gene	L	K_{AB}	K_{AC}	K_{BC}	$K_{AC} - K_{BC}$
Apolipoprotein E	201	7.4	61.3	59.5	1.8 ± 5.3
Actin α	249	17.9	58.2	59.1	−0.9 ± 4.8
Actin β	233	19.7	50.1	45.1	5.0 ± 4.6
Thy-1 antigen	116	19.3	51.8	57.3	−5.5 ± 6.9
Lactate dehydrogenase A	219	30.9	80.4	80.3	0.1 ± 8.2
Glycoprotein hormone, α subunit	58	30.8	97.7	84.3	13.4 ± 18.5
Insulin-like growth factor II	130	4.8	37.0	40.9	−3.9 ± 2.8
Atrial natriuretic factor	107	20.4	69.7	57.4	12.3 ± 8.3
Growth hormone	124	14.1	80.9	79.2	1.7 ± 7.7
Thyroglobulin β	90	25.7	77.4	92.7	−15.3 ± 12.9
Proopiomelanocortin	154	21.4	61.5	52.7	8.8 ± 6.5
Aldolase A	184	15.4	57.5	63.3	−5.8 ± 5.3
Creatine kinase M	251	17.2	48.6	52.2	−3.6 ± 4.3
Metallothionein II	35	19.0	45.5	36.7	8.8 ± 10.2
Total	2,187	19.0	59.8	59.4	0.4 ± 1.5

From Li et al. (1987a).

[a] L = Number of sites compared. K_{ij} = number of substitutions per 100 sites between species i and j. Humans are the reference species (C) in all cases except creatine kinase M, for which a rabbit sequence was used.

Nearly equal rates in mice and rats

Table 5 shows a comparison of the rates of synonymous substitution in mice and rats using the relative-rate test. In all cases, species A is the mouse and species B is the rat. Therefore, a positive sign for the value of $K_{AC} - K_{BC}$ means that the rate in mice is higher than that in rats, whereas a negative sign indicates that the opposite is true.

Because of data limitations, we have used either human or rabbit sequences as references instead of sequences derived from more closely related species to mice and rats, such as hamsters or guinea pigs. As a consequence, the estimate of $K_{AC} - K_{BC}$ is subject to large statistical errors (Table 5). Nevertheless, it is quite clear that the substitution rates in mice and rats are nearly equal. In other words, when all sequences are considered together, the rate difference is close to 0. The same conclusion holds for the rates of nonsynonymous substitution in these two species (Li et al. 1987a).

Lower rates in humans than in monkeys

Based on immunological distance and protein sequence data, Goodman (1961) and colleagues (Goodman et al. 1971) suggested that a rate slowdown occurred in hominoids (humans and apes) after their separation from the Old World monkeys. Wilson et al. (1977), however, contended that the slowdown is an artifact, owing to the use of an erroneous estimate of the ape–human divergence time. They conducted relative-rate tests using both immunological distance data and protein sequence data and concluded that there was no evidence for a hominoid slowdown.

DNA sequence data provide a better resolution of the above controversy. In Table 6, the relative-rate test is used to compare the rate of nucleotide

Table 6. Differences in the number of nucleotide substitutions per 100 sites ($K_{AC} - K_{BC}$) between the Old World monkey lineage (A) and the human lineage (B).[a]

Sequence	No. of sites	K_{AB}	$K_{AC} - K_{BC}$
η-globin pseudogene	2,000	7.4	2.1 ± 0.7**
SYNONYMOUS SITES			
β-globin	71	8.9	2.8 ± 5.6
Apolipoprotein A-I	158	7.9	−5.3 ± 4.8
Erythropoietin	145	11.2	5.1 ± 5.9
α1-antitrypsin	140	10.9	6.7 ± 6.8
Insulin	84	18.6	−7.5 ± 7.2
INTRONS			
δ-globin	601	4.7	3.4 ± 1.4
UNTRANSLATED AND FLANKING REGIONS			
β-globin	179	4.6	1.2 ± 1.7
δ-globin	172	8.8	6.1 ± 3.2
Total	3,550	6.7	2.3 ± 0.6**

From Li et al. (1987a).

[a] The reference species used were owl monkey (η-globin pseudogene), lemur (β- and δ-globins), mouse or rat (erythropoetin, apolipoprotein A-I, and α1-antitrypsin), and dog (insulin).

**Significantly different from 0 at the 1% level.

substitution in the human lineage with that in the Old World monkey lineage. In all tests, lineage B is the human lineage, while lineage A is the monkey lineage. Lineage C is the reference species (see table legend). Therefore, a positive sign for the rate difference ($K_{AC} - K_{BC}$) means that the human lineage has evolved more slowly and a negative sign means the opposite.

We note that there are only two negative signs out of the nine sequences used. The difference in rate is highly significant even when only the η pseudogene is used. When all sequences are considered together, $K_{AC} - K_{BC} = 2.3\%$, and $K_{AB} = 6.7\%$. Therefore, the K value (K_{OA}) in the Old World monkey lineage is $(6.7\% + 2.3\%)/2 = 4.5\%$, and that in the human lineage (K_{OB}) is only $6.7 - 4.5 = 2.2\%$, indicating that the Old World monkey lineage evolved $4.5/2.2 \approx 2$ times faster than the human lineage.

Higher rates in rodents than in primates

Wu and Li (1985) applied the relative-rate test to compare the rates of substitution in the rodent and human lineages, using either the artiodactyl or the carnivore lineage as a reference. They concluded that the rate of synonymous substitution is about two times higher in the rodent lineage than in the human lineage. Note, however, that the estimated difference in rate refers to the long-term average (i.e., from the time of the rodent–primate split to the present). Since at the time of divergence and shortly thereafter the rates of substitution in the two lineages would have been similar, the rate difference must have increased with time. Therefore, the rate difference

Table 7. Rates of synonymous substitution per site per year in primates and rodents.[a]

Species pair	Number of sites	Percent divergence	Substitution rate ($\times 10^9$)
Human vs. chimpanzee	921	1.9	1.3 (0.9–1.9)[b]
Human vs. Old World monkeys	998	11.0	2.2 (1.8–2.8)
Mouse vs. rat	3,886	23.7	7.9 (3.9–11.8)

From Li et al. (1987a).

[a] The divergence times used are 7 (5–10) million years ago for the human-chimpanzee split, 25 (20–30) million years ago for the human–Old World monkey split, and 15 (10–30) million years ago for the mouse-rat split.

[b] Values in parentheses are the range of rate estimates obtained from the upper and lower estimates of divergence times.

value estimated above is probably an underestimate. To know the rate difference in more recent times, we need to estimate the substitution rates among rodents and among primates separately.

Table 7 shows a comparison of substitution rates in primates and rodents. The average rate between human and chimpanzee sequences is 1.3×10^{-9} substitutions per site per year, while that between human and Old World monkey sequences is 2.2×10^{-9} substitutions per site per year, if we assume that the human–chimpanzee split occurred 7 million years ago and that the human–Old World monkey split occurred 25 million years ago. This result is consistent with the conclusion that the Old World monkey lineage has evolved two times faster than the human lineage. The average rate between mouse and rat sequences is 7.9×10^{-9} substitutions per site per year if we assume that the mouse–rat split occurred 15 million years ago. Therefore, the rate in rodents could be four to six times higher than that in higher primates. Although there are uncertainties about the divergence times used, it is clear that rodent sequences evolved much faster than primate sequences.

Causes of variation in substitution rates among evolutionary lineages

The higher substitution rates in monkeys than in humans and the higher rates in rodents than in primates may be explained by the so-called **generation-time effect** (Kohne 1970). The generation time in rodents is much shorter than that in man, so the number of germ-line DNA replications per year could be many times higher in rodents than in man if the number of germ-line replications per generation is not very different in these organisms. Since mutations accumulate mostly during the process of DNA replication, the more cycles of replication there are, the more mutational errors will occur. This factor may largely explain the higher substitution rates in rodents than in humans. Similarly, monkeys have a shorter generation time than humans and so might be expected to have a higher substitution rate.

The differences in substitution rates could also be partly due to differences in the efficiency of the DNA repair system (Britten 1986). There is limited data indicating that rodents have a less efficient DNA repair system than humans and, consequently, accumulate more mutations per replication cycle.

The above results should not be taken as evidence that no molecular clock exists. We note that differences in rates of substitution are observed between organisms with very different generation times. When organisms with similar generation times such as mice and rats are compared, the rate-constancy holds fairly well (see page 82). Thus, although there is no global clock for the mammals, local clocks may exist for many groups of relatively closely related species.

RATES OF SUBSTITUTION IN ORGANELLE DNA

In comparison with nuclear genomes, organelle genomes are much smaller and easier to investigate experimentally. Moreover, the discovery of an exceptionally high rate of substitution in the mammalian mitochondrial genome (Brown et al. 1979) stimulated further interest in the evolution of organelle DNA.

The mammalian mitochondrial genome consists of a circular, double-stranded DNA about 15,000–17,000 base pairs (bp) long, approximately 1/10,000 of the smallest animal nuclear genome. It contains only unique (i.e., nonrepetitive) sequences: 13 protein-coding genes, two rRNA genes, 22 tRNA genes, and a control region that contains sites for replication and transcription initiation. The genome is structurally very stable, as is evident from the small variation in genome size among mammalian species.

In sharp contrast, plant mitochondrial genomes exhibit much more structural variability. They undergo frequent rearrangements, duplications, and deletions (Palmer 1985). For this reason, the genome size varies from 40,000 bp to 2,500,000 bp. The mitochondrial genome in plants can be linear or circular, and in many cases the genetic information is divided into separate DNA molecules, referred to as subgenomic circles. The coding content of plant mitochondria has not been fully determined; however, we do know that there are three rRNA-specifying genes, an undetermined number of tRNA genes, and about 15–30 protein-coding genes, some of which have been identified. (Structural genes may appear in multiple copies in plant mitochondrial genomes.) At the present time, there is no indication of qualitative variation in the coding content of plant mitochondrial genomes, despite the immense variability in genome size.

The chloroplast genome in vascular plants is circular and varies in size from about 120,000 to 220,000 bp, with an average size of 150,000 bp (Palmer 1985). Despite this large variation in size, the genome is known to be structurally stable. The chloroplast genome of the tobacco plant, *Nicotiana tabacum* has been completely sequenced (Shinozaki et al. 1986). It is a circular molecule, 155,844 bp long. It contains 37 tRNA genes (eight of which contain single introns), eight rRNA genes, and 45 protein-coding genes (five of which contain single introns and two of which contain two introns). Both strands are used for coding. The chloroplast genome of *Nicotiana tabacum* also contains 59 additional open reading frames of unknown function, two of which are interrupted by introns.

The synonymous rate of substitution in mammalian mitochondrial genes has been estimated to be 5.7×10^{-8} substitutions per synonymous site per year (Brown et al. 1982). This is about 10 times the value for synonymous substitutions in nuclear protein-coding genes. The rate of nonsynonymous substitution varies greatly among the 13 protein-coding genes, but it is always much higher than the average nonsynonymous rate for nuclear genes. The reason for these high rates of substitution in mammalian mitochondria seems to be a high rate of mutation relative to the nuclear rate. The high mutation rate is due to (a) a low fidelity of the DNA-replication process in mitochondria, (b) absence of repair or the existence of a very inefficient repair mechanism, and (c) a high concentration of mutagens (e.g., superoxide radicals, O_2^-) resulting from the metabolic functions performed by the mitochondria. On the other hand, the stringency of purifying selection against nonsynonymous mutations seems to be of the same order of magnitude as that operating on nuclear genes.

Early studies based on a few gene sequences or on restriction-enzyme mapping have suggested that chloroplast genes have lower rates of nucleotide substitution than mammalian nuclear genes (Curtis and Clegg 1984; Palmer 1985) and that plant mitochondrial DNA evolves slowly in terms of nucleotide substitution, though it undergoes frequent sequence rearrangements (Palmer and Hebron 1987). These results have recently been confirmed by more extensive analyses of DNA sequences (Wolfe et al. 1987, 1989b).

Table 8 shows a comparison of the substitution rates in the three genomes of higher plants. The average numbers of substitutions per nonsynonymous site (K_A) in the chloroplast and mitochondrial genomes are similar, but the average number of substitutions per synonymous site (K_S) in the chloroplast genome is almost three times that in the mitochondrial genome for the

Table 8. Comparison of the rates of nucleotide substitution in plant chloroplast, mitochondrial, and nuclear genes.[a]

Genomes	K_S	L_S	K_A	L_A
COMPARISON BETWEEN MONOCOT AND DICOT SPECIES				
Chloroplast genes	0.58 ± 0.02	4,177	0.05 ± 0.00	14,421
Mitochondrial genes	0.21 ± 0.01	1,219	0.04 ± 0.00	4,380
COMPARISON BETWEEN MAIZE AND WHEAT OR BARLEY				
Nuclear genes	0.71 ± 0.04	1,475	0.06 ± 0.00	5,098
Chloroplast genes	0.17 ± 0.01	2,068	0.01 ± 0.00	7,001
Mitochondrial genes	0.03 ± 0.01	413	0.01 ± 0.00	1,526

From Wolfe et al. (1987, 1989b).

[a] K_S, number of substitutions per synonymous site; K_A, number of substitutions per nonsynonymous site; L_S, number of synonymous sites; and L_A, number of nonsynonymous sites.

comparison between monocot and dicot species, and it is six times higher for the comparisons between maize and wheat or barley. In the following, the former ratio will be used because it is based on a larger data set. The average synonymous substitution rate in plant nuclear genes is about four times that in chloroplast genes. Thus, the synonymous substitution rates in plant mitochondrial, chloroplast, and nuclear genes are in the approximate ratio of 1:3:12.

If we take the divergence time between maize and wheat to be 50–70 million years (Stebbins 1981; Chao et al. 1984), the nuclear data in Table 8 indicate an average synonymous rate of 5.1–7.1 $\times$ 10^{-9} substitutions per site per year. This is similar to the rates of synonymous substitution seen in mammalian nuclear genes (Table 1).

Interestingly, the rate of nucleotide substitution does not correlate well with the rate of structural changes in the genome of organelles. In mammals, the mitochondrial DNA evolves very rapidly in terms of nucleotide substitutions, but the spatial arrangement of genes and the size of the genome are fairly constant among species. In contrast, the mitochondrial genome of plants undergoes frequent structural changes, but the rate of nucleotide substitution is extremely low. In chloroplast DNA, both the rates of nucleotide substitution and structural evolution are very low. The lack of correlation between the rates of substitution and the rates of structural evolution suggests that the two processes occur independently.

PATTERN OF NUCLEOTIDE SUBSTITUTION IN PSEUDOGENES

Since point mutation is one of the most important factors in the evolution of DNA sequences, molecular evolutionists have long been interested in determining the **pattern of spontaneous mutation** (e.g., Beale and Lehmann 1965; Zuckerkandl et al. 1971). This pattern can serve as a standard for inferring how far the observed frequencies of interchange between nucleotides in any given DNA sequence have deviated from the values expected under no selection, i.e., under **selective neutrality.**

One way to study the pattern of point mutation is to examine the pattern of substitution in regions of DNA that are subject to no selective constraints. Pseudogenes are particularly useful in this respect. Since they are devoid of function, all mutations occurring in pseudogenes are selectively neutral and become fixed in the population with equal probability. Thus, the pattern of nucleotide substitution in pseudogenes are expected to reflect the pattern of spontaneous point mutation.

Figure 7 shows a simple method for inferring the nucleotide substitutions in a pseudogene sequence (Gojobori et al. 1982; Li et al. 1984). Sequence 1

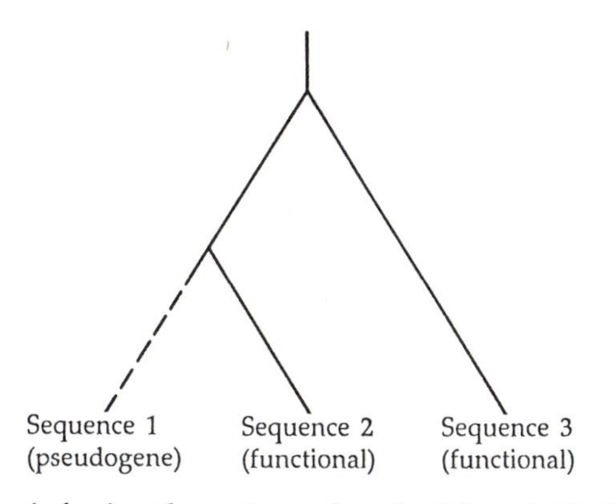

Figure 7. A tree for inferring the pattern of nucleotide substitution in a pseudogene sequence. The dashed line implies "nonfunctional."

is a pseudogene, sequence 2 is its functional counterpart from the same species, and sequence 3 is a functional sequence that has diverged before the emergence of the pseudogene. Suppose that, at a certain nucleotide site, sequences 1 and 2 have A and G, respectively. Then, we can assume that the nucleotide in the pseudogene sequence has changed from G to A if sequence 3 has G, but that the nucleotide in sequence 2 has changed from A to G if sequence 3 has A. However, if sequence 3 has T or C, then we cannot decide the direction of change, and in this case the site is excluded from the comparison. Since the rate of substitution is usually much higher in pseudogenes than that in the homologous functional gene, differences in the nucleotide sequence between a gene and a pseudogene are expected to have occurred in the pseudogene in the majority of cases.

The matrix in Table 9 represents the combined pattern of substitution inferred from 13 mammalian pseudogene sequences. Each entry, f_{ij}, in the matrix represents the expected number of base changes from i to j among every 100 substitutions in a random sequence (i.e., in a sequence in which the four bases are equally frequent). For example, $f_{AT} = 4.7$, i.e., 4.7% of all substitutions are from A to T.

We note that the direction of mutation is nonrandom. For example, A changes more often to G than to either T or C. The four elements from the upper right corner to the lower left corner are the f_{ij} values for transitions, while the other eight elements represent transversions. All transitions, and in particular C → T and G → A, occur more often than transversions. The sum of the relative frequencies of transitions is 59.2% (54.4% if CG dinucleotides are excluded; see below). We note that under random mutation the expected proportion of transitions is only 33%, for there are only four types of transitions but eight types of transversions. The observed proportion is almost twice the value expected under random mutation.

Table 9. Pattern of substitution in pseudogenes.[a]

From	To: A	T	C	G	Row totals
A	—	4.7 ± 1.3 (5.3 ± 1.4)	5.0 ± 0.7 (5.6 ± 0.8)	9.4 ± 1.3 (10.3 ± 1.4)	19.1 (21.2)
T	4.4 ± 1.1 (4.8 ± 1.1)	—	8.2 ± 1.3 (9.2 ± 1.3)	3.3 ± 1.2 (3.6 ± 1.3)	15.9 (17.6)
C	6.5 ± 1.1 (7.1 ± 1.3)	21.0 ± 2.1 (18.2 ± 2.3)	—	4.2 ± 0.5 (4.2 ± 0.6)	31.7 (29.5)
G	20.7 ± 2.2 (18.6 ± 1.9)	7.2 ± 1.1 (7.7 ± 1.3)	5.3 ± 1.0 (5.5 ± 1.3)	—	33.2 (31.8)
Column totals:	31.6 (30.5)	32.9 (38.4)	18.5 (20.3)	16.9 (18.1)	

From Gojobori et al. (1982) and Li et al. (1984).

[a] Table entries are the inferred percentages (f_{ij}) of base changes from i to j based on 13 mammalian pseudogene sequences. Values in parentheses were obtained by excluding all CG dinucleotides from comparison.

We also note that some nucleotides are more mutable than others. In the last column of Table 9, we list the relative frequencies of all mutations from A, T, C, and G to any other nucleotide. Were all the four nucleotides equally mutable, we would expect a value of 25% in each of the column's elements. In practice, we see that G mutates with a relative frequency of 33.2% (i.e., G is a highly mutable nucleotide) while T mutates with a relative frequency of 15.9% (i.e., it is not as mutable). In the bottom row of Table 9, we list the relative frequencies of all mutations that result in A, T, C, or G. We note that 64.5% of all mutations result in either A or T, while the random expectation is 50%. Since there is a tendency for C and G to change frequently to A or T, and since A and T are not as mutable as C and G, pseudogenes are expected to become rich in A and T. This should also be true for other noncoding regions that are subject to no functional constraint. Indeed, noncoding regions are generally found to be AT-rich.

The results in Table 9 are expressed in terms of the sense strand, i.e., the untranscribed strand. Thus, a change from G to A actually means that a G:C pair is replaced by an A:T pair. This can occur as a result of either a G mutating to A in the sense strand or a C to T mutation in the complementary strand. Similarly, a change from C to T can occur as a result of either a C mutating to T in one strand or a G mutating to A in the other. If there is no difference in the pattern of mutation between the two strands, we should have $f_{GA} = f_{CT}$. Similarly, we should obtain $f_{AG} = f_{TC}$, $f_{AT} = f_{TA}$, $f_{AC} = f_{TG}$, $f_{CA} = f_{GT}$, and $f_{CG} = f_{GC}$. These equalities hold only approximately, and indeed there may be minor asymmetries in the mutation pattern between the two strands. The asymmetries could be due to differences in the replication mechanism between the leading strand and the lagging strand during DNA replication (Wu and Maeda 1987).

It is known that, in addition to base mispairing, the transition from C to T can also arise from conversion of methylated C residues to T residues upon deamination (Coulondre et al. 1978; Razin and Riggs 1980). The effect will elevate the frequencies of C:G → T:A and G:C → A:T; i.e., f_{CT} and f_{GA}. Since about 90% of methylated C residues in vertebrate DNA occur at 5′—CG—3′ dinucleotides (Razin and Riggs, 1980), this effect should be expressed mainly as changes of the CG dinucleotide to TG or CA. After a gene becomes a pseudogene, such changes would no longer be subject to any functional constraint and can therefore contribute significantly to C → T and G → A transitions if the frequency of CG is relatively high before **silencing** of the gene (i.e., loss of function) occurs. The substitution pattern obtained by excluding all nucleotide sites where the CG dinucleotides appear to have occurred in the ancestral sequences of these pseudogenes is given in parentheses in Table 9. This pattern is probably more suitable for predicting the pattern of mutations in a sequence that has not been subject to functional constraints for a long time (e.g., some parts of an intron), because in such a sequence there would exist few CG dinucleotides. The pattern obtained after excluding the CG dinucleotides is somewhat different from that obtained otherwise. In particular, the differences among the relative frequencies of the four transitions become somewhat less conspicuous, and the relative frequencies of the transversions become slightly higher, except for G → C and C → G.

NONRANDOM USAGE OF SYNONYMOUS CODONS

Because of the degeneracy of the genetic code, most of the 20 amino acids are encoded by more than one codon (Chapter 1). Since synonymous mutations do not cause any change in amino acid sequence and since natural selection was thought to operate predominantly at the protein level, synonymous mutations were proposed as candidates for selectively neutral mutations (Kimura 1968b; King and Jukes 1969). However, if all synonymous mutations are indeed selectively neutral, the synonymous codons for an amino acid should be used with more or less equal frequency. As DNA sequence data accumulated, however, it became evident that the usage of synonymous codons is distinctly nonrandom in both prokaryotic and eukaryotic genes (Grantham et al. 1980). In fact, in many yeast and *Escherichia coli* genes the bias in usage is highly conspicuous. For example, 21 of the 23 leucine residues in the *E. coli* outer membrane protein II (*ompA*) are encoded by the codon CUG, though there exist five other codons for leucine. Such a bias cannot be explained by nonrandom mutation. How to explain the widespread phenomenon of nonrandom codon usage became a controversial issue, to which, fortunately, some clear answers seem to be emerging.

An observation that has been helpful for understanding the phenomenon of nonrandom usage is that genes in an organism or in related species generally show the same pattern of choices among synonymous codons (Grantham et al. 1980). Thus, mammalian, *E. coli*, and yeast genes fall into distinct classes of codon usage. Grantham et al. (1980) therefore proposed the **genome hypothesis**. According to this hypothesis, the genes in any given genome use the same coding strategy with respect to choices among synonymous codons, i.e., the bias in codon usage is species-specific. The genome hypothesis turns out to be true in general, though there is conside ble heterogeneity in codon usage among genes in a genome (see below).

Studies of codon usage in *E. coli* and yeast have greatly increased our understanding of the factors that affect the choice of synonymous codons. Post et al. (1979) found that *E. coli* ribosomal-protein genes preferentially use synonymous codons that are recognized by the most-abundant tRNA species. They suggested that the preference resulted from natural selection because using a codon that is translated by an abundant tRNA species will increase translational efficiency and accuracy. Their finding prompted Ikemura (1981, 1982) to gather data on the relative abundances of tRNA species in *E. coli* and the yeast *Saccharomyces cerevisiae*. He showed that, in both species, a positive correlation exists between the relative frequencies of the synonymous codons in a gene and the relative abundances of their cognate tRNA species. The correlation is very strong for highly expressed genes. For instance, in *E. coli* the most abundant of the four leucine tRNAs is $tRNA_1^{Leu}$, which recognizes the CUG codon, and the *ompA* gene uses predominately this codon for leucine (see above).

Figure 8 shows schematically the correspondence between the frequencies of the six leucine codons and the relative abundances of their cognate tRNAs. In *E. coli*, $tRNA_1^{Leu}$ is the most abundant leucine tRNA species, and indeed in highly expressed genes, CUG (the codon recognized by this tRNA) is much more frequently used than the other five codons. On the other hand, in yeast, the most abundant leucine tRNA species is $tRNA_3^{Leu}$, and the codon recognized by this tRNA (UUG) is the predominant codon. In contrast, in genes with low levels of expression, the correspondence between tRNA abundance and the use of the respective codon is much weaker in both species (Figure 8).

The importance of translational efficiency in determining the codon usage

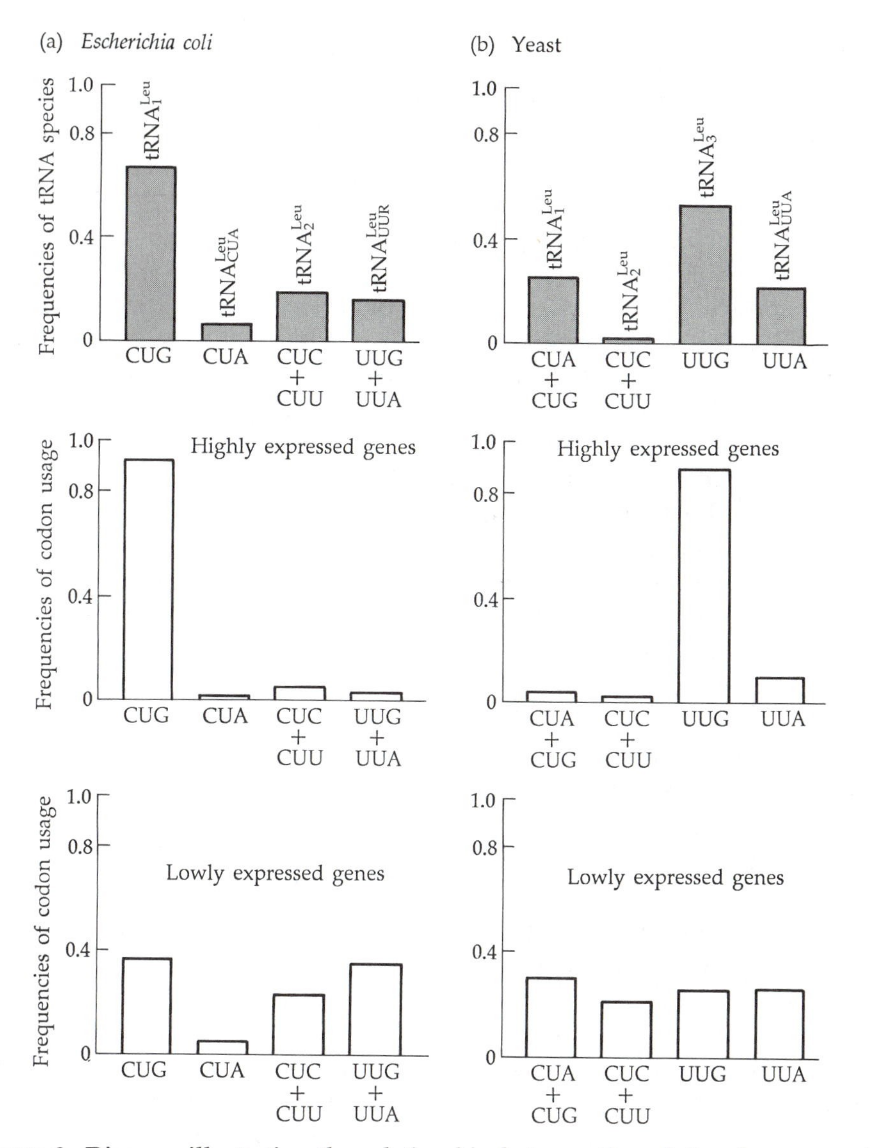

Figure 8. Diagram illustrating the relationship between the relative frequency of codon usage for leucine (open bars) and the relative abundance of the corresponding cognate tRNA species (solid bars) in (a) *Escherichia coli* and (b) *Sacharomyces cerevisiae*. The plus signs (e.g., between codons CUC and CUU for *E. coli*) indicate that each of these pairs of codons is recognized by a single tRNA species (e.g., $tRNA_2^{Leu}$ for CUC and CUU in *E. coli*).

pattern in highly expressed genes is further supported by the following observation (Ikemura 1981). It is known that codon–anticodon pairing involves **wobbling** at the third position. For example, U in the first position of anticodons can pair with both A and G. Similarly, G can pair with both C and U. On the other hand, C in the first anticodon position can only pair with G at the third position of codons and A can only pair with U. Wobbling is also made possible by the fact that some tRNAs contain modified bases at the first anticodon position, and these can recognize more than one codon. For example, inosine (a modified adenine) can pair with any of the three bases U, C, and A. Interestingly, most tRNAs that can recognize more than one codon exhibit differential preferences for one of them. For example, 4-thiouridine (S^4U) in the wobble position of an anticodon can recognize both A and G in the wobble position; however it has a marked preference for A-terminated codons over G-terminated ones. Such a preference should be reflected in highly expressed genes. The two condons for lysine in *E. coli* are recognized by a tRNA molecule that has S^4U in the wobble position of the anticodon, and, indeed, in the *E. coli ompA* gene, 15 of the 19 lysine codons are AAA, and only four are AAG.

Table 10 shows part of an extensive compilation of codon usage by Sharp et al. (1988). For each group of synonymous codons, if the usage is equal, the relative frequency of each codon should be 1. This is clearly not so in the majority of cases. Moreover, in both *E. coli* and yeast the codon-usage bias is much stronger in highly expressed genes than in lowly expressed ones. A simple explanation for this difference is that in highly expressed genes, selection for translation efficiency and accuracy is strong, so that codon-usage bias is pronounced. In lowly expressed genes, on the other hand, selection is relatively weak so that the usage pattern is mainly affected by mutation pressure and random genetic drift, and therefore, is less skewed (Sharp and Li 1986).

In conclusion, in *E. coli* and yeast, the choice of synonymous codons is constrained by tRNA availability and other factors related to translational efficiency. These constraints will result in purifying selection, thus slowing down the rate of synonymous substitution (Ikemura 1981; Kimura 1983). In fact, it has been shown that the rate of synonymous substitution in enterobacterial genes is negatively correlated with the degree of codon-usage bias (Sharp and Li 1986). Therefore, the phenomenon of nonrandom usage of synonymous codons may not be taken as evidence against the neutral theory of molecular evolution, since it can be explained in terms of the principle that stronger selective constraints result in lower rates of evolution (see page 74 and Kimura 1983).

Table 10 also shows that in *Drosophila* the codon usage bias is much stronger in highly expressed genes than in lowly expressed ones, thus indicating that selection for translational efficiency also plays an important role in determining the choice of synonymous codons in this organism.

In many human genes, codons tend to end in either G or C (i.e., to have a high GC content at the third position), whereas other genes have a low

Table 10. Codon usage in four species.[a]

Amino acid	Codon	*Escherichia coli*		*Saccharomyces cerevisiae*		*Drosophila melanogaster*		Human	
		High	Low	High	Low	High	Low	G + C	A + T
Leu	UUA	0.06	1.24	0.49	1.49	0.03	0.62	0.05	0.99
	UUG	0.07	0.87	5.34	1.48	0.69	1.05	0.31	1.01
	CUU	0.13	0.72	0.02	0.73	0.25	0.80	0.20	1.26
	CUC	0.17	0.65	0.00	0.51	0.72	0.90	1.42	0.80
	CUA	0.04	0.31	0.15	0.95	0.06	0.60	0.15	0.57
	CUG	5.54	2.20	0.02	0.84	4.25	2.04	3.88	1.38
Val	GUU	2.41	1.09	2.07	1.13	0.56	0.74	0.09	1.32
	GUC	0.08	0.99	1.91	0.76	1.59	0.93	1.03	0.69
	GUA	1.12	0.63	0.00	1.18	0.06	0.53	0.11	0.80
	GUG	0.40	1.29	0.02	0.93	1.79	1.80	2.78	1.19
Ile	AUU	0.48	1.38	1.26	1.29	0.74	1.27	0.45	1.60
	AUC	2.51	1.12	1.74	0.66	2.26	0.95	2.43	0.76
	AUA	0.01	0.50	0.00	1.05	0.00	0.78	0.12	0.64
Phe	UUU	0.34	1.33	0.19	1.38	0.12	0.86	0.27	1.20
	UUC	1.66	0.67	1.81	0.62	1.88	1.14	1.73	0.80
Met	AUG	1.00	1.00	1.00	1.00	1.00	1.00	1.00	1.00

From Sharp et al. (1988).

[a] For each group of synonymous codons, the sum of the relative frequencies equals the number of codons in the group. For example, there are six codons for leucine, and so the sum of the relative frequencies for these six codons should be 6. Under equal usage, the relative frequencies for each codon in a group should be 1, and so the degree of deviation from one indicates the degree of bias in usage. "High" and "low" denote genes with high and low levels of expression. For humans, "G + C" means high-GC regions, and "A + T" means high-AT regions.

third-position GC content. However, there are several reasons why this bias may not be related to the level of gene expression. First, the α- and β-globin genes have different GC contents at the third position of codons (high and low, respectively), although they are both expressed in the same tissues (erythrocytes) in approximately equal quantities and, therefore, should have the same level of expression. Second, in chicken genes, the frequency with which codons are used does not correlate well with tRNA availability (Ouenzar et al. 1988), though this observation may not be directly applicable to human genes. Finally, the GC content at the third codon position is strongly correlated with the GC level in both flanking regions and in introns (Chapter 8; Bernardi and Bernardi 1985; Aota and Ikemura 1986). For example, the α-globin gene is high in GC and resides in a high-GC region, whereas the β-globin gene is low in GC and resides in a low-GC region (Chapter 8; Bernardi et al. 1985). Thus, it appears that the codon-usage bias in a human gene is largely determined by the GC content in the region that contains the gene. As will be discussed in Chapter 8, whether the GC content in a region is determined by natural selection or mutational bias is still a controversial issue. However, since the GC content at the third position of codons in a gene tends to be higher than in its surrounding regions (Chapter 8; Aota and Ikemura 1986), it is possible that the codon-usage pattern in human genes is affected to some extent by natural selection. Further studies are needed to shed more light on the factors that affect codon usage in humans.

PROBLEMS

1. Why is the denominator of Equation 4.1 $2T$ rather than T?

2. Figure 9 shows the DNA sequences of the first and second exons of the θ1-globin gene from the olive baboon (*Papio anubis*) and the orangutan (*Pongo pygmaeus*). Compute the number of substitutions per site separately for each of the three codon positions by using the one-parameter model. Which position evolves fastest? Why?

3. Figure 10 shows the DNA sequences of the first intron in the θ1-globin gene from the olive baboon and the orangutan. Compute the number of substitutions by using (a) the one-parameter model and (b) the two-parameter model. Do the two estimates differ from one another? In comparison with the results you obtained in Problem 2, does this intron evolve faster or slower than the three codon positions in the exons?

4. Figure 11 shows part of the DNA sequences of the first intron in the nucleolin gene from the hamster, the rat and the mouse. By using the relative-rate test with the hamster sequence as a reference, determine whether there is a difference in the rates of substitution between the rat and the mouse lineages.

5. Two strains of the AIDS virus, denoted as WMJ1 and WMJ2, were isolated from a two-year-old child on October 3, 1984 and January 15, 1985 (Hahn et al. 1986). The child was presumed to have been infected only once (perinatally by her mother). The number of synonymous substitutions per synonymous site in the envelope (*env*) gene between these two isolates was 0.0164 (Li et al. 1988). (a) Obtain a maximum estimate of the rate of synonymous substitution by assuming that WMJ2 evolved directly from WMJ1 and that the two sequences split on October 3, 1984. (b) Obtain a minimum estimate of the rate of synonymous substitution for the gene by assuming that the two strains started diverging from each other at the time of infection and had evolved independently for two years. How much faster are these rates of substitution than the synonymous rate of substitution for the average mammalian gene (Table 1)?

b: ATG GCG CTG TCC GCG GAG GAC CGG GCGGCT GTG CGC GCC CTG

o: ATG GCG CTG TCC GCG GAG GAC CGG GCGCTG GTG CGT GCC CTG

b: TGG AAG AAA CTG GGA AGC AAT GTT GGCGTC TAT GCT ACT GAG

o: TGG AAG AAG CTG GGC AGC AAC GTC GGCGTC TAC ACG ACA GAG

b: GCC CTG GAG AGG ACC TTC CTG GCT TTCCCC GCC ACG AAG ACC

o: GCC CTG GAG AGG ACC TTC CTG GCC TTCCCC GCA ACG AAG ACC

b: TAC TTC TCC CAC CTA GAC CTG AGC CCCGGC TCC GCC CAG GTT

o: TAC TTC TCC CAC CTG GAC CTG AGC CCCGGC TCC TCA CAG GTC

b: AGA GCA CAC GGC CAG AAG GTG GCG GACGCG CTG AGC CTC GCC

o: AGA GCC CAC GGC CAG AAG GTG GCG GACGCG CTG AGC CTC GCC

b: GTG GAG CGC CTA GAC GAC CTA CCC CGCGCG CTG TCC GCT CTG

o: GTG GAG CGC CTG GAC GAC CTA CCC CACGCG CTG TCC GCG CTG

b: AGC CAT CTG CAC GCT TGC CAG CTG CGAGTG GAC CCA GCT AAC

o: AGC CAC CTG CAC GCG TGC CAG CTG CGAGTG GAC CCG GCC AGC

b: TTC CCG

o: TTC CAG

Figure 9. DNA sequences of exons 1 and 2 in the θ1-globin gene from the olive baboon (b) and the orangutan (o). Data from Shaw et al. (1987) and Marks et al. (1986).

b: TGCGGCGAGGCTGGGCGCCCCGCCCTCCGGGGCCCTGCCTCCCCAAGCC

o: TGCGGCGAGGCTGGGCGCCCCGCCCCC-AGGGCCCTCCCTCCCCAAGCC

b: CCCCGGACGCGCCTCACCGCCGTTCCTCTCGCAG

o: CCCCGGACTCGCCTCACCCACGTTCCTCTCGCAG

Figure 10. DNA sequence of the first intron in the θ1-globin gene from olive baboon (b) and orangutan (o). A single gap is marked by –. Data from Shaw et al. (1987) and Marks et al. (1986).

```
m:  GTAAGAGGCCTGGCGCGCCGACGCGGACGACTAGGCCTGCTTTCGGAGGG
r:  GTAATAGGCCTGACGCGCGAACACGGACGACTAGGCCTGCTTTCTGAGAG
h:  GTGAGAGGCCTCGCGCGCGCCGACGGACGGACGGGCCTGCTTTCTGAGGG

m:  GCGCGCGCGCCGTCGCGGAGGGGAGGAGGGCTTGCGCGCAATCCCGGGCG
r:  GCGCGCGCGCCGTCGCGGAGGGGAGGAGGGCCTGCGCACAGTCCCGGGCG
h:  GCGCGCGCGCGGTCGCTCAGGGGAGGAGGGCCTGCGCGCAATCCCGGGCG

m:  CGTTCGAGGGCGCCAGCTGGGGAAGTCTCGCGCGACTAGCGGGAGGTCTC
r:  CGTTCGAGGGCGCATGCTGGGGAAGTCTCGCGCGACTAGCGGAGGGTCTC
h:  CGTTCGAGGGCGCATGCTGGGGAAGTCTCGCGCGACTAGCGGAGGGTCTC
```

Figure 11. Partial DNA sequences of the first intron in the nucleolin gene from mice (m), rats (r) and hamsters (h). Gaps (deletions and insertions) have been omitted. Data from Bourbon et al. (1988).

6. Find a complete cDNA or gene sequence from each of *Escherichia coli*, yeast, and humans. For each gene, compile a table of codon usage (i.e., the number of times each codon is used in the gene). Is the codon usage biased? In what respect? Is the pattern of codon usage similar in the three genes? If not, what are the differences? By using a χ^2 test, determine whether in each gene the deviation from equal codon usage in the valine codon family is statistically significant.

FURTHER READINGS

Ikemura, T. 1985. Codon usage and tRNA content in unicellular and multicellular organisms. Mol. Biol. Evol. 2:13–34.

Kimura, M. 1983. *The Neutral Theory of Molecular Evolution*. Cambridge University Press, Cambridge.

MacIntyre, R. J. (ed.). 1985. *Molecular Evolutionary Genetics*. Plenum, New York.

Nei, M. 1987. *Molecular Evolutionary Genetics*. Columbia University Press, New York.

Sharp, P. M. and W.-H. Li. 1986. An evolutionary perspective on synonymous codon usage in unicellular organisms. J. Mol. Evol. 24: 28–38.

Steinhauer, D. A. and J. J. Holland. 1987. Rapid evolution of RNA viruses. Annu. Rev. Microbiol. 41: 409–433.

5

MOLECULAR PHYLOGENY

Molecular phylogeny is the study of evolutionary relationships among organisms by using techniques of molecular biology. It is one of the areas of molecular evolution that have generated much interest in the last decade, mainly because in many cases phylogenetic relationships are difficult to assess any other way. The purpose of this chapter is to explain how to reconstruct a phylogenetic tree from molecular data and to give some examples in which the molecular approach has been able to provide a much clearer resolution of long-standing phylogenetic issues than was possible with the traditional approaches.

IMPACT OF MOLECULAR DATA ON PHYLOGENETIC STUDIES

The study of molecular phylogeny began at the turn of the century, even before Mendel's laws were rediscovered in 1900. Immunochemical studies showed that serological cross-reactions were stronger for closely related organisms than for distantly related ones. The evolutionary implications of these findings were used by Nuttall (1904) to infer phylogenetic relationships among various groups of animals. For example, he determined that man's closest relatives were the apes, followed, in order of relatedness, by the Old World monkeys, the New World monkeys, and the prosimians.

Since the late 1950's, various techniques have been developed in molecular biology, and this started the extensive use of molecular data in phylogenetic studies. In particular, in the 1960's and 1970's the study of molecular phylogeny by using protein sequence data progressed tremendously. Less expensive and more expedient methods, such as protein electrophoresis, DNA–DNA hybridization, and immunological methods, though less accurate than protein sequencing, were extensively used to study the phylogenetic relationships among populations or closely related species. The application of these methods also stimulated the development of measures of genetic distance and tree-making methods (see Fitch and Margoliash 1967; Nei 1975; Felsenstein 1988).

The rapid accumulation of DNA sequence data since the late 1970's has already had a great impact on molecular phylogeny. DNA sequence data are not only more abundant, but also easier to analyze than protein sequence data. Thus, they have been used, on the one hand, to infer the phylogenetic relationships among such closely related species as humans and apes (see page 119), and on the other hand, to study very ancient evolutionary occurrences, such as the origin of mitochondria and chloroplasts (see page 126) and the divergence of phyla and kingdoms (Woese 1987). In the future, DNA sequencing is likely to resolve many of the long-standing problems in phylogenetic studies, such as the evolutionary relationships among bacteria and unicellular eukaryotes (Sogin et al. 1986, 1989), which could not be resolved by any other method of traditional evolutionary inquiry. Indeed, molecular data have proved so powerful in the study of evolutionary history that we may eventually be able to reconstruct a fairly complete phylogeny of the major groups of the living world.

Of course, we should not abandon traditional means of evolutionary inquiry, such as morphology, anatomy, physiology, and paleontology. Rather, different approaches provide complementary data. We note that taxonomy is based mainly on morphological and anatomical data and that paleontological information is the only data that can provide a time frame for evolutionary study.

PHYLOGENETIC TREES

All life forms on earth, both extant and extinct, share a common origin, and their ancestries can be traced back to one or a few organisms that lived approximately 4 billion years ago. Consequently, all animals, plants, and bacteria are related by descent to each other. Closely related organisms are descended from more recent common ancestors than are distantly related ones. The objectives of phylogenetic studies are (1) to reconstruct the correct genealogical ties between organisms and (2) to estimate the time of divergence between organisms since they last shared a common ancestor.

In phylogenetic studies, the evolutionary relationships among a group of organisms are illustrated by means of a **phylogenetic tree**. A phylogenetic tree is a graph composed of nodes and branches, in which only one branch connects any two adjacent nodes (Figure 1). The **nodes** represent the taxonomic units, and the **branches** define the relationships among the units in terms of descent and ancestry. The branching pattern of a tree is called the **topology**. The **branch length** usually represents the number of changes that have occurred in that branch. The taxonomic units represented by the nodes can be species, populations, individuals, or genes.

When dealing with phylogenetic trees, we distinguish between **external nodes** and **internal nodes**. For example, in Figure 1a nodes A, B, C, D, and E are external, whereas all others are internal. External nodes represent the extant taxonomic units under comparison and are referred to as **operational taxonomic units (OTUs)**. Internal nodes represent ancestral units.

Figure 1 illustrates two common ways of drawing a phylogenetic tree. In Figure 1a, the branches are **unscaled**; their lengths are not proportional to the number of changes, which are indicated on the branches. This presentation allows us to line up the extant OTUs and also to place the nodes representing divergence events on a time scale when the times of divergence are known or have been estimated. In Figure 1b, the branches are **scaled,** and their lengths are proportional to the numbers of changes.

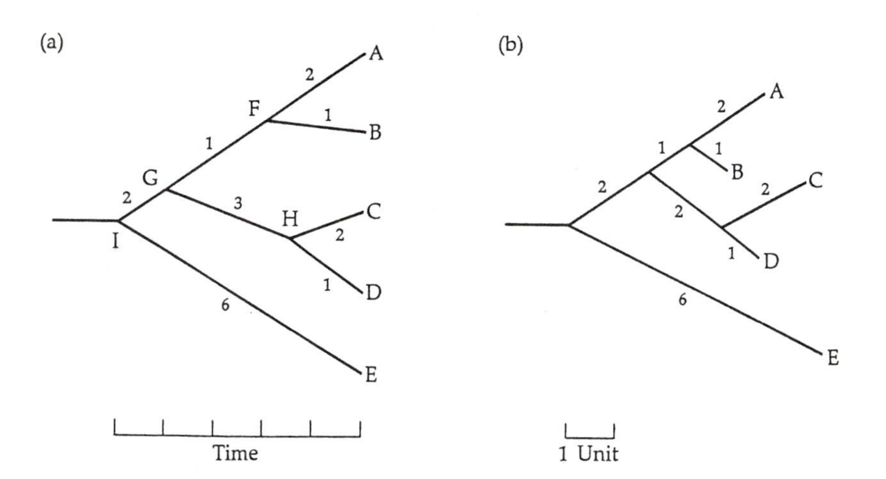

Figure 1. Two alternative representations of a phylogenetic tree for five OTUs. (a) Unscaled branches: extant OTUs are lined up and nodes are positioned proportionally to times of divergence. (b) Scaled branches: lengths of branches are proportional to the numbers of molecular changes.

A tree is said to be **additive** if the distance between any two OTUs is equal to the sum of the lengths of all the branches connecting them. For example, if additivity holds, the distance between OTUs A and C in Figure 1a should be equal to 2 + 1 + 3 + 2 = 8. The distance between two OTUs is calculated directly from molecular data (e.g., DNA sequences) while the branch lengths are estimated from the distances between OTUs according to certain rules (see page 114). Additivity usually does not hold if multiple substitutions have occurred at any nucleotide sites (Figure 5 in Chapter 3).

A node is **bifurcating** if it has only two immediate descendant lineages, but **multifurcating** if it has more than two immediate descendant lineages. For simplicity, we shall only consider bifurcating trees like those in Figure 1.

Rooted and unrooted trees

Phylogenetic trees can be either **rooted** or **unrooted** (Figure 2). In a rooted tree there exists a particular node, called the root (R in Figure 2a), from which a unique path leads to any other node. The direction of each path corresponds to evolutionary time, and the root is the common ancestor of all the OTUs under study. An unrooted tree is a tree that only specifies the relationships among the OTUs and does not define the evolutionary path (Figure 2b).

For three species, there are three different possible rooted trees but only one unrooted tree (Figure 3). The number of bifurcating rooted trees (N_R) for n OTUs is given by

$$N_R = \frac{(2n-3)!}{2^{n-2}(n-2)!} \tag{5.1}$$

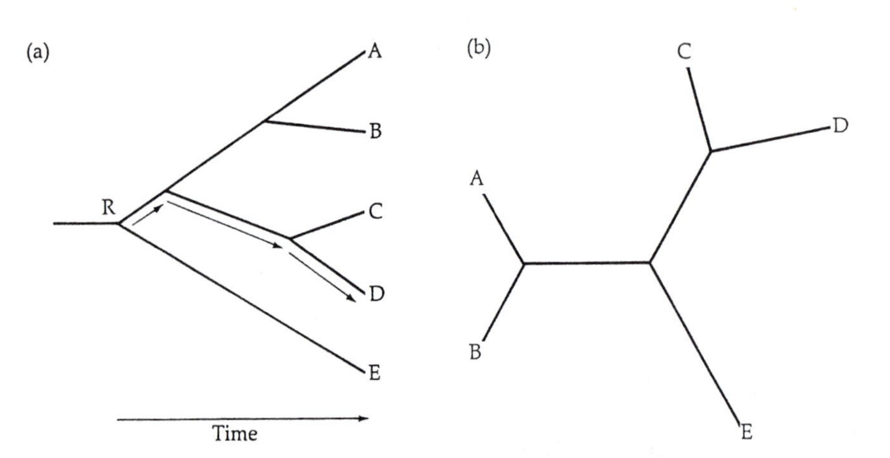

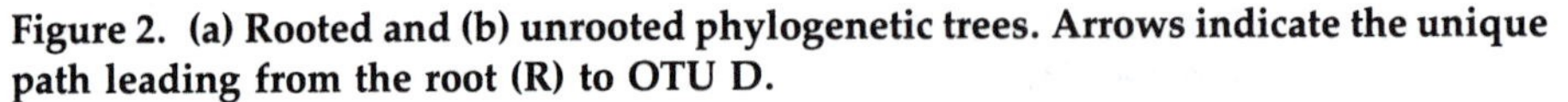
Figure 2. (a) Rooted and (b) unrooted phylogenetic trees. Arrows indicate the unique path leading from the root (R) to OTU D.

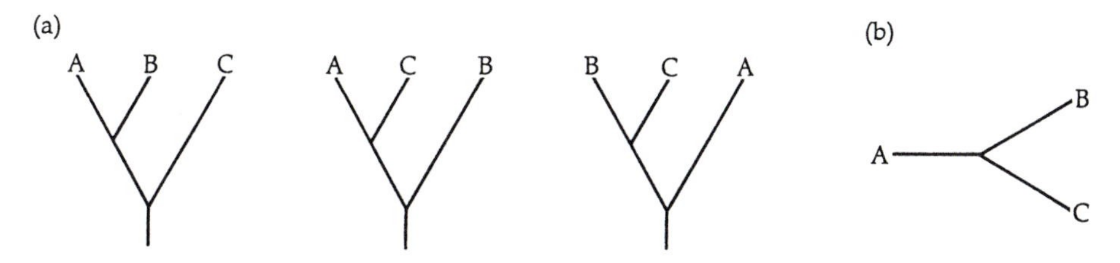

Figure 3. From three OTUs it is possible to construct three different rooted trees (a), but only one unrooted tree (b).

for $n \geq 2$. The number of bifurcating unrooted trees (N_U) for $n \geq 3$ is

$$N_U = \frac{(2n - 5)!}{2^{n-3}(n - 3)!} \tag{5.2}$$

Note that the number of possible unrooted trees for n OTUs is equal to the number of possible rooted trees for $n - 1$ OTUs. The numbers of possible rooted and unrooted trees for up to 10 OTUs are given in Table 1. We see that both N_U and N_R increase very rapidly with n, and for 10 OTUs there are already more than 2 million bifurcating unrooted trees and close to 35 million rooted ones. Since only one of these trees represents correctly the true evolutionary relationships among the OTUs, it is usually very difficult to infer the true phylogenetic tree when n is large.

Table 1. Possible numbers of rooted and unrooted trees for 1–10 OTUs.

Number of OTUs	Number of rooted trees	Number of unrooted trees
2	1	1
3	3	1
4	15	3
5	105	15
6	954	105
7	10,395	954
8	135,135	10,395
9	2,027,025	135,135
10	34,459,425	2,027,025

From Felsenstein (1978).

True and inferred trees

The sequence of speciation events that has led to the formation of any group of OTUs is historically unique. Thus, only one of all the possible trees that can be built with a given number of OTUs represents the true evolutionary history. Such a phylogenetic tree is called a **true tree.** A tree that is obtained by using a certain set of data and a certain method of tree reconstruction is called an **inferred tree**. An inferred tree may or may not be identical with the true tree.

Gene trees and species trees

A phylogenetic tree that represents the evolutionary pathways of a group of species is called a **species tree**. When a phylogenetic tree is constructed from one gene from each species, the inferred tree is a **gene tree** (Nei 1987). It can differ from the species tree in two respects. First, the divergence of two genes sampled from two different species can predate the divergence of the two species (Figure 4). This will result in an overestimate of the branch length but will not present a serious problem if we are concerned with long-term evolution, in which the component of divergence due to genetic polymorphism within species may be ignored.

The second problem with gene trees is that the branching pattern of a gene tree (i.e., its topology) may be different from that of the species. Figure

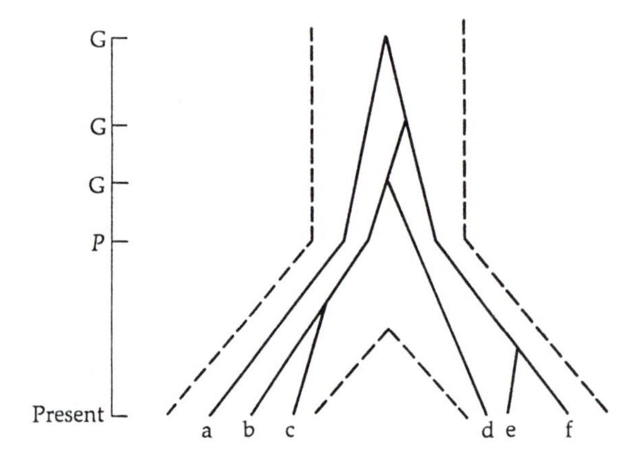

Figure 4. Diagram showing that gene splitting (G) usually occurs earlier than population splitting (P) if the population is genetically polymorphic at time P. The evolutionary history of gene splitting resulting in the six alleles denoted *a–f* is shown in solid lines, and population splitting is shown in broken lines. From Nei (1987).

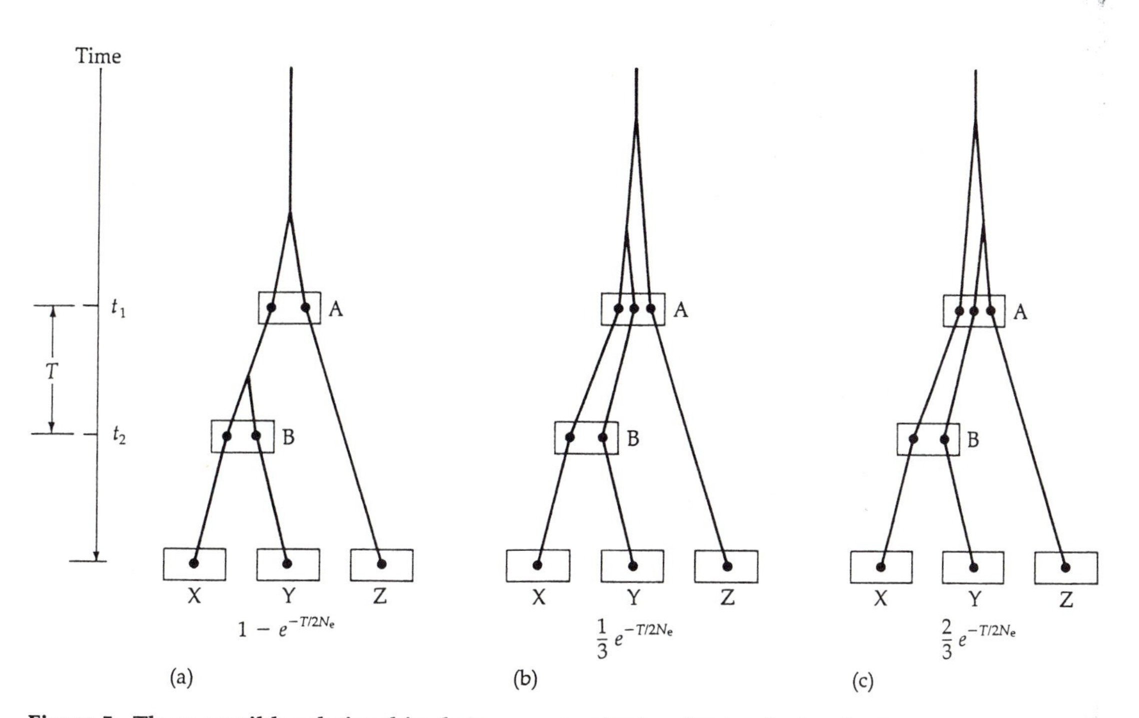

Figure 5. Three possible relationships between a species tree (rectangles) and a gene tree (dots). In (a) and (b), the topologies of the species trees are identical to those of the gene trees. Note that in (a) the time of divergence between the genes is roughly equal to the time of divergence between the populations. In (b), on the other hand, the time of divergence between genes X and Y greatly predates the time of divergence between the respective populations. The topology of the gene tree in (c) is different from that of the species tree. For neutral alleles, the probability of occurrence of each tree is given underneath the tree. t_1 is the time at which the first speciation event occurred and t_2 is the time at which the second speciation event occurred. $T = t_1 - t_2$, and N_e is the effective population size. From Nei (1987).

5 shows three different possible relationships between the two trees. The topologies of gene trees (a) and (b) are identical with those of the corresponding species trees (e.g., X and Y form a cluster). Gene tree (c), however, is different from the true species tree, since now Y and Z are sister groups. The probability of obtaining the erroneous tree (c) is quite high when the time interval between the first and second species splitting ($t_1 - t_0$) is short, as is probably true in the case of the phylogenetic relationships among humans, chimpanzees, and gorillas. To avoid this type of error, one needs to use many genes in the reconstruction of the phylogeny. A large amount of data is also required to avoid stochastic errors, which can occur because nucleotide substitutions occur randomly, so that, for instance, lineage Z in Figure 5a may have by chance accumulated fewer substitutions than lineages X and Y, despite the fact that it has branched off earlier in time.

METHODS OF TREE RECONSTRUCTION

Numerous tree-making methods have been proposed in the literature. For a detailed treatment, readers may consult Sneath and Sokal (1973), Nei (1987), and Felsenstein (1988). Here we describe four methods that have been frequently used in phylogenetic studies. For simplicity, we consider nucleotide sequence data, but the methods described are equally applicable to other types of molecular data such as amino acid sequence data.

The methods described below can be classified into two types: **distance matrix methods** and **maximum parsimony methods**. In the distance matrix methods, evolutionary distances (usually the number of nucleotide or amino acid substitutions separating two taxonomic units) are computed for all pairs of taxa, and a phylogenetic tree is constructed by using an algorithm based on some functional relationships among the distance values. In maximum parsimony methods, character states (e.g., the nucleotide or amino acid at a site) are used, and the shortest pathway leading to these character states is chosen as the phylogenetic tree.

Unweighted pair group method with arithmetic mean (UPGMA)

The **unweighted pair group method with arithmetic mean** (**UPGMA**) is the simplest method for tree reconstruction. It was originally developed for constructing taxonomic phenograms, i.e., trees that reflect the phenotypic similarities between OTUs (Sokal and Michener 1958, and see page 113), but it can also be used to construct phylogenetic trees if the rates of evolution are approximately constant among the different lineages so that an approximately linear relation exists between evolutionary distance and divergence

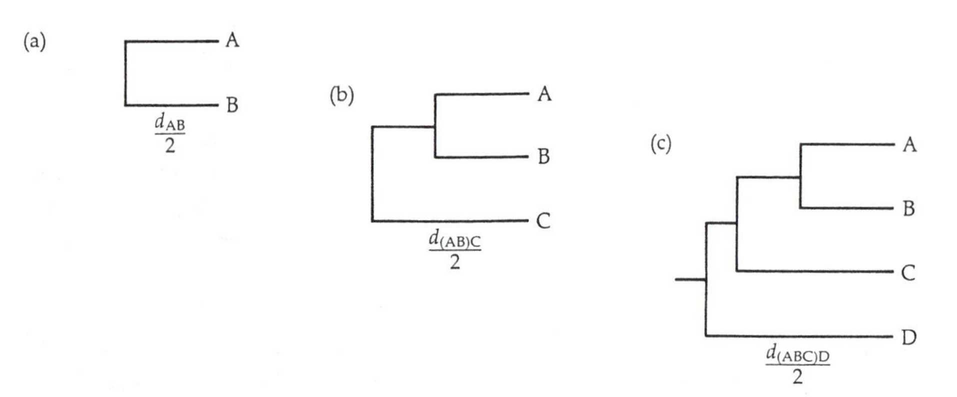

Figure 6. Diagram illustrating the stepwise construction of a phylogenetic tree for four OTUs according to the UPGMA method (see text).

time (Nei 1975). For such a relation to hold, linear distance measures such as the number of nucleotide (or amino acid) substitutions should be used.

The UPGMA method employs a sequential clustering algorithm, in which local topological relationships are identified in order of similarity, and the phylogenetic tree is built in a stepwise manner. In other words, we first identify from among all the OTUs the two OTUs that are most similar to each other and treat these as a new single OTU. Such an OTU is referred to as a **composite OTU.** Subsequently, from among the new group of OTUs we identify the pair with the highest similarity, and so on, until we are left with only two OTUs.

To illustrate the method let us consider a case of four OTUs. The pairwise evolutionary distances, such as Jukes and Cantor's (1969) estimates (Chapter 3), are given by the following matrix:

	OTU		
OTU	A	B	C
B	d_{AB}		
C	d_{AC}	d_{BC}	
D	d_{AD}	d_{BD}	d_{CD}

In this matrix, d_{ij} stands for the distance between OTUs i and j. The first two OTUs to be clustered are the ones with the smallest distance. Let us assume that d_{AB} has the smallest value. Consequently, OTUs A and B are the first to be clustered, and the branching point is positioned at a distance of $d_{AB}/2$ substitutions (Figure 6a).

Following the first clustering, A and B are considered as a single composite OTU, and a new distance matrix is computed:

	OTU	
OTU	(AB)	C
C	$d_{(AB)C}$	
D	$d_{(AB)D}$	d_{CD}

In this matrix, $d_{(AB)C} = (d_{AC} + d_{BC})/2$, and $d_{(AB)D} = (d_{AD} + d_{BD})/2$. In other words, the distance between a simple OTU and a composite OTU is the average of the distances between the simple OTU and the constituent simple OTUs of the composite OTU. If $d_{(AB)C}$ turns out to be the smallest distance in the new matrix, then OTU C will be joined to the composite OTU (AB) with a branching node at $d_{(AB)C}/2$ (Figure 6b).

The final step consists of clustering the last OTU, D, with the composite OTU (ABC). The root of the entire tree is positioned at $d_{(ABC)D}/2 = [(d_{AD} + d_{BD} + d_{CD})/3]/2$. The final tree inferred by using the UPGMA method is shown in Figure 6c.

In the UPGMA method, the distance between two composite OTUs is computed as the arithmetic mean of the distances between the constituent OTUs in each composite OTU. For example, the distance between a composite OTU (*ij*) and a composite OTU (*mn*) is

$$d_{(ij)(mn)} = (d_{im} + d_{in} + d_{jm} + d_{jn})/4 \quad \textbf{(5.3)}$$

In the case of composite OTUs (*ijk*) and (*mn*), the distance is

$$d_{(ijk)(mn)} = (d_{im} + d_{in} + d_{jm} + d_{jn} + d_{km} + d_{kn})/6 \quad \textbf{(5.4)}$$

Transformed distance method

If the assumption of rate constancy among lineages does not hold, UPGMA may give an erroneous topology. For example, suppose that the phylogenetic tree in Figure 7a is the true tree. The pairwise evolutionary distances are given by the following matrix:

	OTU		
OTU	A	B	C
B	8		
C	7	9	
D	12	14	11

By using the UPGMA method, we obtain an inferred tree that differs from the true tree in its branching pattern (Figure 7b). For example, OTUs A and

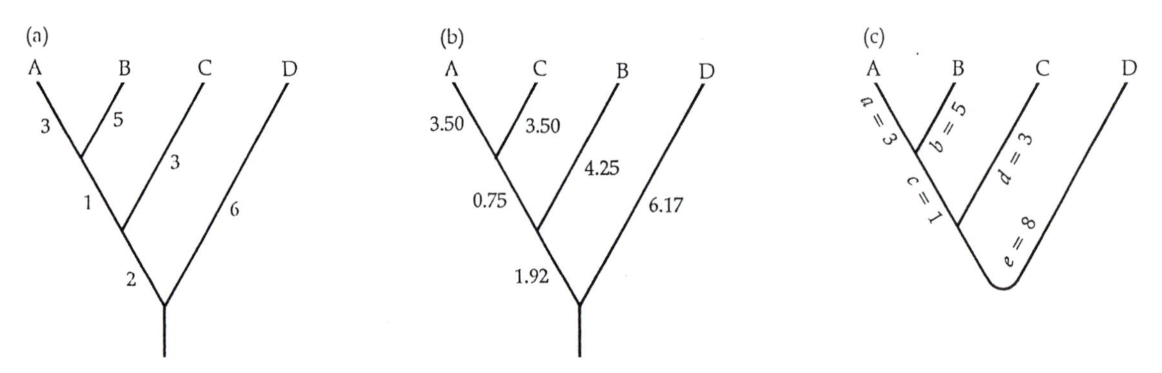

Figure 7. (a) The true phylogenetic tree. (b) The erroneous phylogenetic tree reconstructed by using the UPGMA method, which does not take into account the possibility of unequal substitution rates along different branches. (c) The tree inferred by the transformed distance method. The root must be between OTU D and the node of the common ancestor of OTUs A, B, and C, but its exact location cannot be determined.

C are grouped together, whereas in the true tree, A and B are sister OTUs. (Note that additivity does not hold in this case. For example, the true distance between A and B is 8, whereas the sum of the lengths of the estimated branches connecting A and B is 3.50 + 0.75 + 4.25 = 8.50.)

The topological errors might be remedied, however, by using a correction called the **transformed distance method** (Farris 1977; Klotz et al. 1979). In brief, this method uses an **outgroup** as reference to make corrections for unequal rates of evolution among the lineages under study and then applies UPGMA to the new distance matrix to infer the topology of the tree. An outgroup is an OTU for which we have external knowledge, such as taxonomic or paleontological information, that clearly shows it to have diverged from the common ancestor prior to all the other OTUs under consideration.

In the present case, let us assume that taxon D is an outgroup to all other taxa. D can then be used as a reference to transform the distances by the following equation:

$$d'_{ij} = [(d_{ij} - d_{i\mathrm{D}} - d_{j\mathrm{D}})/2] + \bar{d}_{\mathrm{D}} \qquad \textbf{(5.5)}$$

where d'_{ij} is the transformed distance, i = A, B, or C, and $\bar{d}_{\mathrm{D}} = (d_{\mathrm{AD}} + d_{\mathrm{BD}} + d_{\mathrm{CD}})/3$. The term $\bar{d}_{\mathrm{D}}$ was introduced to assure that all d'_{ij} values are positive. This is done because in practice a distance can never be negative. For the general case of n OTUs (not including the outgroup), $\bar{d}_{\mathrm{D}} = \Sigma d_{i\mathrm{D}}/n$.

In our example, $\bar{d}_{\mathrm{D}} = 37/3$, and the new distance matrix for taxa A, B, and C is:

	OTU	
OTU	A	B
B	10/3	
C	13/3	13/3

Since d'_{AB} has the smallest value, A and B are the first to be clustered together and, subsequently, C is added to the tree. By definition, the outgroup OTU, D, determines the root of the tree and is the last to be added. This gives the correct topology (Figure 7c). In the above example, we considered only three taxa with one outgroup but the method can be easily extended to more taxa and/or more outgroups.

In many instances, it is impossible to decide a priori which of the taxa under consideration is an outgroup. To overcome this difficulty, a two-stage approach has been proposed (Li 1981). In the first step, one infers the root of the tree by using the UPGMA method. After that, the taxa on one side of the root are used as references (outgroups) for making corrections for the unequal rates of evolution among the lineages on the other side of the root, and vice versa. In our example, this approach also identifies the correct tree.

Neighbors relation methods

In an unrooted bifurcating tree, two OTUs are said to be **neighbors** if they are connected through a single internal node. For example, in Figure 8a, A and B are neighbors and so are C and D. In comparison, in Figure 8b, neither A and C nor B and C are neighbors. However, if we combine OTUs A and B into one composite OTU, then the composite OTU (AB) and the simple OTU C become a new pair of neighbors.

Let us now assume that the tree shown in Figure 8a is the true tree. Then, if additivity holds, we should have

$$d_{AC} + d_{BD} = d_{AD} + d_{BC} = a + b + c + d + 2x = d_{AB} + d_{CD} + 2x \quad \textbf{(5.6)}$$

where x is the length of the internal branch. Therefore, the following two conditions hold:

$$d_{AB} + d_{CD} < d_{AC} + d_{BD} \quad \textbf{(5.7a)}$$

and

$$d_{AB} + d_{CD} < d_{AD} + d_{BC} \quad \textbf{(5.7b)}$$

These two conditions are collectively known as the **four-point condition.** They may hold even if additivity holds only approximately.

Conversely, for four OTUs with unknown phylogenetic relationships, the above two conditions can be used to identify the neighbors (A and B; C and D). Once the two pairs of neighbors are determined, so is the topology of the phylogenetic tree.

Sattath and Tversky (1977) proposed the following method for dealing with more than four OTUs. First, compute a distance matrix as in the case of UPGMA. For every possible quadruple, say OTUs i, j, m, and n, compute $d_{ij} + d_{mn}$, $d_{im} + d_{jn}$, and $d_{in} + d_{jm}$. Suppose that the first sum is the smallest; then, assign a score of 1 to both the pair i and j and the pair m and n. Pairs i and m, j and n, and j and m are assigned a score of 0. If, on the other hand,

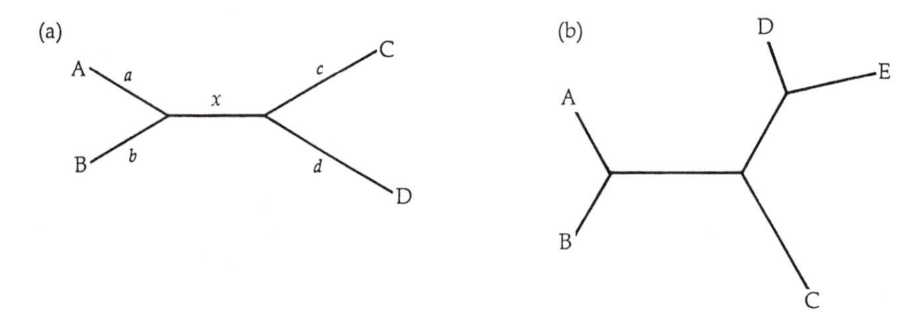

Figure 8. Bifurcating unrooted trees with (a) four OTUs, and (b) five OTUs.

$d_{im} + d_{jn}$ has the smallest value, then we assign the pair i and m and the pair j and n each a score of 1 and assign the other four possible pairs a score of 0. When all the possible quadruples are considered, the pair with the highest total score is selected as the first pair of neighbors and treated as a single OTU. Next, compute a new distance matrix as in the case of UPGMA and then repeat the same process to select the second pair of neighbors. This process is continued until all OTUs are clustered. A detailed illustration of this method will be given when we consider the phylogeny of apes and humans (see page 119).

Another method based on the neighbors relation concept has been proposed by Fitch (1981). Saitou and Nei (1987) proposed a method, called the **neighbor-joining method**, which sequentially identifies neighbor pairs that minimize the total length of the tree.

Maximum parsimony methods

The principle of maximum parsimony or minimum evolution involves the identification of a tree that requires the smallest number of evolutionary changes to explain the differences observed among the OTUs under study. Such a tree is called a **maximum parsimony tree**. Often more than one tree with the same minimum number of changes are found, so that no unique tree can be inferred.

The method discussed below was first developed for amino acid sequence data (Eck and Dayhoff 1966) and was later modified for use on nucleotide sequences (Fitch 1977).

We start with the definition of **informative sites**. A nucleotide site is phylogenetically informative only if it favors some trees over the others. To illustrate the distinction between informative and noninformative sites, consider the following four hypothetical sequences.

	Site								
Sequence	1	2	3	4	5	6	7	8	9
1	A	A	G	A	C	T	G	C	A
2	A	G	C	C	C	T	G	C	G
3	A	G	A	T	T	T	C	C	A
4	A	G	A	G	T	T	C	C	G
					*		*		*

There are three possible unrooted trees for four OTUs (Figure 9). Site 1 is not informative because all sequences at this site have A, so that no change is required in any of the three possible trees. At site 2, sequence 1 has A while all other sequences have G, and so a simple assumption is that the

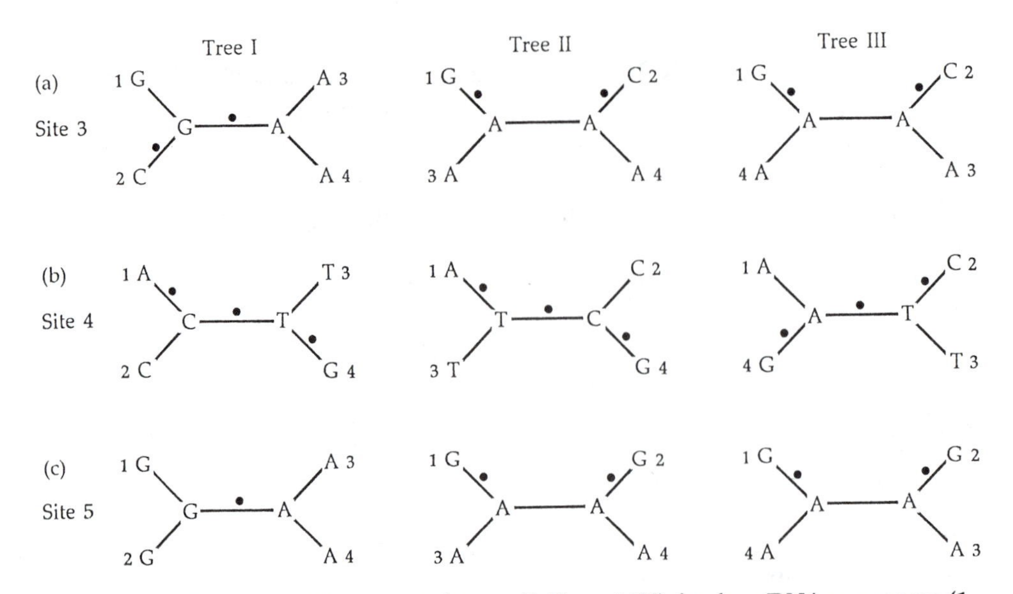

Figure 9. Three possible unrooted trees (I, II, and III) for four DNA sequences (1, 2, 3, and 4) that have been used to choose the most parsimonious tree (see text). The terminal nodes indicate the nucleotide type at homologous positions in the extant species. Each dot on a branch means a substitution is inferred on that branch. Note that the nucleotides at the two internal nodes of each tree represent one possible reconstruction from among several alternatives. For example, the nucleotide at both the internal nodes of tree III(c) (bottom right) can be G instead of A. In this case, the two substitutions will be positioned on the branches leading to species 3 and 4. However, the minimum number of required substitutions remains the same.

nucleotide has changed from G to A in the lineage leading to sequence 1. Thus, this site is also not informative, because each of the three possible trees requires 1 change. As shown in Figure 9a, for site 3 each of the three possible trees requires 2 changes and so it is also not informative. Note that if we assume that the nucleotide at the node connecting OTUs 1 and 2 in tree I in Figure 9a is C (or A) instead of G, the number of changes required for the tree remains 2. Figure 9b shows that for site 4 each of the three trees requires 3 changes and thus site 4 is also noninformative. For site 5, tree I requires only 1 change, whereas trees II and III require 2 changes each (Figure 9c). Therefore, this site is informative.

From these examples, we see that, as far as molecular data are concerned, a site is informative only when there are at least two different kinds of nucleotides at the site, each of which is represented in at least two of the sequences under study. In the above example, informative sites are indicated by an asterisk (*).

To infer a maximum parsimony tree, we first identify all the informative sites. Next, for each possible tree we calculate the minimum number of substitutions at each informative site. In the above example, for sites 5, 7, and 9, tree I requires 1, 1, and 2 changes respectively, tree II requires 2, 2, and 1 changes, and tree III requires 2, 2, and 2 changes. In the final step, we sum the number of changes over all the informative sites for each tree and choose the tree associated with the smallest number of substitutions. In our case, tree I is chosen because it requires the smallest number of changes (4) at the informative sites.

In the case of four OTUs, an informative site favors only one of the three possible alternative trees. For example, site 5 favors tree I over trees II and III, and is said to support tree I. It is easy to see that the tree supported by the largest number of informative sites is the maximum parsimony tree. For instance, in the above example, tree I is supported by 2 sites, tree II by one site, and tree III by none.

PHENETICS VERSUS CLADISTICS

A long-standing controversy in taxonomy has been the often acrimonious dispute between "cladists" and "pheneticists." The term **cladistics** can be defined as the study of the pathways of evolution. In other words, cladists are interested in such questions as: how many branches there are among a group of organisms; which branch connects to which other branch; and what is the branching sequence (Sneath and Sokal 1973). A tree-like network that expresses such ancestor–descendant relationships is called a **cladogram**. To put it another way, a cladogram refers to the topology of a rooted phylogenetic tree.

On the other hand, **phenetics** is the study of relationships among a group of organisms on the basis of the degree of similarity between them, be that similarity molecular, phenotypic, or anatomical. A tree-like network expressing phenetic relationships is called a **phenogram**. While a phenogram may serve as an indicator of cladistic relationships, it is not necessarily identical to the cladogram. If there is a linear relationship between the time of divergence and the degree of genetic (or morphological) divergence, the two types of trees may become identical to each other.

Among the methods discussed above, the maximum parsimony method is a typical representative of the cladistic approach, whereas the UPGMA method is a typical phenetic method. The other methods, however, cannot be classified easily according to the above criteria. For example, the transformed distance method and the neighbors relation method have often been said to be phenetic methods, but this is not an accurate description. Although these methods use similarity (or dissimilarity, i.e., distance) measures, they do not assume a direct connection between similarity and evolutionary relationship, nor are they intended to infer phenetic relationships.

In molecular phylogeny, a better classification of methods would be to distinguish between **distance** and **character-state approaches**. Methods belonging to the former approach are based on distance measures, such as the number of nucleotide or amino acid substitutions, while methods belonging to the latter approach rely on the state of the character, such as the nucleotide or amino acid at a particular site, or the presence or absence of a deletion or an insertion at a certain DNA location. According to this classification, the UPGMA method, the transformed distance method, and the neighbors relation method are distance methods, while the maximum parsimony method is a character-state method.

It has often been argued that character-state methods are more powerful than distance methods, because the raw data is a string of character states (e.g., the nucleotide sequence) and in transforming character-state data into distance matrices some information is lost. We note, however, that while the maximum parsimony method indeed uses the raw data, it usually uses only a small fraction of the available data. For instance, in the example on page 111, only three sites are used while six sites are excluded from the analysis. For this reason, this method is often less efficient than some distance matrix methods (e.g., see Saitou and Nei 1986). Of course, if the number of informative sites is large, the maximum parsimony method is generally very effective.

Finally, it should be noted that an inferred tree often contains topological errors, regardless of the method used. To obtain a correct tree, a large amount of data is usually required.

ESTIMATION OF BRANCH LENGTHS

We have not discussed how to estimate branch lengths, except in the case of the UPGMA method. We shall now deal with this problem, assuming that the tree topology has already been inferred. We consider only trees inferred by distance matrix methods. For the maximum parsimony method, the problem is more complicated (see Fitch 1971). The method discussed below is that of Fitch and Margoliash (1967).

First, let us consider the simplest case, i.e., an unrooted tree with three OTUs (A, B, and C) and a single node (Figure 10a). Let x, y, and z be the lengths of the branches leading to A, B, and C, respectively. It is easy to see that the following equations hold:

$$d_{AB} = x + y \quad \textbf{(5.8a)}$$

$$d_{AC} = x + z \quad \textbf{(5.8b)}$$

$$d_{BC} = y + z \quad \textbf{(5.8c)}$$

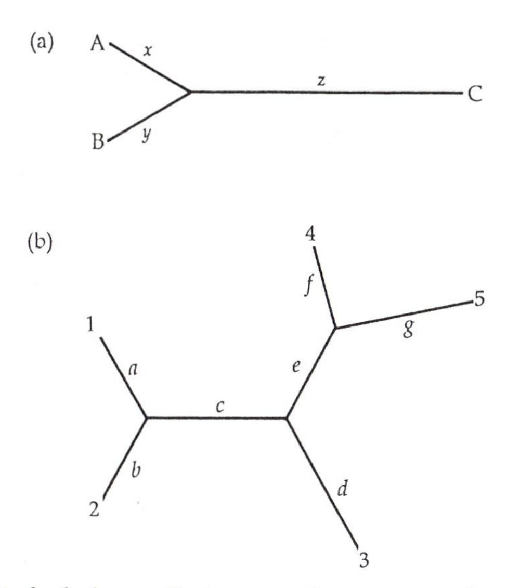

Figure 10. Unrooted phylogenetic trees used to compute branch lengths by Fitch and Margoliash's (1967) method. (a) A tree with three OTUs. (b) A tree with five OTUs.

From these equations, we obtain the following solutions:

$$x = (d_{AB} + d_{AC} - d_{BC})/2 \quad \textbf{(5.9a)}$$

$$y = (d_{AB} + d_{BC} - d_{AC})/2 \quad \textbf{(5.9b)}$$

$$z = (d_{AC} + d_{BC} - d_{AB})/2 \quad \textbf{(5.9c)}$$

Let us now deal with the case of more than three OTUs. For simplicity, let us assume that there are five OTUs (1, 2, 3, 4, and 5) and that the topology and the branch lengths are as in Figure 10b. Suppose that OTUs 1 and 2 were the first OTUs to be clustered together in the tree reconstruction process. We then use A and B to denote OTUs 1 and 2, respectively, and put all the other OTUs (3, 4, and 5) into a composite OTU denoted as C. By this arrangement, we can apply Equations 5.9a–5.9c to estimate the lengths of the branches leading to A, B, and C, except that now $d_{AC} = d_{1(345)} = (d_{13} + d_{14} + d_{15})/3$, and $d_{BC} = d_{2(345)} = (d_{23} + d_{24} + d_{25})/3$. Then we have $a = x$ and $b = y$. OTUs 1 and 2 are subsequently considered as a single composite OTU. In the next step, suppose that the composite OTU (12) and the simple OTU 3 were the next pair to be joined together. Then, we denote OTUs (12) and 3 by A and B, respectively, and put the other OTUs (i.e., 4 and 5) into the new composite OTU C. In the same manner as above, we obtain x, y, and z. Note that $d = y$ and $c + (a + b)/2 = x$. From the values for a and b, which have been obtained previously, we can calculate c. The process is continued until all branch lengths are obtained.

Note that sometimes an estimated branch length can be negative. Since the true length can never be negative, it is better to replace such an estimate by 0.

As an example of using the above method, let us compute the branch lengths of the tree in Figure 7c. For convenience, we again present the distance matrix that was used to infer the topology of this tree. To avoid confusion with the notation in Equation 5.9 we rename OTUs A, B, C, and D as OTUs 1, 2, 3, and 4, respectively.

	OTU		
OTU	1	2	3
2	8		
3	7	9	
4	12	14	11

Since OTUs 1 and 2 were clustered first, we first compute the lengths (a and b) of the branches leading to these two OTUs by putting OTUs 3 and 4 into a composite OTU C. We then have $d_{AB} = d_{12} = 8$, $d_{AC} = (d_{13} + d_{14})/2 = (7 + 12)/2 = 9.5$ and $d_{BC} = (d_{23} + d_{24})/2 = 11.5$. From Equations 5.9a–5.9c, we have $a = x = (8 + 9.5 - 11.5)/2 = 3$ and $b = y = (8 + 11.5 - 9.5)/2 = 5$. Next we treat OTUs 1 and 2 as a single OTU (12) and denote it by A. Since we are left with only three OTUs, we denote OTU 3 by B and OTU 4 by C. We then have $d_{AB} = d_{(12)3} = (d_{13} + d_{23})/2 = (7 + 9)/2 = 8$, $d_{AC} = d_{(12)4} = (d_{14} + d_{24})/2 = (12 + 14)/2 = 13$, and $d_{BC} = d_{34} = 11$. Therefore, from Equations 5.9a–5.9c we have $x = (8 + 13 - 11)/2 = 5$, and $d = y = (8 + 11 - 13)/2 = 3$, and $e = z = (13 + 11 - 8)/2 = 8$. We note from Figure 7c that $(a + b)/2 + c = x$, and so $c = 1$. This completes the computation. Note, however, that since we do not know the exact location of the root, we cannot estimate the length of the branch connecting the root and OTU D but can only estimate the length from the common ancestral node of OTUs A, B, and C through the root to OTU D, i.e., $e = 8$.

ROOTING UNROOTED TREES

The majority of tree-making methods yield unrooted trees. To root an unrooted tree, we usually need an outgroup (i.e., an OTU for which external information, such as paleontological evidence, clearly indicates that it has branched off earlier than the taxa under study). The root is then placed between the outgroup and the node connecting it to the other OTUs.

While we must be certain that the outgroup did indeed diverge prior to all other taxa, it is advisable not to choose an outgroup that is too distantly related to the taxa in question, because in such cases it is difficult to obtain reliable estimates of the distances between the outgroup and the other taxa. For example, in reconstructing the phylogenetic relationships among a group of placental mammals, we may use a marsupial as an outgroup. Birds may serve as reliable outgroups only if the DNA sequences used have been highly conserved in evolution. Plants or fungi would have clearly qualified as outgroups in this example; however, by being only very distantly related to the mammals, their use as outgroups may result in serious topological errors. The use of more than one outgroup generally improves the estimate of the tree topology, provided again that they are not too distant from the rest of the taxa. The outgroup must also not be phylogenetically too close to the other OTUs, because in this case we cannot be certain that it is a true outgroup, i.e., that it diverged from the other OTUs prior to their divergence from one another.

In the absence of an outgroup, we may position the root by assuming that the rate of evolution has been approximately uniform over all the branches. Under this assumption we put the root at the midpoint of the longest pathway between two OTUs.

ESTIMATION OF SPECIES-DIVERGENCE TIMES

Because the paleontological record is far from complete, we are often ignorant of the dates of divergence between species. DNA sequence data can be of great help in this respect. Let us assume that the rate of evolution for a DNA sequence is known from a previous study to be r substitutions per site per year. To obtain the divergence time (T) between species A and B, we compare the sequences from both species and compute the number of substitutions per site (K). As shown in Chapter 4, we have the following equation:

$$r = K/(2T) \tag{5.10}$$

Therefore, T is estimated as

$$T = K/(2r) \tag{5.11}$$

As noted in Chapter 4, the rate of nucleotide substitution obtained from one group of organisms may not be applicable to another group of organisms. To avoid this problem, we may estimate the substitution rate by adding a third species, C, whose divergence time (T_1) from the species pair A and B is known (Figure 11). Let K_{ij} be the number of nucleotide substitutions per site between species i and j. Then, the rate of nucleotide substitution is estimated by

$$r = \frac{K_{AC} + K_{BC}}{2(2T_1)} \tag{5.12}$$

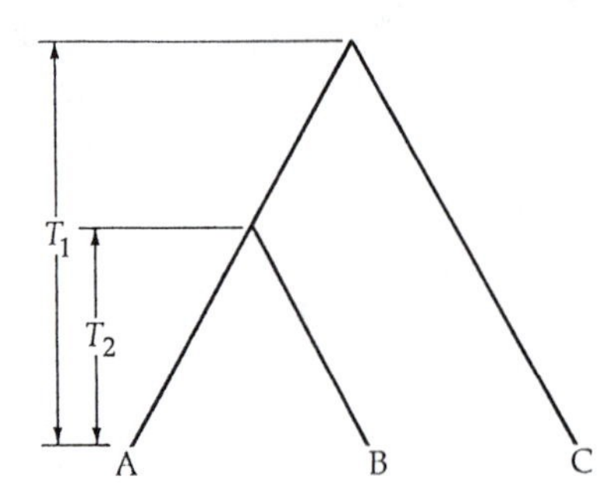

Figure 11. Model tree for estimating times of divergence (see text).

The unknown divergence time between species A and B (T_2) is estimated by

$$T_2 = \frac{K_{AB}}{2r} = \frac{2K_{AB}T_1}{K_{AC} + K_{BC}} \qquad \textbf{(5.13)}$$

Conversely, in the case that T_2 is known but T_1 is not, T_1 is given by

$$T_1 = \frac{(K_{AC} + K_{BC})T_2}{2K_{AB}} \qquad \textbf{(5.14)}$$

The above formulation assumes rate constancy. As discussed in the previous chapter, this assumption often does not hold, and so the estimated divergence time should be treated with caution. Li and Tanimura (1987) have proposed a method that can reduce the effects of unequal rates of substitution on divergence-time estimates.

We have also noted earlier that the divergence time between two sequences may predate the divergence between the species from which the sequences were obtained. However, this error is usually not very serious if we are concerned with long-term divergence events, say on the order of several million years or longer. One should also note that estimates of divergence time are usually subject to large stochastic errors. To reduce such errors, many sequences should be used in the estimation.

CLADES

The purpose of phylogenetic studies is to establish the evolutionary relationships among different species. In particular, we are interested in the identification of natural **clades.** A clade is defined as a group of species that share a common ancestor, which is not shared by any other species outside the clade.

Figure 12 shows the evolutionary relationships among three classes of

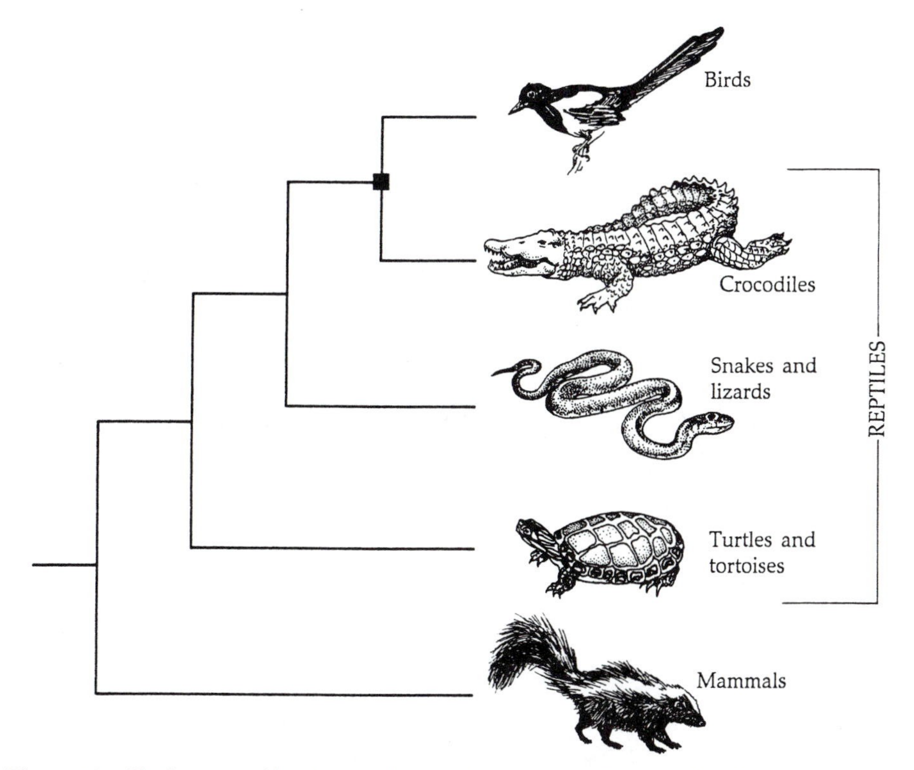

Figure 12. Cladogram of birds, reptiles, and mammals. The reptiles do not constitute a natural clade, since they share ancestors with the birds, which are not included in the Reptilia. Birds and crocodiles, on the other hand, constitute a natural clade (Archosauria) since they share a common ancestor (black box) not shared by any other organism.

vertebrates: birds, reptiles, and mammals. We see that the classical taxonomic assignment of reptiles to a separate class does not fit the definition of a clade, since the three groups of reptiles share a common ancestor with another group, the birds, which is not included within the definition of the class Reptilia. Birds and crocodiles, on the other hand, do constitute a natural clade, the Archosauria, since they share a common ancestor not shared by any extant organism other than birds and crocodiles. Similarly, all birds and all reptiles taken together constitute a natural clade.

PHYLOGENY OF HUMANS AND APES

The issue of man's closest evolutionary relatives has always intrigued biologists. Darwin (1871), for instance, claimed that the African apes, the chimpanzee (*Pan*) and the gorilla (*Gorilla*), are man's closest relatives and, hence,

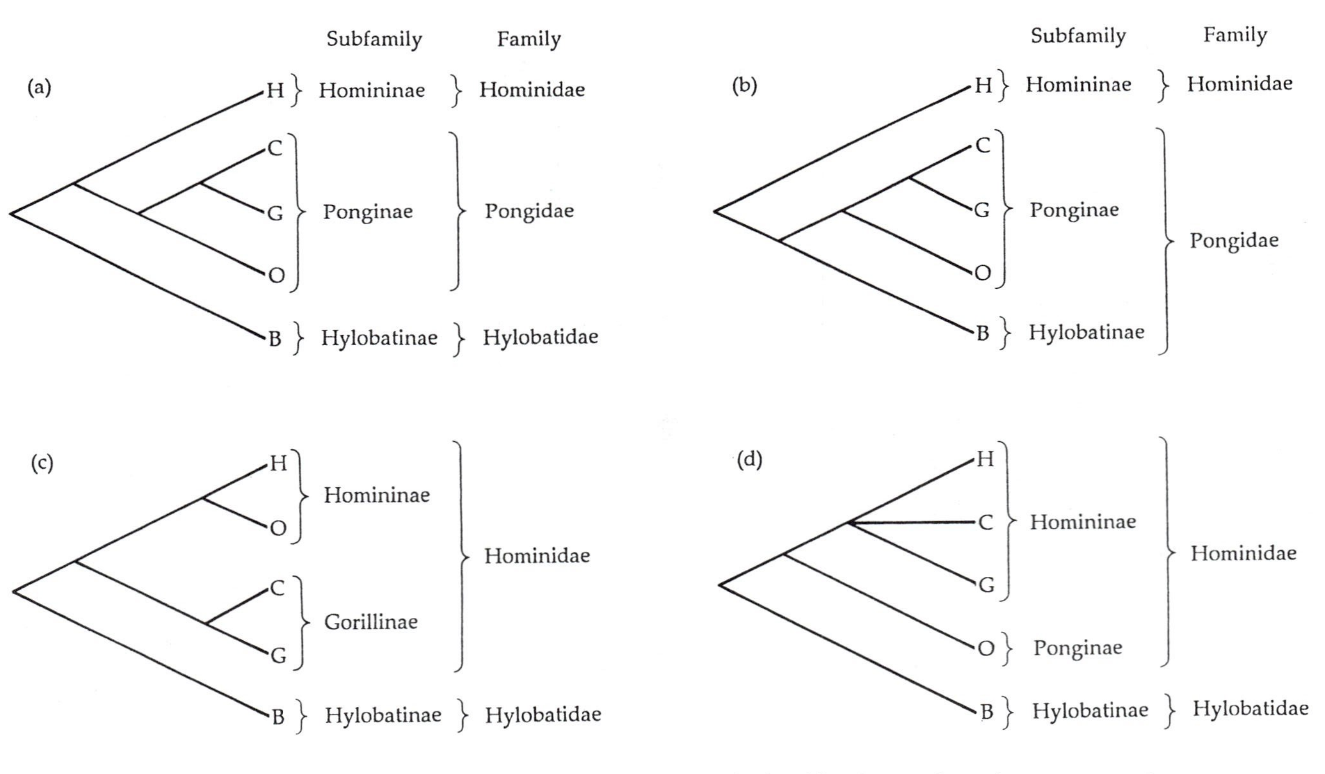

Figure 13. Four alternative phylogenies and classifications of modern apes and humans (Hominoidea). Traditional classifications setting humans apart are shown in (a) and (b). The clustering of humans with the orangutan is shown in (c). Cumulative molecular as well as morphological evidence favors the classification in (d).

he concluded that man's evolutionary origins were to be found in Africa. Darwin's view fell into disfavor for various reasons, and for a long time taxonomists believed that the genus *Homo* was only distantly related to the apes, and thus *Homo* was given a family of its own, Hominidae. Chimpanzees, gorillas, and orangutans (*Pongo*) were usually placed in a separate family, the Pongidae (Figure 13a). The gibbons (*Hylobates*) were either classified separately or with the Pongidae (Figure 13b; see Simpson 1961). Goodman (1963) correctly recognized that this systematic arrangement was anthropocentric, because humans represent "a new grade of phylogenetic development, one which is 'higher' than the pongids and all other preceding grades." However, placing the various apes in one family and humans in another implies that the apes share a more recent common ancestry with each other than with humans. When *Homo* was put in the same clade with a living ape, it was usually with the Asian ape, the orangutan (Figure 13c; Schultz 1963).

By using a serological precipitation method, Goodman (1962) was able to demonstrate that humans, chimpanzees, and gorillas constitute a natural clade (Figure 13d), with orangutans and gibbons having diverged from the other apes at much earlier dates. From microcomplement fixation data, Sarich and Wilson (1967) estimated the divergence time between humans and chimpanzees or gorillas to be as recent as 5 million years ago, rather than a minimum date of 15 million years ago as was commonly accepted by paleontologists at that time.

However, serological, electrophoretic, and amino acid sequences could not resolve the evolutionary relationships among humans and the African apes, and the so-called human–gorilla–chimpanzee trichotomy remained unsolved and has continued to be an extremely controversial issue (Figure 14). In the

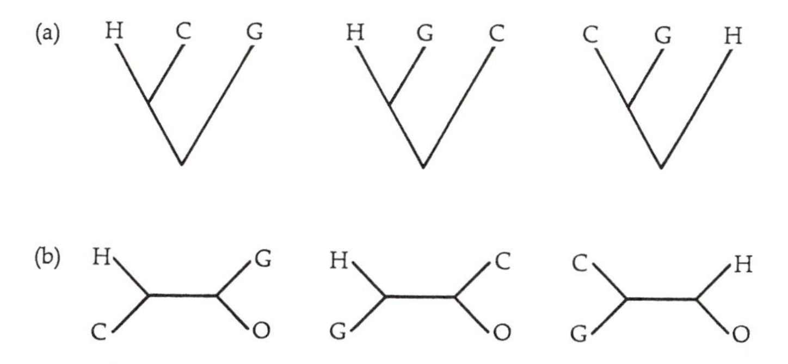

Figure 14. (a) Three possible rooted trees for gorillas, chimpanzees and humans. (b) Comparable unrooted trees with the orangutan as an outgroup. Species abbreviations: C, chimpanzee (*Pan troglodytes*); H, human (*Homo sapiens*); G, gorilla (*Gorilla gorilla*); and O, orangutan (*Pongo pygmaeus*).

Table 2. **Mean (below diagonal) and standard error (above diagonal) of the number of nucleotide substitutions per 100 sites between OTUs.**[a]

OTU	Human	Chimpanzee	Gorilla	Orangutan	Rhesus monkey
Human		0.17	0.18	0.25	0.41
Chimpanzee	1.45		0.18	0.25	0.42
Gorilla	1.51	1.57		0.26	0.41
Orangutan	2.98	2.94	3.04		0.40
Rhesus monkey	7.51	7.55	7.39	7.10	

From Li et al. (1987b).

[a] The sequence data used are 5.3 kb of noncoding DNA, which is made up of two separate regions: (1) the η-globin locus (2.2 kb) described by Koop et al. (1986b) and (2) 3.1 kb of the η-δ globin intergenic region sequenced by Maeda et al. (1983, 1988).

following, we shall use DNA sequence data obtained by M. Goodman and colleagues (see Miyamoto et al. 1987) and by Maeda et al. (1988) to show that the molecular evidence supports the human–chimpanzee clade and, at the same time, to illustrate the tree-making methods discussed in the previous sections.

Table 2 shows the number of nucleotide substitutions per 100 sites between each pair of the following OTUs: humans (H), chimpanzees (C), gorillas (G), orangutans (O), and rhesus monkeys (R). Let us first apply the UPGMA method to these distances. The distance between humans and chimpanzees is the shortest (d_{HC} = 1.45). Therefore, we join these two OTUs first and place the node at 1.45/2 = 0.73 (Figure 15a). We then compute the distances between the composite OTU (HC) and each of the other species, and obtain a new distance matrix:

OTU	(HC)	G	O
G	1.54		
O	2.96	3.04	
R	7.53	7.39	7.10

Since (HC) and G are separated by the shortest distance, they are the next to be joined together, and the connecting node is placed at 1.54/2 = 0.77. Continuing the process, we obtain the tree in Figure 15a. We note that the estimated branching node for H and C is very close to that for (HC) and G. In fact, the distance between the two nodes is smaller than all the standard

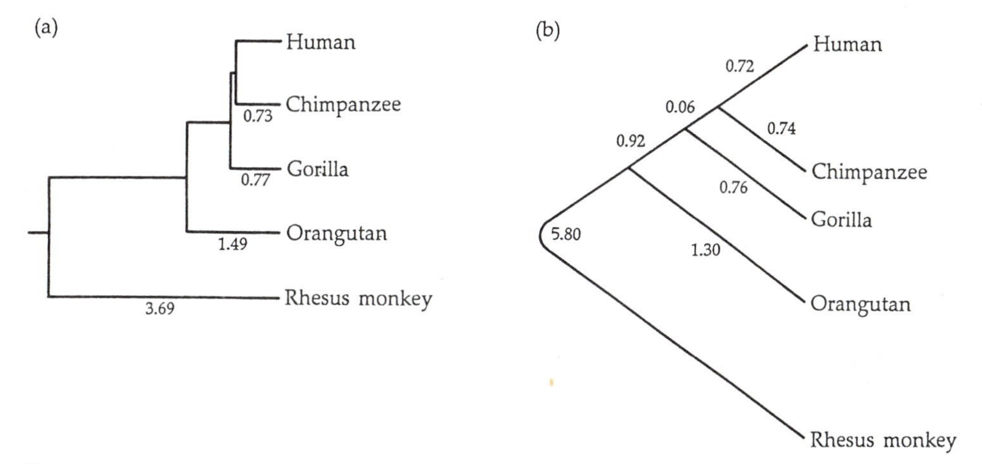

Figure 15. Phylogenetic tree for humans, chimpanzees, gorillas, orangutans, and rhesus monkeys inferred from the UPGMA method (a) and from Sattath and Tversky's neighbors relation method (b).

errors for the estimates of the pairwise distances among H, C, and G (Table 2). Thus, although the data suggest that man's closest living relatives are the chimpanzees, the data do not provide a conclusive resolution of the branching order. On the other hand, it is unequivocal that the orangutan is an outgroup to the human–chimpanzee–gorilla clade.

Next, we use Sattath and Tversky's neighbors relation method. We consider four OTUs at one time. Since there are five OTUs, there are $5!/[4!(5 - 4)!] = 5$ possible quadruples. We start with OTUs H, C, G, and O and compute the following sums of distances (data from Table 2): $d_{HC} + d_{GO} = 1.45 + 3.04 = 4.49$, $d_{HG} + d_{CO} = 4.45$, and $d_{HO} + d_{CG} = 4.55$. Since the second sum is the smallest, we choose H and G as one pair of neighbors and C and O as the other (Table 3). Similarly, we consider the four other possible quadruples. The results are shown in Table 3. Noting from the bottom of Table 3 that (OR) has the highest neighbors-relation score among all neighbor pairs, we choose (OR) as the first pair of neighbors. Treating this pair as a single OTU, we obtain the following new distance matrix:

	OTU		
OTU	H	C	G
C	1.45		
G	1.51	1.57	
(OR)	5.25	5.25	5.22

As only four OTUs are left, it is easy to see that $d_{HC} + d_{G(OR)} = 6.67 < d_{HG}$

Table 3. Neighbors-relation scores obtained from the distance matrix in Table 1.

OTUs compared[a]	Sum of pairwise distances	Neighbor pairs chosen
H,C,G,O	$d_{HC} + d_{GO} = 4.49$	(HG), (CO)
	$d_{HG} + d_{CO} = 4.45$	
	$d_{HO} + d_{CG} = 4.55$	
H,C,G,R	$d_{HC} + d_{GR} = 8.84$	(HC), (GR)
	$d_{HG} + d_{CR} = 9.06$	
	$d_{HR} + d_{CG} = 9.08$	
H,C,O,R	$d_{HC} + d_{OR} = 8.55$	(HC), (OR)
	$d_{HO} + d_{CR} = 10.53$	
	$d_{HR} + d_{CO} = 10.45$	
H,G,O,R	$d_{HG} + d_{OR} = 8.61$	(HG), (OR)
	$d_{HO} + d_{GR} = 10.37$	
	$d_{HR} + d_{GO} = 10.55$	
C,G,O,R	$d_{CG} + d_{OR} = 8.67$	(CG), (OR)
	$d_{CO} + d_{GR} = 10.33$	
	$d_{CR} + d_{GO} = 11.59$	
Total scores:	(HC) = 2, (HG) = 2, (HO) = 0, (HR) = 0, (CG) = 1, (CO) = 1, (CR) = 0, (GO) = 0, (GR) = 1, (OR) = 3	

[a] H, human; C, chimpanzee; G, gorilla; O, orangutan; and R, rhesus monkey.

$+ d_{C(OR)} = 6.76 < d_{H(OR)} + d_{CG} = 6.82$. Therefore, we choose H and C as one pair of neighbors and G and (OR) as the other. The final tree obtained by this method is shown in Figure 15b. The topology of this tree is identical to that in Figure 15a. Note, however, that in this method O and R rather than H and C were the first pair to be joined to each other. This is because, in an unrooted tree, O and R are in fact neighbors. The branch lengths in Figure 15b were estimated by the method presented earlier (page 117).

Finally, let us consider the maximum parsimony method. For simplicity, let us consider only humans, chimpanzees, gorillas and orangutans. Table 4 shows the informative sites for the 10.2-kb region including the η-globin pseudogene and its surrounding regions (Koop et al. 1986a; Miyamoto et al. 1987; Maeda et al. 1988). For each site, the hypothesis supported is given in the last column. If we consider base changes only, then there are 15 informative sites, of which eight support the human–chimpanzee clade (hypothesis I), four support the chimpanzee–gorilla clade (hypothesis II), and three support the human–gorilla clade (hypothesis III). Moreover, all the four informative sites involving a gap support the human–chimpanzee clade. Therefore, the human–chimpanzee clade is chosen as the most likely representation of the true phylogeny. In a more detailed analysis, Williams and

Goodman (1989) have shown that support for the human–chimpanzee clade is statistically significant at the 1% level.

Some other types of molecular data also support the clustering of humans and chimpanzees. For example, DNA–DNA hybridization data (Sibley and Ahlquist 1984; Caccone and Powell 1989) distinguish unambiguously among the three alternatives in Figure 14 in favor of the phylogeny that clusters humans and chimpanzees in a clade. Thus, the closest extant relatives of humans are the two chimpanzee species, followed by the gorilla, the orangutan, and the nine gibbon species.

Table 4. Informative sites among human, chimpanzee, gorilla, and orangutan sequences.

Site[a]	Sequence				Hypothesis supported[b]
	Human	Chimpanzee	Gorilla	Orangutan	
DATA FROM MIYAMOTO ET AL. (1987)					
34	A	G	A	G	III
560	C	C	A	A	I
1287	*[c]	*	T	T	I
1338	G	G	A	A	I
3057–3060	****	****	TAAT	TAAT	I
3272	T	T	*	*	I
4473	C	C	T	T	I
5153	A	C	C	A	II
5156	A	G	G	A	II
5480	G	G	T	T	I
6368	C	T	C	T	III
6808	C	T	T	C	II
6971	G	G	T	T	I
DATA FROM MAEDA ET AL. (1988)					
127–132	******	******	AATATA	AATATA	I
1472	G	G	A	A	I
2131	A	A	G	G	I
2224	A	G	A	G	III
2341	G	C	G	C	III
2635	G	G	A	A	I

Modified from Williams and Goodman (1989).

[a] Site numbers correspond to those given in the original sources. The total length of the sequence used is 10.2 kb, about twice that used in Table 2.

[b] Hypotheses: I, human and chimpanzee in one clade; II, chimpanzee and gorilla in one clade; and III, human and gorilla in one clade.

[c] Each asterisk denotes the deletion of a nucleotide at the site.

Table 5. Molecular characters that distinguish the genomes of both chloroplasts and prokaryotes from the nuclear genome of eukaryotes.

1. Histoneless DNA
2. 120,000–150,000 base pairs in size
3. Circular genome
4. Rifampicin-sensitivity of transcription
5. Inhibition of ribosomes by streptomycin, chloramphenicol, spectromycin, and paromonycin
6. Insensitivity of translation to cycloheximide
7. Translation starts with formylmethionine
8. Polyadenylation of mRNA absent or very short
9. Prokaryotic promoter structure

ENDOSYMBIOTIC ORIGIN OF MITOCHONDRIA AND CHLOROPLASTS

There are essentially two types of theories to explain the existence of separate nuclear, mitochondrial, and chloroplast genomes in eukaryotes. Theories in the first category (e.g., Cavalier-Smith 1975) stipulate that the genomes of organelles have autogenous origins and are descended from nuclear genes by direct filiation, whereby part of the nuclear genome became incorporated into a membrane-enclosed organelle and subsequently assumed a quasi-independent existence. In contrast, the endosymbiotic theories (e.g., Margulis 1981) claim that the origin of extranuclear DNA is exogenous. According to this proposal, the ancestors of eukaryotic organisms engulfed prokaryotes, which were subsequently retained because of a mutually beneficial or symbiotic relationship. With time, the endosymbionts were streamlined by means of loss of genes and became obligatory symbionts (i.e., organisms incapable of an independent existence outside their host).

The molecular evidence is now overwhelmingly in favor of the endosym-

Figure 16. (a) Eubacterial–chloroplast–mitochondrial portion of an unrooted tree inferred from small-subunit rRNA sequences. Note that while the chloroplastic sequences from both green algae and higher plants are monophyletic, the mitochondrial sequences are polyphyletic. (b) An unrooted tree inferred from nuclear small-subunit rRNA sequences. Step 1 (indicated by a circle) is the early symbiosis postulated to have given rise to the mitochondrial genomes of most eukaryotes while step 2 is the late symbiosis postulated by Gray et al. (1989) to have contributed the rRNA genes of higher-plant mitochondria. The length of each branch is proportional to the number of substitutions between its two endpoints, as indicated by the scales. Modified from Gray et al. (1989).

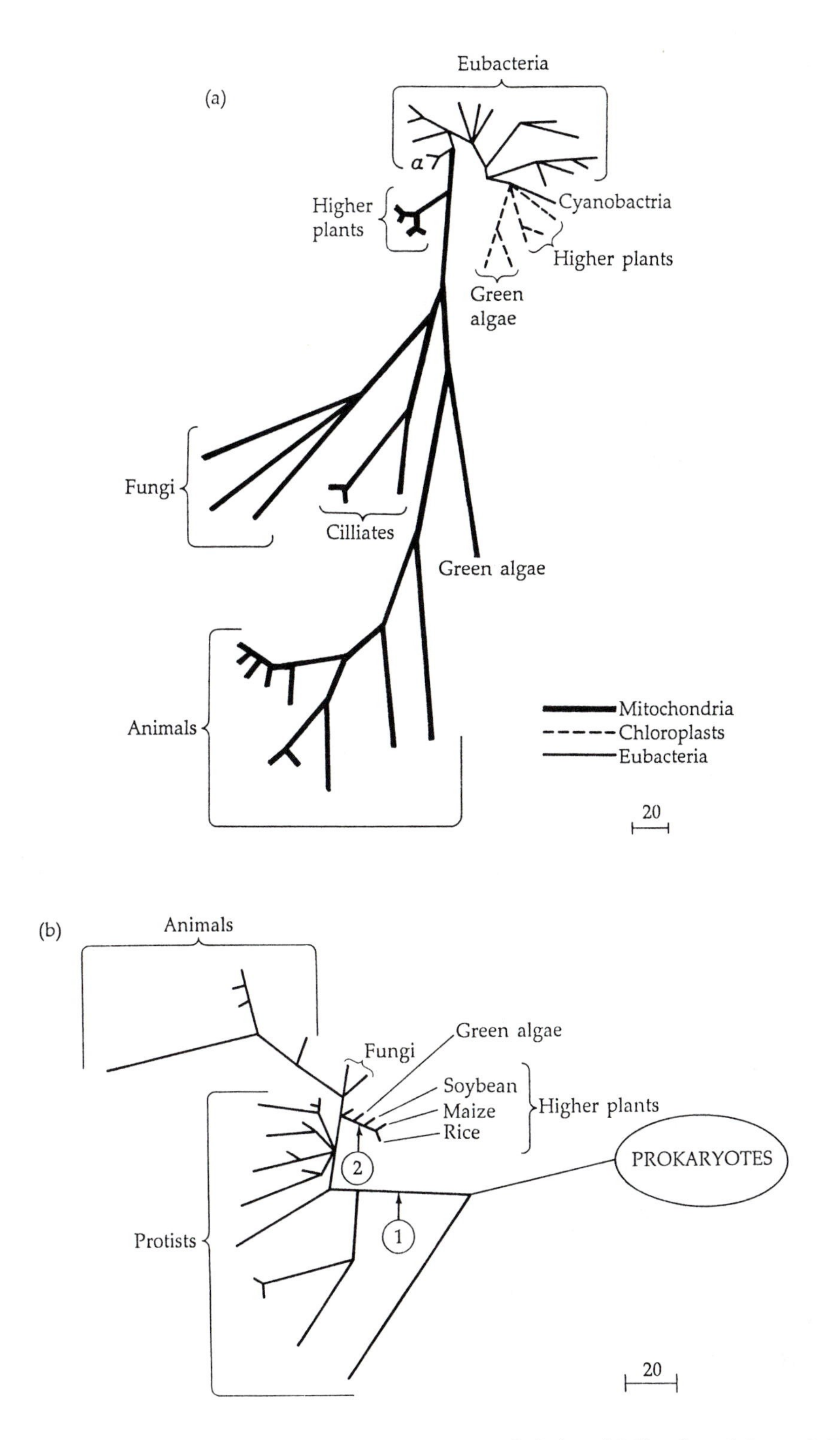
(a)
Eubacteria
a
Higher plants
Cyanobactria
Higher plants
Green algae
Fungi
Cilliates
Green algae
Animals
Mitochondria
Chloroplasts
Eubacteria
20
(b)
Animals
Green algae
Fungi
Soybean
Maize
Rice
Higher plants
PROKARYOTES
2
1
Protists
20

biotic theory. A list of biochemical characters that distinguish the genomes of both chloroplasts and prokaryotes from the nuclear genome of eukaryotes is given in Table 5. The ultimate support, however, came from rRNA sequence data. Because of their low rates of substitution, rRNA sequences have proved to be very useful for addressing questions concerning very ancient evolutionary divergence events.

Schwarz and Kössel (1980) showed that the nucleotide sequence of the 16S rRNA gene from the chloroplast of maize (*Zea mays*) has regions with a strong similarity to those in the 16S rRNA from the bacterium *Escherichia coli*. The degree of similarity between nuclear and chloroplast rRNA sequences was much lower. A detailed analysis of 16S rRNA sequences from the photosynthetic cyanobacteria (Giovannovi et al. 1988) supports the proposal that green chloroplasts were derived from a group of photosynthetic bacteria called cyanobacteria (Bonen et al. 1979) (see Figure 16a).

Phylogenetic analyses of rRNA sequences suggest that mitochondria were derived from the α subdivision of the purple bacteria (Figure 16a; Cedergren et al. 1988). However, the nuclear rRNA sequences used indicate that higher plants branched off at about the same time as animals and fungi (Figure 16b), consistent with the traditional view, whereas the mitochondrial rRNA sequences indicate that higher plants cluster very near the root of the purple bacteria, but are separated from fungi, green algae, and animals (Figure 16a). This is contrary to the traditional view, which groups higher plants and green algae together in one clade. For this reason, Gray et al. (1989) suggested that the rRNA genes in the mitochondria of higher plants are of a more recent evolutionary origin than the rRNA genes in other mitochondria. Whether or not this hypothesis will be substantiated by further evidence remains to be seen.

MOLECULAR PALEONTOLOGY

It is now possible to sequence segments of DNA from unpurified samples derived from micrograms of preserved tissues. The method employed is the polymerase chain reaction (PCR). PCR involves the amplification of unique sequences from a mixture of sequences via the use of two known primers (Figure 17; Saiki et al. 1985, 1988; Scharf et al. 1986; Engelke et al. 1988). Each primer attaches to a complementary piece of DNA, which triggers the binding of a DNA polymerase that then copies the segment. Because each newly made copy can serve as a template for further duplication, the number of copies of the target segment grows exponentially. By using this procedure, it is possible to synthesize many copies of a chosen piece of DNA in the presence of vast excesses of other DNA sequences (Kocher et al. 1989).

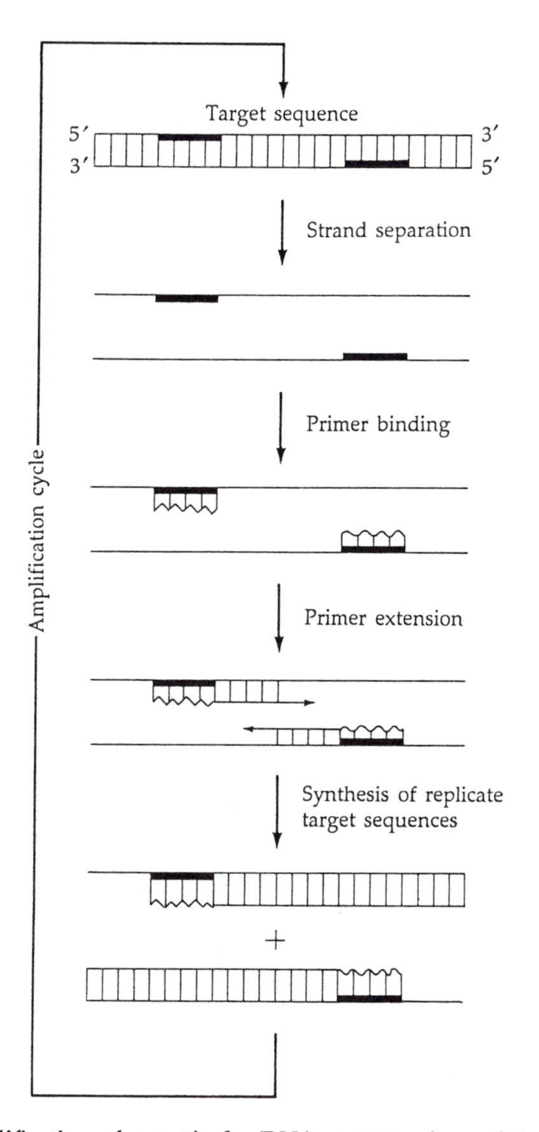

Figure 17. Amplification of a particular DNA sequence in a mixture by the method of the polymerase chain reaction (PCR). The DNA is separated into single-stranded molecules by heating. Two small pieces of synthetic DNA (zigzag and wavy lines), each complementing a specific sequence at one end of the target sequence (black bars), serve as primers. Each primer binds to its complementary sequence on different strands. DNA polymerases then extend the primers by adding nucleotides to them. Within a short time, exact replicas of the target sequence are produced. In subsequent cycles, both the original and the replicate target sequences can serve as templates. In contrast, DNA molecules that do not contain complementary sequences to the primers are not amplified. Therefore, unpurified mixtures of DNA can be used, from which only one type of sequence will be amplified even in the presence of vast excesses of other sequences. For more details, see Mullis (1990).

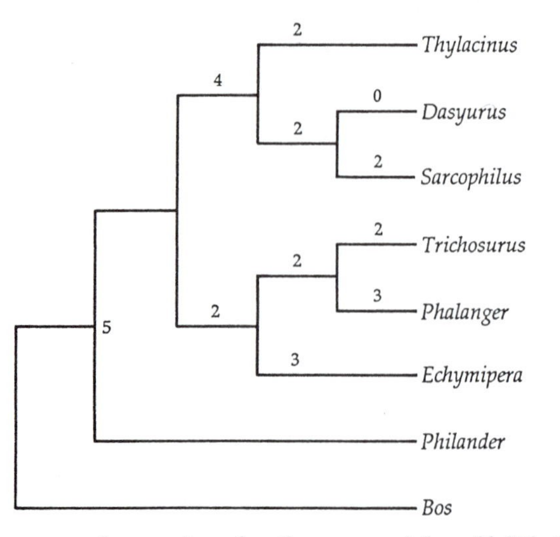

Figure 18. Maximum parsimony tree for the marsupial wolf (*Thylacinus*) and six other marsupials based on mitochondrial 12S rRNA sequences. The tree has been rooted by using the cow (*Bos*) as an outgroup. The numbers on the branches represent the number of substitutions. From Thomas et al. (1989).

Consequently, by employing PCR we can retrieve particular DNA sequences from museum specimens such as preserved organic material (mainly skin and muscle), badly damaged archeological remains, and even bones (Hagelberg et al. 1989).

By using this method it is now possible to establish phylogenetic affiliations of extinct species, such as the quagga and the Australian marsupial wolf, or to determine ancestor–descendant relationships among extinct human populations, such as the bog peoples of the Iron Age in Scandinavia and Egyptian mummies, for which morphological comparisons have yielded ambiguous results.

By using the PCR method, Thomas et al. (1989) were able to sequence and compare 219 nucleotides from the mitochondria of the marsupial wolf, *Thylacinus cynocephalus*, with those from other Australian and South American marsupials as well as with homologous sequences from placental mammals. By using this method, they were able to decide between two claims: (a) that the marsupial wolf is related to a South American group of marsupials and (b) that the marsupial wolf is closely related to other Australian marsupials. From these sequence comparisons, it was concluded that *Thylacinus* is closely related to two other Australian marsupials, the nearly extinct Tasmanian devil (*Sarcophilus harrisii*) and the Australian tiger cat (*Dasyurus maculatus*), but only distantly related to a South American marsupial, the opossum (*Philander opossum andersoni*) (Figure 18). Thus, the morphological similarity between *Thylacinus* and South American marsupials is thought to represent an instance of convergent evolution at the morphological level that has no parallels in the mitochondrial DNA.

THE DUSKY SEASIDE SPARROW: A LESSON IN CONSERVATION BIOLOGY

The last dusky seaside sparrow died on June 16, 1987, in a zoo at Walt Disney World, near Orlando, Florida. Dusky seaside sparrows were discovered in 1872, and their melanic appearance led to their being classified as a distinct subspecies (*Ammodramus maritimus nigrescens*). The geographical distribution of *A. m. nigrescens* was confined to salt marshes in Brevard County, Florida (Figure 19), and by 1980 only six individuals, all males, could be found in nature. Obviously, the population was doomed, and an artificial breeding program was launched as a last-ditch attempt to preserve genes of the subspecies.

In such a case, the conservation program involves the mating of the males

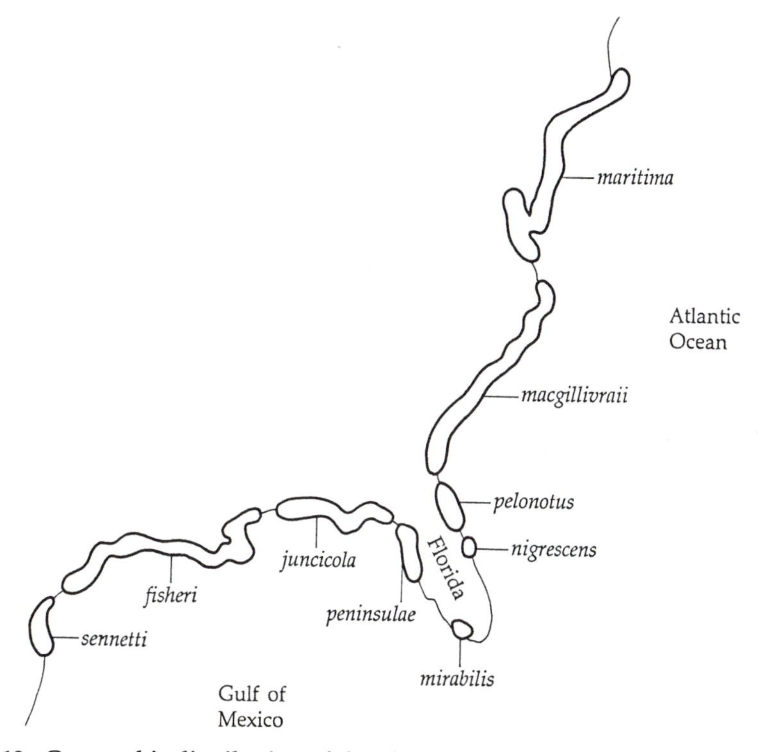

Figure 19. Geographic distribution of the nine taxonomically recognized subspecies of the seaside sparrow (*Ammodramus maritimus*). From Avise and Nelson (1989).

from the nearly extinct subspecies with females from the closest subspecies available. The female hybrids of the first generation are then backcrossed to the males, their offspring are again backcrossed to the original males, and the process is continued for as long as the males live. The crux of such an experiment is to decide from which population to choose the females, i.e., which subspecies is phylogenetically closest to the endangered one.

In the case of *A. maritimus*, there were eight recognized subspecies from which to choose. The geographical ranges of these species are shown in Figure 19. On the basis of morphological and behavioral characters, it was decided that the closest subspecies to *A. m. nigrescens* is Scott's seaside sparrow (*A. m. peninsulae*), which inhabits Florida's Gulf shores. As a consequence of this decision, several males of the *nigrescens* subspecies were mated with females of the *peninsulae* subspecies. Two successful backcrosses were accomplished and the resulting population has since been kept inbred with the view of someday releasing the "reconstructed" subspecies into its original habitat.

In order to find out whether or not the choice of females was correct, Avise and Nelson (1989) compared the restriction-enzyme pattern of mitochondrial DNA from the last pure *A. m. nigrescens* specimen with that of 39 individual birds belonging to five of the eight extant subspecies of *A. maritimus*. They chose mitochondrial DNA for several reasons. First, mitochondrial DNA in vertebrates is known to evolve very rapidly (Chapter 4), and hence it can provide a high resolution for distinguishing between closely related organisms. Second, mitochondria are maternally inherited, and thus complications due to allelic segregation do not arise. Finally, because of the maternal mode of transmission, mitochondrial DNA from the last male dusky sparrow had not been transmitted to the hybrids in the restorative breeding program. Thus, unlike nuclear genes, some of which survive in the hybrids, the mitochondrial genes of the dusky seaside sparrow are truly extinct.

From the restriction-enzyme patterns, Avise and Nelson (1989) reconstructed the evolutionary relationships between several subspecies of *A. maritimus* by the UPGMA and maximum parsimony methods (Figure 20). As seen from the figure, the Atlantic Coast populations, including *A. m. nigrescens*, are nearly indistinguishable from one another. The same is true for the three Gulf Coast subspecies in this study. In comparison, the Atlantic Coast subspecies are quite distinct from the Gulf Coast subspecies. The number of nucleotide substitutions per site between the two groups has been estimated to be about 1%. If mitochondrial DNA in sparrows evolves at about the same rate as mitochondrial DNA in mammals and other birds (i.e., 2–4% sequence divergence per million years), then these two groups of populations have separated from each other some 250,000–500,000 years ago. While these estimates of absolute age of divergence must remain qualified due to uncer-

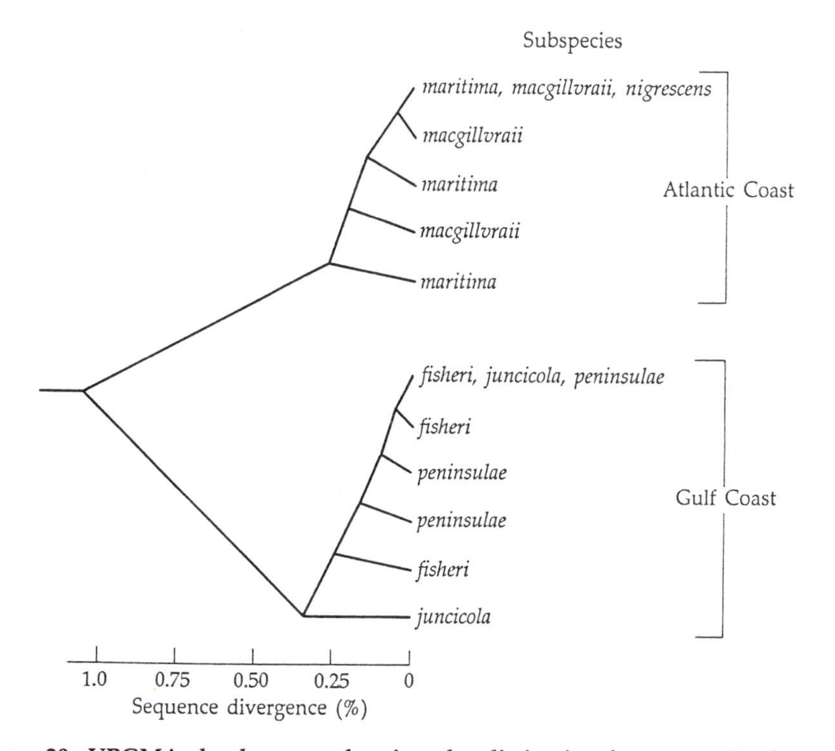

Figure 20. UPGMA dendrogram showing the distinction between mitochondrial DNA genotypes of the Atlantic Coast versus Gulf Coast populations of seaside sparrows. By using maximum parsimony methods many equally parsimonious trees were obtained, including one identical to the tree shown. All alternative trees involved minor branch rearrangements within either the Atlantic clade or the Gulf clade, while the distinction between the two groups remained unaltered. The multiple appearance of the same subspecies name in the dendrogram indicates that different individuals belonging to the same subspecies exhibit different restriction-enzyme patterns. Conversely, the appearance of several subspecies names at the end of a single branch indicates that individuals classified as different subspecies on morphological and zoogeographical grounds exhibit identical restriction-enzyme patterns for the enzymes used. From Avise and Nelson (1989).

tainties in calibration, they do agree well with the dates obtained for the falling of the sea level that exposed the Florida peninsula, which serves as a reproductive barrier between the populations on the two sides of the peninsula.

Most importantly, Avise and Nelson's (1989) molecular study showed that, while the *A. m. nigrescens* subspecies is molecularly indistinguishable from the two other Atlantic subspecies (i.e., *A. m. maritima* and *A. m. macgillivraii*), it is quite different from the Gulf subspecies, such as *A. m. peninsulae*, whose females had been chosen for the breeding program. In conclusion, the salvation program of the dusky seaside sparrow may have rested on an erroneous phylogenetic premise. Instead of reconstructing an extinct subspecies, the program may have created a new one. Thus, knowledge of phylogenetic relationships is essential in making rational decisions for the conservation of biotic diversity. A faulty taxonomy may turn well-intentioned efforts into fiascoes.

PROBLEMS

1. Draw all the possible rooted and unrooted trees for four OTUs: A, B, C, and D.

2. Identify all informative sites in the following five hypothetical sequences:

	1	2	3	4	5	6	7	8
(a)	A	T	G	A	C	T	A	A
(b)	G	T	G	A	T	T	G	A
(c)	A	C	G	G	A	T	A	A
(d)	A	T	G	C	A	T	T	A
(e)	A	C	G	C	A	T	C	A

3. Construct phylogenetic trees by using (a) the UPGMA method, (b) the transformed distance method, and (c) Sattath and Tversky's neighbors relation method to the following distance matrix. In the case of the transformed distance method, assume OTU E to be a known outgroup to all the other OTUs.

	OTU			
OTU	A	B	C	D
B	3			
C	8	7		
D	7	6	3	
E	11	10	13	12

4. The tree in Figure 15b was obtained by the Sattath and Tversky's neighbors relation method (see pages 110–111 and 123–124). Verify the branch lengths shown on the tree by following the method outlined on pages 114–116.

5. In classical entomological taxonomy, the divergence between Hemimetabola (e.g., Orthoptera) and Holometabola (e.g., Diptera, Lepidoptera, Homoptera) is supposed to be very ancient. (a) By using the six 5S rRNA sequences shown in Figure 21, construct a phylogenetic tree by using the UPGMA method. (b) By using five of the sequences in Figure 21 (a, b, c, e, and f) and the maximum parsimony method, construct an unrooted phylogenetic tree. Place the root on the branch connecting the outgroup, *Artemia salina* (f), with the other OTUs. Do the two trees agree with classical taxonomy? Do the two reconstruction methods result in identical topologies? If not, what may be the reasons for the difference?

(a) GCCAACGTCCATACCACGTTGAAAGCACCGGTTCTCGTCCGATCACCGAAGTTAAGCAGC

(b) GGCAACGACCATACCACGTTGAATACACCAGTTCTCGTCCGATCACTGAAGTTAAGCAAC

(c) GCCAACGTCCATACCACGTTGAAAACACCGGTTCTCGTCCGATCACCGAAGTCAAGCAAC

(d) GCCAACGTCCATACCACGTTGAAAACACCGGTTCTCGTCCGATCACCGAAGTTAAGCAAC

(e) GCCAACGACCATACCACGCTGAATACATCGGTTCTCGTCCGATCACCGAAATTAAGCAGC

(f) ACCAACGGCCATACCACGTTGAAAGTACCCAGTCTCGTCAGATCCTGGAAGTCACACAAC

(a) GTCGGGCGCGGTTAGTACTTGGATGGGTGACCGCCTGGGAACCCCGCGTGACGTTGGCA

(b) GTCGGGCGTAGTTAGTACTTGGATGGGTGACCGCTTGGGAACACTACGTGCCGTTGGCA

(c) GTCGGGCGTAGTCAGTACTTGGATGGGTGACCGCCTGGGAACACTACGTGATGTTGGCT

(d) GTCGGGCGCGGTCAGTACTTGGATGGGTGACCACCTGGGAACACCGCGTGCCGTTGGCT

(e) GTCGGGCGCGGTTAGTACTTAGATGGGGGACCGCTTGGGAACACCGCGTGTTGTTGGCC

(f) GTCGGGCCCGGTCAGTACTTGGATGGGTGACCGCCTGGGAACACCGGGTGCTGTTGGCA

Figure 21. DNA sequences of 5S rRNA genes from six arthropods. (a) Locust (*Acheta domesticus*, Orthoptera). (b) Aphid (*Acyrthosiphon magnoliae*, Homoptera). (c) Moth (*Bombyx mori*, Lepidoptera). (d) Moth (*Philosamia cynthia ricini*, Lepidoptera). (e) Fruit fly (*Drosophila melanogaster*, Diptera). (f) Crustacean (*Artemia salina*, Crustacea). Data from Kawata and Ishikawa (1982), Morton and Sprague (1982), Gu et al. (1982), Bagshaw et al. (1987), Cave et al. (1987), and Samson and Wegnez (1988).

FURTHER READINGS

Felsenstein, J. 1988. Phylogenies from molecular sequences: Inference and reliability. Annu. Rev. Genet. 22: 521–565.

Goodman, M. (ed.). 1982. *Macromolecular Sequences in Systematic and Evolutionary Biology*. Plenum, New York.

Hillis, D. M. and C. Moritz (eds.). 1990. *Molecular Systematics*. Sinauer Associates, Sunderland, MA.

Margulis, L. 1981. *Symbiosis in Cell Evolution*. Freeman, San Francisco.

Nei, M. 1987. *Molecular Evolutionary Genetics*. Columbia University Press, New York.

Sneath, P. H. A. and R. R. Sokal. 1973. *Numerical Taxonomy*. Freeman, San Francisco.

Woese, C. R. 1987. Bacterial evolution. Microbiol. Rev. 51: 221–271.

6

EVOLUTION BY GENE DUPLICATION AND EXON SHUFFLING

The importance of gene duplication in evolution was first noted by Haldane (1932) and Muller (1935). They suggested that a redundant duplicate of a gene may acquire divergent mutations and eventually emerge as a new gene. Using molecular, biochemical, and cytological evidence, Ohno (1970) took this suggestion to the extreme, arguing that gene duplication is the only means by which a new gene can arise. Although other means of creating new functions are now known (see page 157), Ohno's view remains largely valid.

The discovery of split genes prompted Gilbert (1978) to suggest that recombination in introns provides a mechanism for the exchange of exon sequences between genes. Many examples of such exon exchanges have been found, indicating that this mechanism has played a significant role in the evolution of eukaryotic genes with new functions.

TYPES OF DNA DUPLICATION

An increase in the number of copies of a DNA segment can be brought about by several types of **DNA duplication**. These are usually classified according to the extent of the genomic region involved. The following types of duplication are recognized: (1) **partial** or **internal gene duplication**, (2) **complete gene duplication**, (3) **partial chromosomal duplication**, (4) **aneuploidy** or **chromosomal duplication**, and (5) **polyploidy** or **genome duplication**. The first four categories are also referred to as **regional duplications**, because they do not affect the entire haploid set of chromosomes. Ohno (1970) has argued that genome duplication has generally been more important than regional duplication, because in the latter case only parts of the regulatory system of structural genes may be duplicated, and such an imbalance may disrupt the normal function of the duplicated genes. However, as discussed below, regional duplications have apparently played a very important role in evolution.

DNA duplication has long been recognized as an important factor in the evolution of genome size (see Ohno 1970). In particular, the duplication of the entire genome or a major part of it, such as a chromosome, may result in a sudden substantial increase in genome size. Genome duplication events have been registered repeatedly during the evolution of different groups of organisms, most notably in plants, bony fishes, and amphibians. Evolutionary pathways resulting in genome enlargement will be discussed in Chapter 8.

DOMAINS AND EXONS

A protein **domain** is a well-defined region within a protein that either performs a specific function, such as substrate binding, or constitutes a stable, compact structural unit within the protein that can be distinguished from all the other parts. The former is referred to as a **functional domain**, and the latter, a **structural domain** or **module** (Gō and Nosaka 1987). Defining the boundaries of a functional domain is often difficult because functionality is in many cases conferred by amino acid residues that are scattered throughout the polypeptide. A structural module, on the other hand, consists of a continuous stretch of amino acids.

The above distinction is important when considering possible evolutionary mechanisms by which multidomain proteins have come into existence. If a functional domain coincides with a module, its duplication will increase the number of functional segments. In contrast, if functionality is conferred by amino acid residues scattered among different modules, the effects of a duplication of a single module may not be functionally desirable. The internal repeats found in many proteins often correspond to either structural modules or single modular functional domains (Barker et al. 1978).

Theoretically, several possible relationships may be envisioned between the structural domains and the arrangement of exons in the gene (Figure 1).

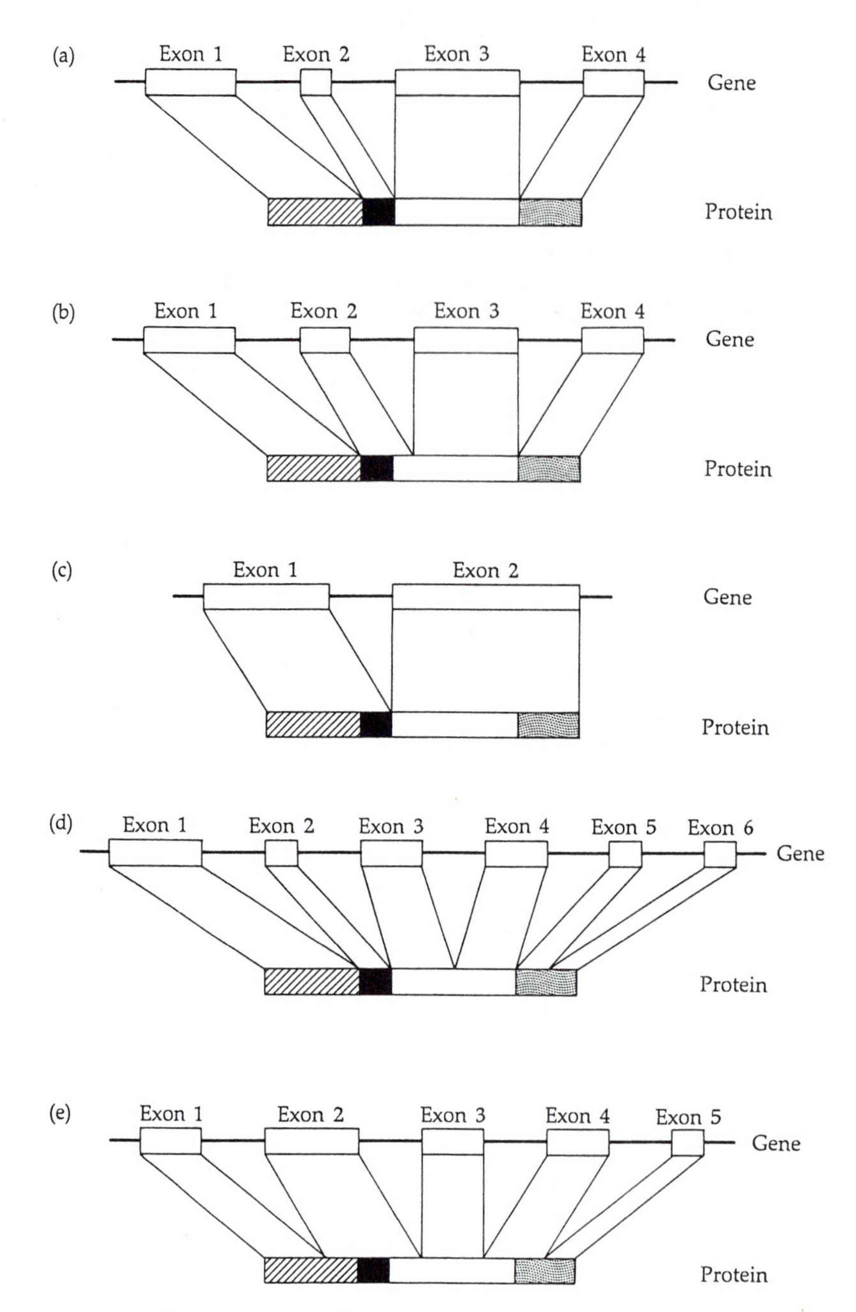

Figure 1. Possible relationships between the arrangement of exons in a gene and the structural domains of the protein it encodes: (a) each exon corresponds exactly to a structural domain; (b) the correspondence is only approximate; (c) an exon encodes two or more domains; (d) a single structural domain is encoded by two or more exons; and (e) lack of correspondence between exons and domains. The four structural domains of the protein are designated by different boxes (hatched, solid black, white, and stippled).

Gō (1981) found that in many globular proteins for which the internal division into structural domains has been determined, a more or less exact correspondence exists between the exons of the gene and the domains (Figure 1a, b). In a few cases, a single module was found to be encoded by more than one exon (Figure 1d). A complete discordance between the modular structure of a protein and the division of its gene into exons (Figure 1e) was not found in her study. In a considerable number of cases, however, several adjacent domains were found to be encoded by the same exon (Figure 1c). Hemoglobin α and β chains, for instance, consist of four domains while their genes consist of only three exons, the second of which encodes two adjacent domains. Gō postulated that a merger occurred between two exons as a result of the loss of an intron. Indeed, the gene for leghemoglobin, the homologous protein in plants, was found to contain an additional intron at exactly the position predicted by the domain structure of globins (following amino acid 68). Thus, during the evolution of the globin family of genes, several lineages lost some or all of their introns (Figure 2).

In the majority of cases, a domain duplication at the protein level indicates that an exon duplication has occurred at the DNA level. It has, thus, been suggested that exon duplication is one of the most important types of internal duplication. Eukaryotic genes generally consist of many exons and introns

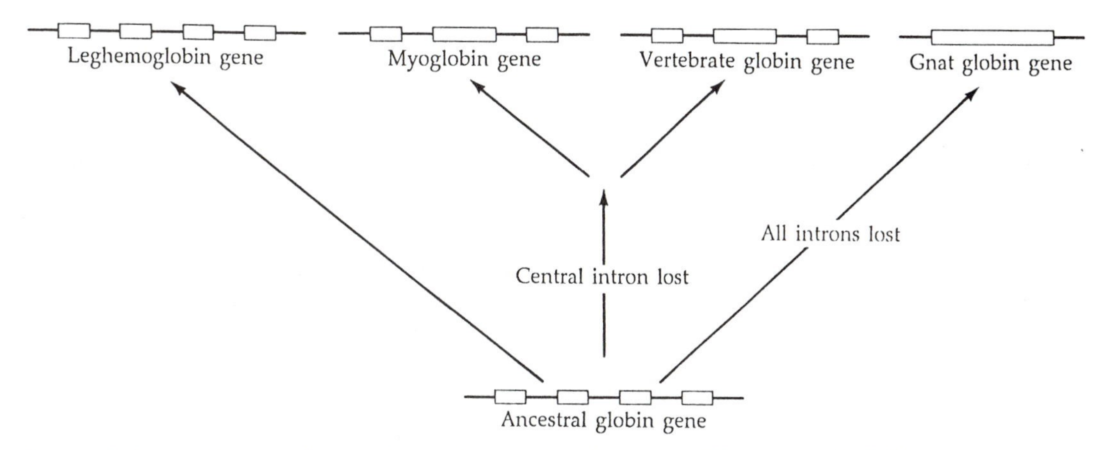

Figure 2. Loss of introns during the evolution of globin genes. The ancestral globin gene had three introns and four exons. The leghemoglobin gene has retained the ancestral structure, while other lineages have lost at least one intron. Note that the introns are not drawn to scale. The two introns of the myoglobin gene in mammals (bovine, human, mouse, pig and seal) are ~4,800 bp and ~3,400 bp in length, while the homologous introns of the globin and leghemoglobin genes are only 108–192 bp and 103–904 bp long, respectively. The middle intron of the leghemoglobin genes is 99–234 bp long (Blanchetot et al. 1983). The globin data are derived from many amphibians, birds and mammals. The leghemoglobin data are derived from three legume species (*Phaseolus vulgaris, Glycine max,* and *Vicia faba*).

(Chapter 1), and neighboring exons are often identical or very similar to one another. These facts suggest that many complex genes in modern organisms have evolved by internal duplication and subsequent modification of primordial genes, which presumably contained one or few exons and could only perform simple biological functions (Li 1983).

DOMAIN DUPLICATION AND GENE ELONGATION

A survey of modern genes in eukaryotes shows that internal duplications have occurred frequently in evolution. This increase in gene size, or **gene elongation**, is one of the most important steps in the evolution of complex genes from simple ones. Theoretically, elongation of genes can also occur by other means. For example, a mutational change converting a stop codon into a sense codon can also elongate the gene (Chapter 1). Similarly, insertion of a foreign DNA segment into one of the exons or the occurrence of a mutation obliterating a splicing site will achieve the same result. These types of molecular changes, however, would most probably disrupt the function of the elongated gene, because the added regions would consist of an almost random array of amino acids. Indeed, in the vast majority of cases, such molecular changes have been found to be associated with pathological manifestations. For instance, the hemoglobin abnormalities Constant Spring and Icaria resulted from mutations turning the stop codon into codons for glutamine and lysine, respectively, thus adding 30 additional residues to the α chains of these variants (Weatherall and Clegg 1979). By contrast, duplication of a structural domain is less likely to be problematic. Indeed, such a duplication can sometimes even enhance the function of the protein produced, for example, by increasing the number of active sites.

Thus, gene elongation during evolution seems to have depended largely on the duplication of domains. In the following section, we shall present an example of internal gene duplications to illustrate the consequences of gene elongation during evolution.

The ovomucoid gene

Ovomucoid is a protein present in the egg white of birds that can inhibit the activity of trypsin, an enzyme that catalyzes the digestion of proteins. The ovomucoid polypeptide can be divided into three functional domains (Figure 3). Each domain is capable of binding one molecule of either trypsin or another serine proteinase. The DNA regions coding for the three functional domains clearly share a common evolutionary origin and are separated from each other by introns (Stein et al. 1980). Each of the three regions consists

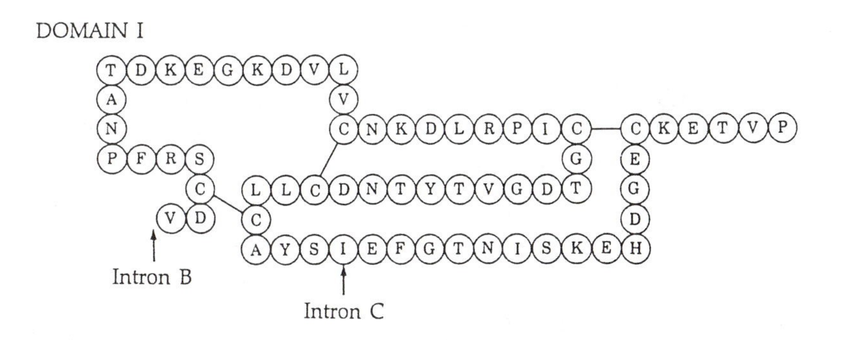

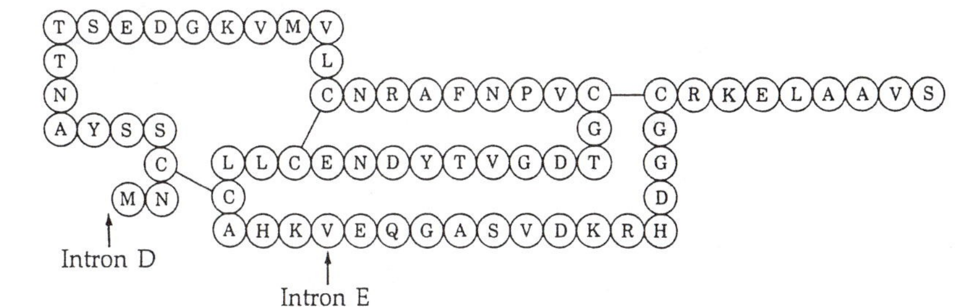

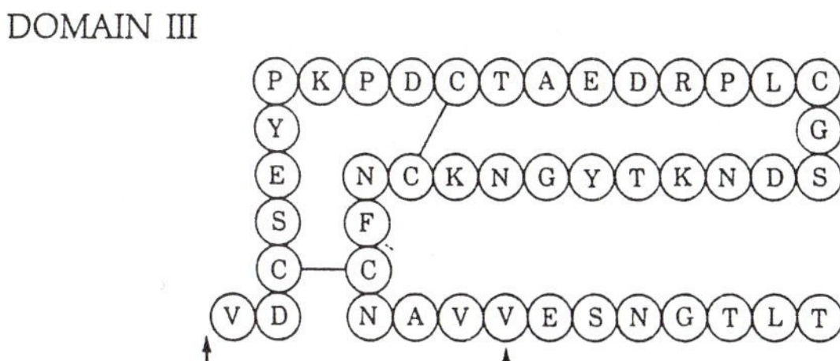

	Percent similarity	
Domains	Amino acids	Nucleic acids
I vs. II	46	66
II vs. III	30	42
I vs. III	33	50

Figure 3. The three functional domains of the secreted ovomucoid and the degree of sequence similarity between the domains at the amino acid and nucleotide levels. From Stein et al. (1980).

of two exons interrupted by an intron, and these two exons exhibit no similarity between them. Thus, the ovomucoid gene appears to have been derived from a primordial single-domain gene by two internal duplications, each of which involved two neighboring exons. Since domains I and II are more similar to each other than either of them is to domain III, they were probably derived from the second duplication, while domain III was the product of the first duplication.

Prevalence of domain duplication

Table 1 presents a list of several genes for which there is evidence of internal duplication during their evolutionary histories. All involve one or more domain duplications, and some of the sequences were derived from multiplications of a primordial sequence, resulting in a repetitive structure that takes up the entire length of the protein. In each of these examples, the duplication event could easily be inferred from protein or DNA sequence similarity. There may be many other complex genes that have also evolved by internal gene duplication, but their duplicated regions may have diverged from each other to such an extent that the sequence homology between them is no longer discernable. In some cases, such as the constant and variable regions of immunoglobulin genes, we can infer common ancestry by comparing the secondary structures of the domains, because the secondary struc-

Table 1. Proteins with internal domain duplications.

Sequence	Length of protein[a]	Length of repeat	Number of repeats	Percent repetition[b]
Immunoglobulin ε-chain C region (humans)	423	108	4	100
Immunoglobulin γ-chain C region (humans)	329	108	3	98
Serum albumin (humans)	584	195	3	100
Parvalbumin (humans)	108	39	2	72
Protease inhibitor, Bowman-Birk type (soybean)	71	28	2	79
Protease inhibitor, submandibular-gland type (rodents)	115	54	2	94
Ferredoxin (*Clostridium pasteurianum*)	55	28	2	100
Plasminogen (humans)	790	79	5	50
Calcium-dependent regulator protein (humans)	148	74	2	100
Tropomyosin α chain (humans)	284	42	7	100

From Barker et al. (1978).

[a] Number of amino acid residues.

[b] The percent of the total length of the protein occupied by repeated sequences.

ture has been preserved better than the amino acid sequence. Thus, internal duplications in proteins are probably much more ubiquitous than the empirical data have indicated.

FORMATION OF GENE FAMILIES AND THE ACQUISITION OF NEW FUNCTIONS

A complete gene duplication produces two identical copies. How they will evolve varies from case to case. The copies may, for instance, retain their original function, enabling the organism to produce a larger quantity of certain RNA species or proteins. Alternatively, one of the copies may be incapacitated by the occurrence of a deleterious mutation and become a functionless pseudogene (see page 147). More importantly, however, gene duplication may result in the emergence of genetic novelties or new genes. This will happen if one of the duplicates retains its original function while the other accumulates molecular changes such that, in time, it can perform a rather different task.

Repeated genes can be divided into two types: variant and invariant repeats. **Invariant repeats** are identical or nearly identical in sequence to one another. In several cases, the repetition of identical sequences can be shown to be correlated with the synthesis of increased quantities of a gene product that is required for the normal function of the organism. Such repetitions are referred to as **dose repetitions**. Dose repetitions are quite common whenever a metabolic need for producing large quantities of specific RNAs or proteins arises (Ohno 1970). Representative examples include the genes for rRNAs and tRNAs, which are required for translation, and the histone genes, which constitute the chief protein component of chromosomes and therefore must be synthesized in large quantities.

Variant repeats consist of copies of a gene that, although similar to each other, differ in their sequence to a lesser or greater extent. Interestingly, variant repeats can sometimes perform markedly different functions. For example, thrombin, which cleaves fibrinogen during the process of blood clotting, and the digestive enzyme trypsin have been derived from a complete gene duplication in the past. Similarly, lactalbumin, a subunit of the enzyme that catalyzes the synthesis of the sugar lactose, and lysozyme, which dissolves certain bacteria by cleaving the polysaccharide component of their cell walls, are related by descent to each other. Differentiation in function usually requires a large number of substitutions. However, under certain circumstances, a novel function may be created through a relatively small number of substitutions (e.g., Betz et al. 1974).

All the genes that belong to a certain group of repeated sequences in a genome are referred to as a **gene family** or **multigene family**. Members of a gene family usually reside in close proximity to each other on the same chromosome. In some cases, some functional or nonfunctional family members may be located on other chromosomes.

When duplicate genes become very different from each other in either function or sequence, it may no longer be convenient to assign them to the same gene family. The term **superfamily** was coined by Dayhoff (1978) in order to delineate between closely related and distantly related proteins. Accordingly, proteins that exhibit at least 50% similarity to each other at the amino acid level are considered to be members of a family, while homologous proteins exhibiting less than 50% similarity are considered to be members of a superfamily. For example, the α-globins and the β-globins are classified into two separate families, and together with myoglobin they form the globin superfamily (see page 151). However, the two terms cannot always be used strictly according to Dayhoff's criteria. For example, human and carp α globin chains exhibit only a 46% sequence similarity, which is below the limit for assignment to the same gene family. For this reason, the classification of proteins into families and superfamilies is determined not only according to sequence resemblance, but also by considering auxiliary evidence pertaining to functional similarity or tissue specificity.

The number of genes within gene families varies widely. Some genes are repeated within the genome a few times and these are referred to as **lowly repetitive.** Others may be repeated hundreds of times within the genome and are called **highly repetitive.** In the following sections, rRNA and tRNA genes will be used to illustrate highly repetitive invariant genes. Lowly repetitive genes will be represented by isozymes and the color-sensitive pigment proteins.

RNA-specifying genes

Table 2 shows the numbers of rRNA and tRNA genes for a variety of organisms. The mitochondrial genome of mammals contains only one copy of both the 12S and the 16S rRNA genes. This is apparently sufficient for the mitochondrial translation system, because the genome contains only 13 protein-coding genes (Chapter 4). The mycoplasmas, which are the smallest self-replicating prokaryotes, contain two sets of rRNA genes. The genome of *Escherichia coli* is 4–5 times larger and it contains seven sets of rRNA genes. The number of rRNA genes in yeast is approximately 140, and the numbers in fruit flies and humans are even larger. *Xenopus laevis* has a larger genome and more rRNA genes than humans. Thus, there exists a strong positive

Table 2. Numbers of rRNA and tRNA genes per haploid genome in various organisms.

Genome source	Number of rRNA genes[a]	Number of tRNA genes	Approximate genome size (bp)
Human mitochondrion	1	22	1.7×10^4
Mycoplasma capricolum	2	ND[b]	1×10^6
Escherichia coli	7	~100	4×10^6
Saccharomyces cerevisiae	~140	320–400	5×10^7
Drosophila melanogaster	130–250	~750	2×10^8
Human	~300	~1,300	3×10^9
Xenopus laevis	400–600	~7,800	8×10^9

From Li (1983).

[a] For rRNA genes, the values refer to the number of complete sets of rRNA genes.

[b] ND = not determined.

correlation between the number of rRNA genes and genome size. This rule also holds for the tRNA genes (Table 2) and other RNA-specifying genes.

Highly repetitive genes, such as the rRNA genes, are generally very similar to each other. One factor responsible for the homogeneity may be purifying selection, because these genes should abide by very specific functional and structural requirements. However homogeneity often extends to regions devoid of any functional significance, and thus the maintenance of homogeneity necessitates that other mechanisms be invoked (see page 162).

Isozymes

In addition to invariant repeats, the genomes of higher organisms contain numerous multigene families whose members have diverged from each other to various extents. Good examples are families of genes coding for isozymes, such as lactate dehydrogenase, aldolase, creatine kinase, and pyruvate kinase. **Isozymes** are enzymes that catalyze the same biochemical reaction but may differ from each other in tissue specificity, developmental regulation, electrophoretic mobility, or biochemical properties. Note that isozymes are encoded by different loci, usually duplicated genes, as opposed to **allozymes,** which are distinct forms of an enzyme encoded by different alleles at a single locus.

Let us consider the two genes encoding the A and B subunits of lactate dehydrogenase (LDH) in vertebrates. These two subunits form five tetrameric isozymes, A_4, A_3B, A_2B_2, AB_3, and B_4, all of which catalyze either the conversion of lactate into pyruvate in the presence of the oxidized coenzyme nicotinamide adenine dinucleotide (NAD^+) or the reverse reaction in the presence of the reduced coenzyme (NADH). It has been suggested that LDH-B_4 and the other isozymes rich in B subunits, which have a high affinity for NAD^+, function as true lactate dehydrogenases in aerobically metabolizing tissues such as the heart, whereas LDH-A_4 and the isozymes rich in A subunits, which have a high affinity for NADH, are especially geared to serve as pyruvate reductases in anaerobically metabolizing tissues such as skeletal muscle (Everse and Kaplan 1975; Nadal-Ginard and Markert 1975). Figure 4 shows the developmental sequence of LDH production in the heart. We see that the more anaerobic the heart is, specifically in the early stages of gestation, the higher the proportion of LDH isozymes rich in A subunits. Thus, the two duplicate genes have become specialized to different tissues and to different developmental stages. As the two subunits are present in almost all the vertebrates studied to date, the duplication that produced the genes for LDH-A and LDH-B probably occurred either before or during the early stages of vertebrate evolution. An interesting feature of LDH is that the two subunits can form heteromultimers, thus further increasing the physiological versatility of the enzyme.

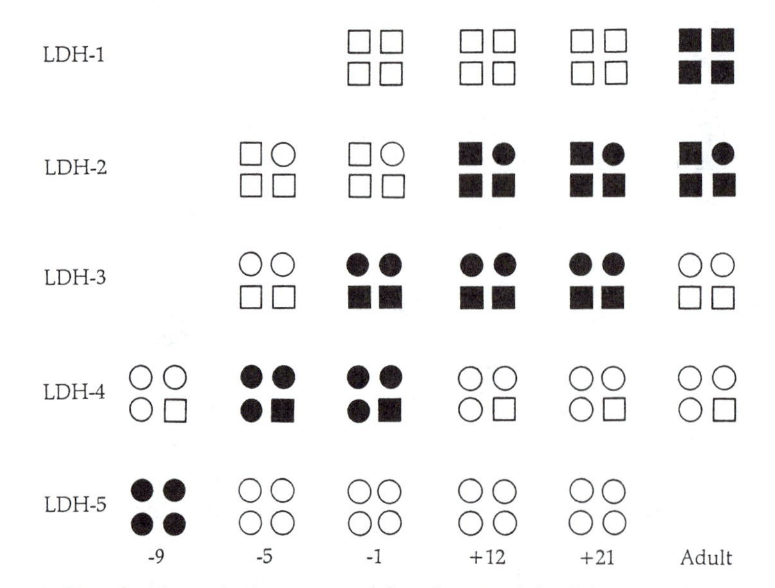

Figure 4. Developmental sequences of five lactate dehydrogenase (LDH) isozymes in the rat heart. Negative and positive numbers denote days before or after birth, respectively. Squares indicate B subunits, circles indicate A subunits. Solid symbols indicate quantitatively predominant forms. Notice the shift from A to B subunits during ontogenesis. Data from Markert and Ursprung (1971).

Color-sensitive pigment proteins

Humans, apes, and Old World monkeys possess three color-sensitive pigment proteins. The blue pigment is encoded by an autosomal gene, while the red and the green pigments are each encoded by an X-linked gene (Nathans et al. 1986). The amino acid sequences of the red and green pigments are 96% identical, but their similarity with the blue pigment is only 43%. The blue-pigment gene and the ancestor of the green- and red-pigment genes diverged about 500 million years ago. In contrast, the close linkage and high homology between the red and green pigments point to a very recent gene duplication. Because New World monkeys have only one X-linked pigment gene, whereas Old World monkeys and humans have two or more, it is assumed that the duplication occurred in the ancestor of Old World monkeys after their divergence from the New World monkeys about 35–40 million years ago. As a consequence of this duplication, humans, apes, and Old World monkeys can distinguish three colors (i.e., they are **trichromatic**), whereas New World monkeys, such as squirrel monkeys, can only distinguish between blue and green or blue and red, but not between green and red (i.e., they are **dichromatic**).

Interestingly, female squirrel monkeys heterozygous for two X-linked alleles are trichromatic (Jacobs and Neitz 1986). On the other hand, males, who carry only one X chromosome, can never achieve true trichromatic vision. Thus, in the case of humans and Old World monkeys, trichromatic vision is achieved by a mechanism akin to isozymes (i.e., two distinct proteins encoded by different loci), while in heterozygous female squirrel monkeys the same end is achieved through the use of two "allozymes" (i.e., two distinct allelic forms at a single locus) (Figure 5). If trichromatic vision confers a selective advantage on its carriers, then the long-term maintenance of two color-sensitive alleles at a locus in New World monkeys has probably been achieved through a form of overdominant selection (Chapter 2).

NONFUNCTIONALIZATION OF DUPLICATE GENES

A redundant duplicate gene is more likely to become nonfunctional than to evolve into a new gene, because deleterious mutations occur far more often than advantageous ones. The nonfunctionalization of a duplicate gene produces a pseudogene. Pseudogenes thus produced are called **unprocessed** pseudogenes, as opposed to processed pseudogenes, which will be discussed in Chapter 7. Table 3 lists the structural defects found in several globin pseudogenes. Most of these unprocessed pseudogenes contain multiple defects such as frameshifts, premature stop codons, and obliterations of

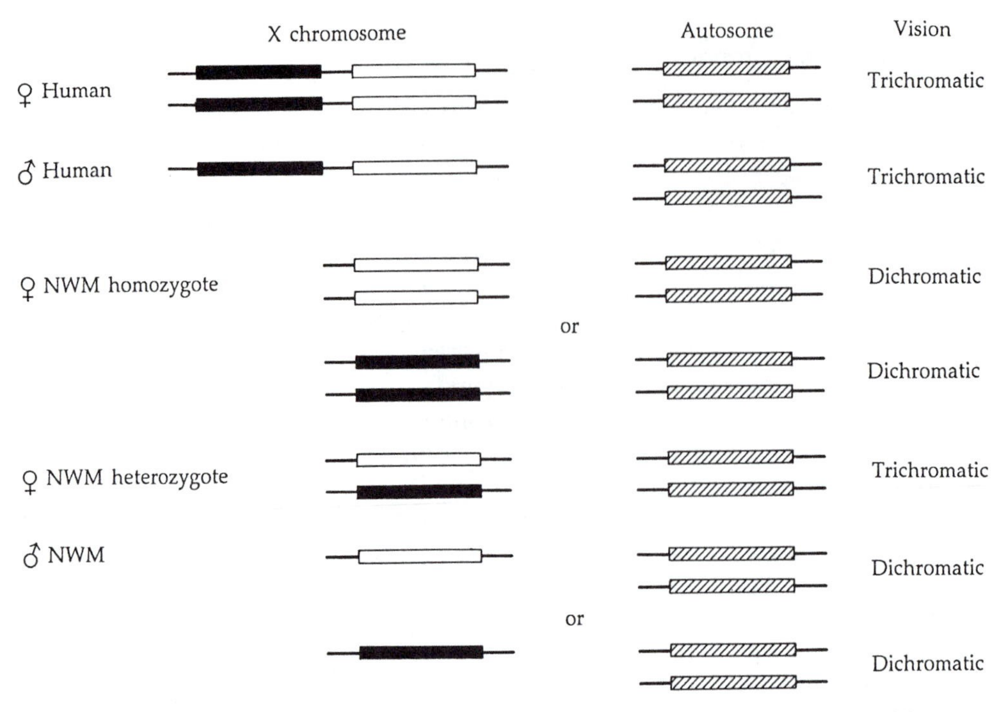

Figure 5. Molecular basis of trichromatic vision in males and females of humans and New World monkeys (NWM). Note that male New World monkeys cannot achieve trichromatic vision. The solid, empty, and hatched boxes denote the green-, red-, and blue-pigment genes, respectively.

Table 3. Defects in globin pseudogenes.[a]

Pseudogene	TATA box	Initiation codon	Frame-shift	Premature stop	Lacks essential amino acid	Splice GT/AG rule	Altered stop codon	Polyadenylation signal: AATAAA
Human $\psi\alpha 1$		+	+	+	+	+	+	+
Human $\psi\zeta 1$				+				
Mouse $\psi\alpha 3$	+		+	+		+		
Mouse $\psi\alpha 4$			+		+			
Mouse $\beta h3$	?	+	+	+	+	+	?	?
Goat $\psi\beta^x$	+		+	+	+	+	+	+
Goat $\psi\beta^z$	+		+	+	+	+	+	+
Rabbit $\psi\beta 2$			+	+	+	+		

From Li (1983).

[a] A plus indicates the existence of a particular type of defect; a question mark indicates the possibility of a defect.

splicing sites or regulatory elements, so that it is difficult to identify the mutation that was the direct cause of gene silencing. In a few cases, identification of the "culprit" is possible. For example, human $\psi\zeta$ contains only a single major defect, a nonsense mutation, which is probably the direct cause of nonfunctionalization. (The notation ψ is used to distinguish a pseudogene from its functional counterpart.) Some pseudogenes, such as $\psi\beta^X$ and $\psi\beta^Z$ in the goat β-globin multigene family, have been derived from the duplication of a preexisting pseudogene.

DATING GENE DUPLICATIONS

Two genes are said to be **paralogous** if they are derived from a duplication event, but **orthologous** if they are derived from a speciation event. For example, in Figure 6, genes α and β were derived from duplication of an ancestral gene and are therefore paralogous, while gene α from species 1 and gene α from species 2 are orthologous, and so are gene β from species 1 and gene β from species 2.

We can estimate the date of duplication, T_D, from sequence data if we know the rate of substitution in genes α and β. The rate of substitution can be estimated from the number of substitutions between the orthologous genes in conjunction with knowledge of the time of divergence, T_S, between species 1 and 2 (Figure 6). We show below how an estimate of T_D can be obtained.

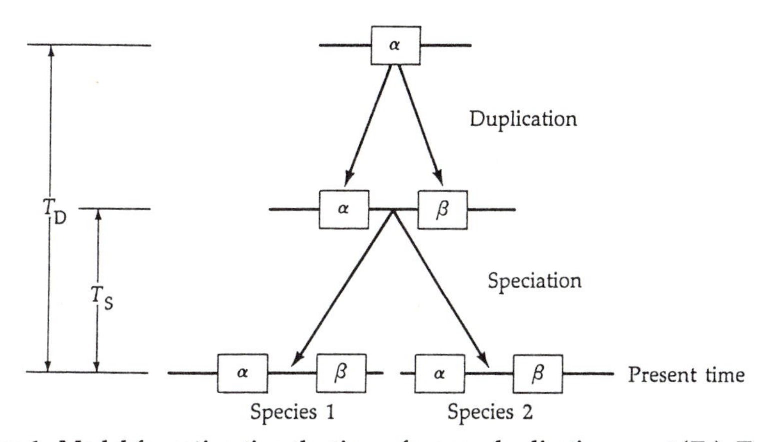

Figure 6. Model for estimating the time of a gene duplication event (T_D). Two genes α and β were derived from a duplication event T_D time units ago in an ancestral species. The species then split into two species, 1 and 2, T_S time units ago. The α genes in species 1 and 2 are orthologous, and so are the β genes, but the α genes are paralogous to the β genes.

For gene α, let K_α be the number of substitutions per site between the two species. Then, the rate of substitution in gene α is estimated by

$$r_\alpha = K_\alpha/(2T_S) \tag{6.1}$$

The rate of substitution in gene β, r_β, can be obtained in a similar manner. The average substitution rate for the two genes is given by

$$r = (r_\alpha + r_\beta)/2 \tag{6.2}$$

To estimate T_D, we need to know the number of substitutions per site between genes α and β ($K_{\alpha\beta}$). This number can be estimated from four pairwise comparisons: (1) gene α from species 1 and gene β from species 2, (2) gene α from species 2 and gene β from species 1, (3) both genes from species 1, and (4) both genes from species 2. From these four estimates we can compute the average value for $K_{\alpha\beta}$ ($\overline{K}_{\alpha\beta}$), from which we can estimate T_D as

$$T_D = \overline{K}_{\alpha\beta}/(2r) \tag{6.3}$$

Note that in the case of protein-coding genes, by using the numbers of synonymous and nonsynonymous substitutions separately, we can obtain two independent estimates of T_D. The average of these two estimates may be used as the final estimate of T_D. However, if the number of substitutions per synonymous site between genes α and β is large, say larger than 1, then the number of synonymous substitutions cannot be estimated accurately, and so synonymous substitutions may not provide a reliable estimate of T_D. In such cases, only the number of nonsynonymous substitutions should be used. Conversely, if the number of substitutions per nonsynonymous site between the paralogous genes is small, then the estimate of the number of nonsynonymous substitutions is subject to a large sampling error, and in such cases, only the number of synonymous substitutions should be used.

In the above, we have assumed rate constancy. This assumption can be tested by the four pairwise comparisons mentioned above. The assumption fails if an approximate equality does not hold among the four comparisons. As will be discussed later (see page 162), problems due to concerted evolutionary events may also arise and complicate the estimation of T_D.

Another method for dating gene-duplication events is to consider the phylogenetic distribution of genes in conjunction with paleontological data pertinent to the divergence date of the species in question. For example, all vertebrates with the exception of jawless fish (Agnatha) encode α and β globin chains. There are two possible explanations for this observation. One is that the duplication event producing the α- and β-globins occurred in the common ancestor of the Agnatha and the other vertebrates, but all the Agnatha species have lost one of the two duplicates. This is possible but not very likely because such a scenario would require that the losses occurred independently in many evolutionary lineages. The other explanation is that the duplication event occurred after the divergence of jawless fish from the ancestor of all other vertebrates but before the radiation of the other vertebrates from each other (450–500 million years ago). This latter explanation is thought to be more plausible, and the duplication date is commonly taken to be 450–500 million years ago (Dayhoff 1972; Dickerson and Geis 1983).

Obviously, the above methods can only provide us with rough estimates of duplication dates, and so all estimates should be taken with caution.

THE GLOBIN SUPERFAMILY OF GENES

The globin superfamily has experienced all the possible evolutionary pathways that can occur in families of repeated sequences: (1) retention of original function, (2) acquisition of new function, and (3) loss of function in some duplicates. In humans, the globin superfamily consists of three families: the myoglobin family, whose single member is located on chromosome 22, the α-globin family on chromosome 16, and the β-globin family on chromosome 11 (Figure 7). Together these three families produce two types of functional proteins: myoglobin and hemoglobin. The two proteins diverged about 600–800 million years ago (see Figure 8; Dayhoff 1972; Doolittle 1987;) and have become specialized in several respects. In terms of tissue specificity, myoglobin became the oxygen-storage protein in muscles, whereas hemoglobin became the oxygen carrier in blood. In terms of quaternary structure, myoglobin retains a monomeric structure, while hemoglobin has become a tetra-

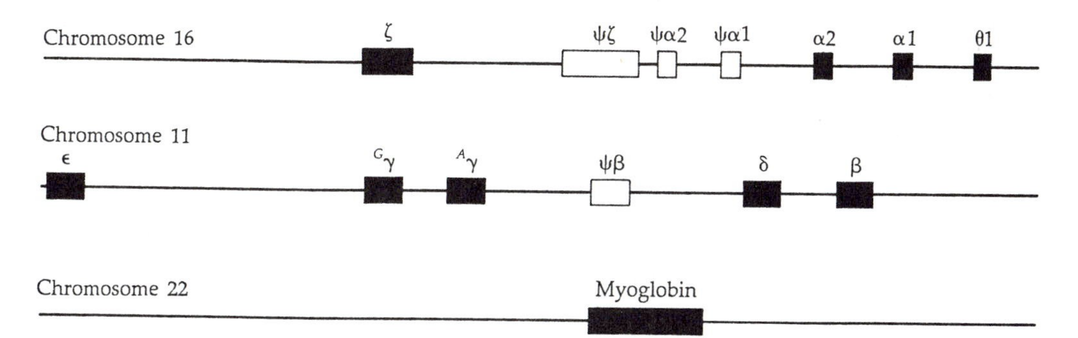

Figure 7. The chromosomal arrangement of the three gene families belonging to the globin superfamily of genes in humans: the α-globin family on chromosome 16, the β-globin family on chromosome 11, and myoglobin on chromosome 22. Solid black boxes denote functional genes, while empty boxes denote pseudogenes.

mer. In terms of function, myoglobin evolved a higher affinity for oxygen than did hemoglobin, while the function of hemoglobin has become much more refined and regulated (see Stryer 1988). Mammalian hemoglobin, for instance, has the capability to regulate its oxygen affinity according to the level of organic phosphate in the blood. Apparently, the heteromeric structure has facilitated the refinement of the function of hemoglobin.

The hemoglobin in humans and the vast majority of vertebrates is made up of two types of chains, one encoded by an α family member, the other by a member of the β family. As discussed above, the α and β families diverged about 450–500 million years ago (Figure 8). Since jawless fish contain only one type of monomeric hemoglobin, polymerization of hemoglobin in vertebrates must have occurred close to the time of the α–β divergence.

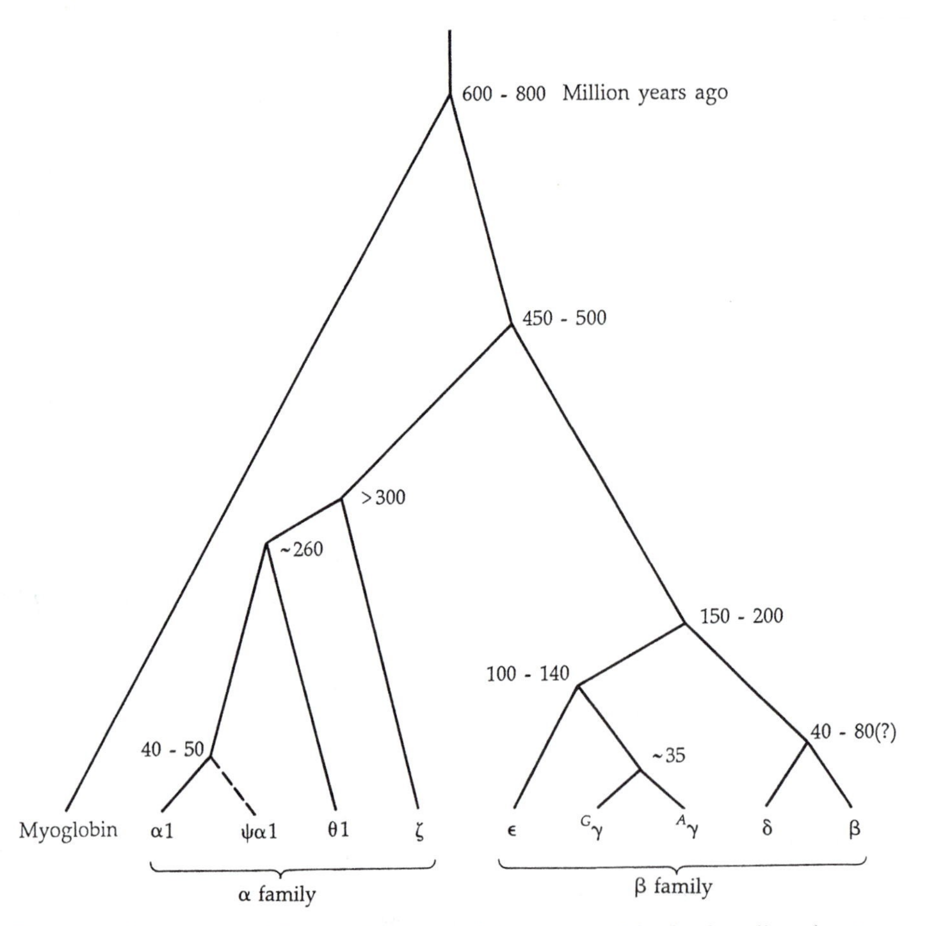

Figure 8. Evolutionary history of human globin genes. The broken line denotes a pseudogene lineage. Only one of the two α-globin genes is shown in the figure because the date of their divergence from each other is uncertain.

In humans, the α family consists of four functional genes: ζ, $\alpha1$, $\alpha2$, and the recently discovered $\theta1$ (Figure 7). It also contains three pseudogenes: $\psi\zeta$, $\psi\alpha1$, and $\psi\alpha2$. The β family consists of five functional genes: ϵ, ${}^G\gamma$ ${}^A\gamma$, β, and δ, and one pseudogene, $\psi\beta$. The two families have diverged in both physiological properties and ontological regulation. In fact, distinct globins appear at different developmental stages: $\zeta_2\epsilon_2$ and $\alpha_2\epsilon_2$ in the embryo, $\alpha_2\gamma_2$ in the fetus, and $\alpha_2\beta_2$ and $\alpha_2\delta_2$ in adults; the time at which $\theta1$ is expressed is not known. Furthermore, differences in oxygen-binding affinity have evolved among these globins. For example, the fetal hemoglobin $\alpha_2\gamma_2$ has a higher oxygen affinity than either adult hemoglobin ($\alpha_2\beta_2$ and $\alpha_2\delta_2$) and can thus function better in the fetus, which exists in a relatively hypoxic (low-oxygen) environment (Wood et al. 1977). This phenomenon exemplifies again the fact that gene duplication can result in refinements of a physiological system.

Among the α family members the embryonic type, ζ, is the most divergent, having branched off more than 300 million years ago (Figure 8). The $\theta1$ globin branched off about 260 million years ago. Because the divergence time between the two α genes is uncertain, only the *$\alpha1$* gene is shown in the figure. The $\alpha1$ and $\alpha2$ genes have almost identical DNA sequences and produce an identical polypeptide. This would seem to indicate a very recent divergence time. However, the similarity could also be the result of concerted evolution (Zimmer et al. 1980), a phenomenon that will be discussed later (see page 167). The two genes are present in humans and all the apes and so could have arisen more than 20 million years ago.

Among the β family members, the adult types (β and δ) and the nonadult types (γ and ϵ) diverged about 155–200 million years ago (Efstratiadis et al. 1980). The ancestor of the two γ genes diverged from the ϵ gene about 100–140 million years ago. The duplication that created ${}^G\gamma$ and ${}^A\gamma$ occurred after the separation of the human lineage from the New World monkey lineage, about 35 million years ago (Shen et al. 1981). The divergence between the δ and β genes was previously estimated to be 40 million years ago (Dayhoff 1972; Efstratiadis et al. 1980), but recent DNA sequence data suggest that it may have occurred even before the eutherian radiation, about 80 million years ago (Hardison and Margot 1984; Goodman et al. 1984). We note from the above discussion that, in both families, there is a good correlation between the time of divergence and the degree of functional or regulatory divergence between genes.

EXON SHUFFLING

There are two types of **exon shuffling**: exon duplication and exon insertion. Exon duplication refers to duplication of one or more exons in a gene and so is a type of internal duplication, which was discussed in the context of gene elongation (see page 140). Exon insertion is the process by which structural or functional domains are exchanged between proteins or inserted into a protein. Both types of shuffling have been used in the evolutionary process of creating new genes. Here, we discuss the insertion of an exon from one gene into another, with the consequent production of mosaic or chimeric proteins (Doolittle 1985; Patthy 1985).

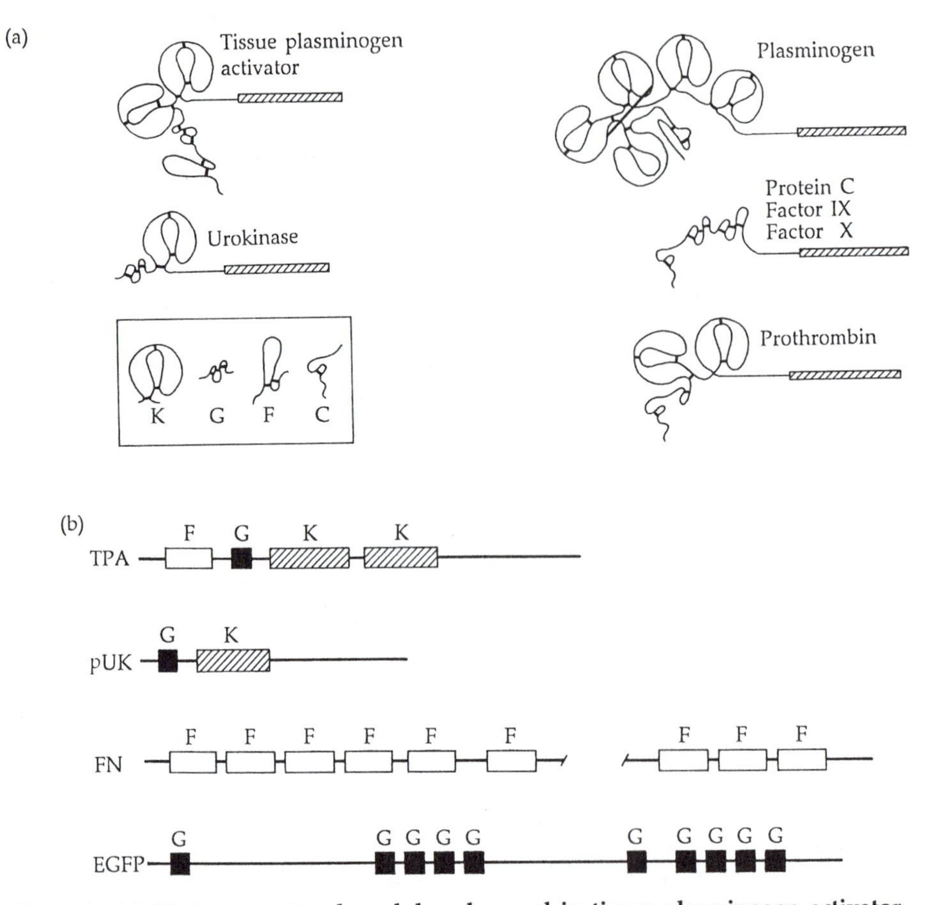

Figure 9. (a) Various structural modules observed in tissue plasminogen activator (TPA) and other proteins involved in blood coagulation and fibrinolysis. The inset shows the structure of the modules in the nonproteinase regions: K, kringle module; G, growth-factor module; F, finger module; C, vitamin K-dependent calcium-binding module. The crosshatched bars represent the proteinase regions homologous to trypsin. Modified from Patthy (1985). (b) The origin of the modules acquired through exon insertion in the tissue plasminogen activator (TPA) protein. pUK, prourokinase; EGFP, epidermal growth factor precursor; FN, fibronectin.

Mosaic proteins

The first mosaic protein to be discovered was tissue plasminogen activator (TPA) (Figure 9). Plasminogen is converted by TPA into its active form, plasmin, which dissolves fibrin, a soluble fibrous protein in blood clots. The conversion of plasminogen into plasmin is greatly accelerated by the presence of fibrin, the substrate of plasmin. Fibrin binds both plasminogen and TPA, thus aligning them for catalysis. This mode of molecular alignment allows plasmin production only in the proximity of fibrin, thus conferring fibrin-specificity to plasmin. By contrast, urokinase (UK), a urinary plasminogen activator, lacks fibrin-specificity. A comparison of the amino acid sequences of TPA and the precursor of UK, prourokinase, showed that TPA contains a 43-residue sequence at its amino-terminal end that has no counterpart in UK (Banyai et al. 1983). This segment can form a finger-like structure (Figure 9a) and is homologous to the finger domains responsible for the fibrin affinity of another protein, fibronectin (FN) (Figure 9b), which is a large glycoprotein present in the plasma and on cell surfaces that promotes cellular adhesion. Deletion of this segment leads to a loss of the fibrin affinity of TPA. The homology of TPA with FN is restricted to this finger domain. Thus, exon shuffling must have been responsible for the acquisition of this domain by TPA from either FN or a similar protein.

TPA also contains a segment homologous to epidermal growth factor (EGF) and the growth-factor-like regions of other proteins, such as Factor IX and Factor X, which are blood-clotting enzymes in the blood-coagulation pathway. In addition, the carboxy-terminal regions of TPA are homologous to the proteinase parts of trypsin and other trypsin-like serine proteinases, which are enzymes that hydrolyze proteins into peptide fragments. Finally, the nonproteinase parts of TPA contain two structures similar to the kringles of plasminogen. (A kringle is a cysteine-rich sequence that contains three internal disulfide bridges and forms a pretzel-like structure resembling the Danish cake bearing this name.) Thus, during its evolution, TPA captured at least four DNA segments from at least three other genes: plasminogen, epidermal growth factor, and fibronectin (Figure 9b). Moreover, the junctions of these acquired units coincide precisely with the borders between exons and introns, thus lending further credibility to the idea that exons have indeed been transferred from one gene to another. For more examples of exon shuffling, see Doolittle (1985) and Patthy (1985).

Phase limitations on exon shuffling

For an exon to be inserted into an intron of a gene without causing a frameshift in the reading frame, the phase limitations of the receiving gene must be respected. In order to understand this constraint, let us consider the

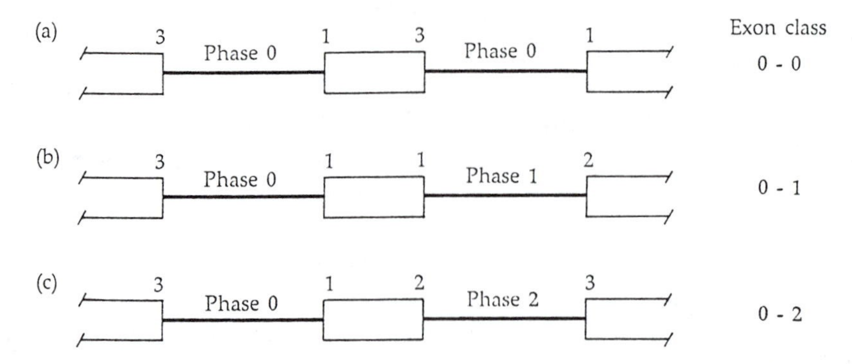

Figure 10. Phases of introns and classes of exons. Exons are represented by boxes. The number at the exon–intron junction indicates the codon position of the last nucleotide of the exon, while the number at the intron–exon junction indicates the codon position of the first nucleotide of the exon. Only three of the nine possible exon classes are shown.

different types of introns in terms of their possible positions relative to the coding regions. Introns residing between coding regions are classified into three types according to the way in which the coding region is interrupted. An intron is of phase 0 if it lies between two codons, of phase 1 if it lies between the first and second nucleotides of a codon, and of phase 2 if it lies between the second and third nucleotides of a codon (Figure 10). Exons are grouped into classes according to the phases of their flanking introns. For

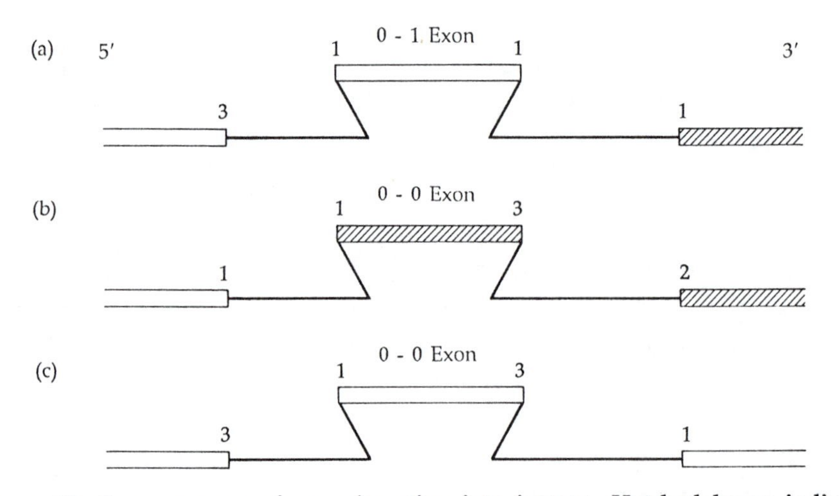

Figure 11. Consequences of exon insertion into introns. Hatched boxes indicate frameshifts. (a) Insertion of a 0-1 asymmetrical exon into a phase-0 intron; (b) insertion of a 0-0 symmetrical exon into a phase-1 intron; (c) insertion of a 0-0 symmetrical exon into a phase-0 intron. The insertions in (a) and (b) are abortive insertions.

example, the middle exon in Figure 10b is flanked by a phase-0 intron at its 5′ end and by a phase-1 intron at its 3′ end and is said to be of class 0-1. An exon that is flanked by introns of the same phase at both ends is called a **symmetrical exon**, otherwise it is **asymmetrical.** For example, the middle exon in Figure 10a is symmetrical. Of the nine possible classes of exons, three are symmetrical (0-0, 1-1, and 2-2) and six are asymmetrical.

Only symmetrical exons can be inserted into introns. For example, in Figure 11a the insertion of a 0-1 exon into a phase-0 intron causes a frameshift in all the subsequent exons. Moreover, insertion of symmetrical exons is also restricted; a 0-0 exon can be inserted only into introns of phase 0, an 1-1 exon can be inserted only into introns of phase 1, and a 2-2 exon can be inserted only into introns of phase 2. For example, Figure 11b shows that the insertion of a 0-0 exon into an intron of phase 1 causes a frameshift in the inserted exon and in all exons on the 3′ side, while Figure 11c shows that the insertion of a 0-0 exon into an intron of phase 0 causes no frameshift.

ALTERNATIVE PATHWAYS FOR PRODUCING NEW FUNCTIONS

In addition to gene duplication and exon shuffling, there are many other mechanisms for producing new genes or polypeptides. Three such mechanisms are considered below.

Overlapping genes

It has been found that a DNA segment can code for more than one gene by using different reading frames. This phenomenon is widespread in viruses, organelles, and bacteria. Figure 12a shows the genetic map of ϕX174, which is a single-stranded DNA bacteriophage. Several overlapping genes are observed. For example, gene *B* is completely contained inside gene *A* while gene *K* overlaps gene *A* on the 5′ end and gene *C* on the 3′ end. A more detailed illustration of the latter case is given in Figure 12b.

Overlapping genes can also arise by the use of the complementary strands of a DNA sequence. For example, the genes specifying $tRNA^{Ile}$ and $tRNA^{Gln}$ in the human mitochondrial genome are located on different strands and there is a three-nucleotide overlap between them that reads 5′—CTA—3′ in the former and 5′—TAG—3′ in the latter (Anderson et al. 1981).

The question arises as to how overlapping genes may come into existence during evolution. To answer this question, we note that open reading frames abound throughout the genome. Therefore, it is possible that potential coding regions of considerable length exist in either a different reading frame of an existing gene or on the complementary strand. Because only three of 64 possible codons are termination codons, even a random DNA sequence might

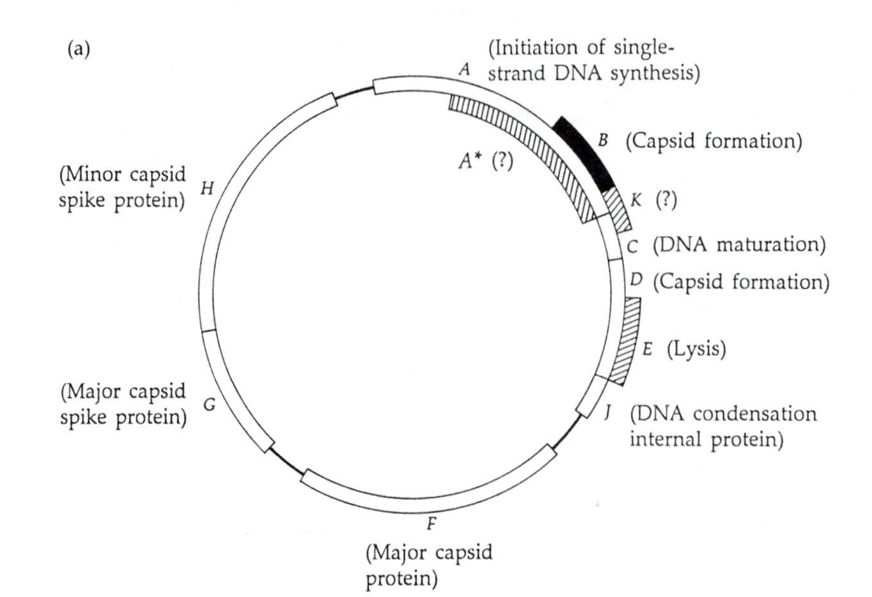

(b)

K protein: M S R K I I L I K Q E L L L L V Y E L
A protein: ...S D E S K N Y L D K A G I T T A C L R I
C protein:
DNA sequence: TCTGATGAGTCGAAAAATTATCTTGATAAAGCAGGAATTACTACTGCTTGTTTACGAATT

K protein: N R S G L L A E N E K I R P I L A Q L E
A protein: K S K W T A G G K *
C protein: M R K F D L S L R S S R
DNA sequence: AAATCGAAGTGGACTGCTGGCGGAAAATGAGAAAATTCGACCTATCCTTGCGCAGCTCGAG

K protein: L L L C D L S P S T N D S V K N *
A protein:
C protein: S S Y F A T F R H Q L T I L S K T D A L...
DNA sequence: AGCTCTTACTTTGCGACCTTTCGCCATCAACTAACGATTCTGTCAAAAACTGACGCGTTG

Figure 12. (a) Genetic map of the circular genome of the single-stranded DNA bacteriophage ϕX174. Note that the *B* protein-coding gene (black) is completely contained within the *A* protein-coding gene and that gene *K* overlaps two genes, *A* and *C*. Modified from Kornberg (1982). (b) Sequence of the *K* gene showing overlap with the 5′ part of the *A* gene and the 3′ of the *C* gene. Asterisks indicate stop codons. (See Table 1 in Chapter 1 for the one-letter abbreviations for amino acids.)

contain open reading frames hundreds of nucleotides long. If by chance such a reading frame contains an initiation codon and a transcription-initiation site, or if such sites are created by mutation, an additional mRNA will be transcribed and subsequently translated into a new protein. Whether the new product has a beneficial function or not is another matter, but if it does, the trait may become fixed in the population.

We also note that the rate of evolution is expected to be slower in stretches of DNA encoding overlapping genes than in similar DNA sequences using only one reading frame. The reason is that the proportion of nondegenerate sites is higher in overlapping genes than in nonoverlapping genes, thus vastly reducing the proportion of synonymous mutations out of the total number of mutations (Miyata and Yasunaga 1978).

Alternative splicing

Alternative splicing of a primary RNA transcript can result in the production of different polypeptides from the same DNA segment. In this case, the distinction between exons and introns is no longer absolute but depends on the mRNA of reference. Many cases of alternative RNA processing have been found in multicellular organisms.

Alternative splicing has often been used as a means of developmental regulation. A very intriguing situation is seen in several genes involved in the process of sex determination in *Drosophila melanogaster*. At least three genes, *sexlethal* (*Sxl*), *transformer* (*tra*), and *doublesex* (*dsx*), are spliced differently in males and females (Figure 13). In the case of *dsx*, the gene has six exons; exons 1, 2, 3, and 4 are used in the female, and exons 1, 2, 3, 5, and 6 are used in the male. In the cases of *Sxl* and *tra*, the products of the alternative splicing in males contain premature termination codons and are therefore nonfunctional. For example, exon 3 in *Sxl* contains an in-frame stop codon, but the mRNA in females does not contain this exon.

A special instance of alternative splicing is illustrated by the case of intron-encoded proteins (Perlman and Butow 1989). In such cases, the intron contains an open reading frame that encodes a protein or part of a protein that is completely different in function from the protein encoded by the flanking exons. In some cases, the open reading frame is an extension of the upstream exon, e.g., intron a14α in the yeast mitochondrial gene *cox I* (Figure 14a). In other cases, the intron includes not only a free-standing protein-coding gene, but also the necessary signals for transcription initiation and termination (Figure 14b). Intron a14α is intriguing, since it encodes an enzyme called maturase that is required for the proper self-splicing of this intron from the pre-mRNA. This maturase also functions as an endonuclease in DNA recombination.

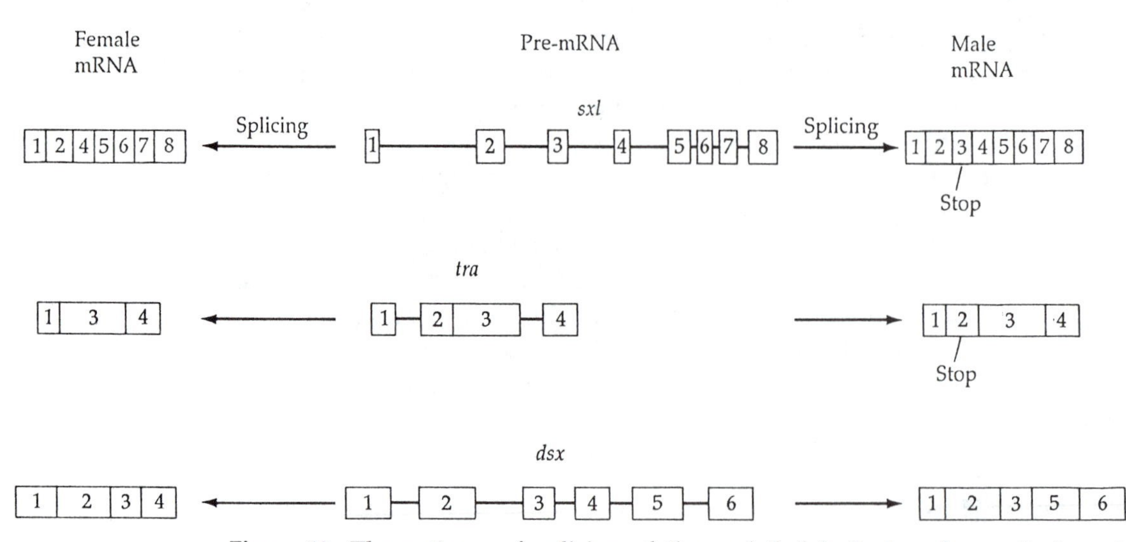

Figure 13. The patterns of splicing of the *sexlethal* (*sxl*), *transformer* (*tra*), and *doublesex* (*dsx*) genes in *Drosophila melanogaster* females (left) and males (right). "Stop" indicates a termination codon that truncates the coding region of the mature mRNA and renders the product nonfunctional. From Baker (1989).

The evolution of alternative splicing requires that an alternative splice-junction site be created de novo. Since splicing signals are usually 5–10 nucleotides long, it is possible that such sites are created with an appreciable frequency by mutation. Indeed, many such examples are known in the literature. For example, Figure 15 illustrates a case in which a synonymous substitution in a glycine codon turned a coding region into a splice junction. In cases of pathological manifestations such as the β^+-thalassemia in Figure 15, the new splice site is usually stronger than the old splice site (i.e., most of the mRNA synthesized after such a mutation occurs is of the altered type).

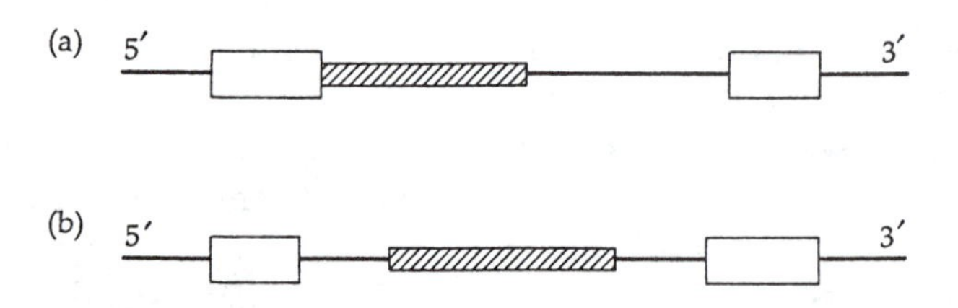

Figure 14. Examples of intron-encoded proteins: (a) an open reading frame (hatched) that is an extension of the upstream exon (empty box) (e.g., intron a14α in the yeast mitochondrial gene *cox I*); (b) a free-standing open reading frame, the transcription initiation and termination signals of which reside within the intron (e.g., the intron of *sun Y* gene in bacteriophage T4). Data from Perlman and Butow (1989).

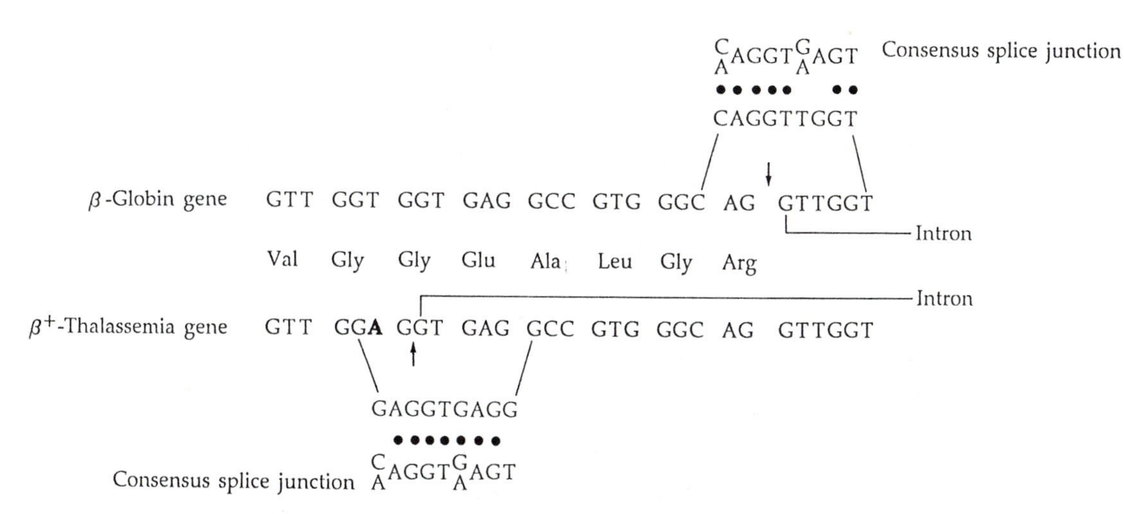

Figure 15. The nucleotide sequences at the border between exon 1 and intron I in the β-globin gene from a normal individual and a patient with β^+-thalassemia. The mutated nucleotide is shown in boldfaced type. The arrows indicate the splicing sites. Each of the splice junctions is compared with the sequence of the consensus splice junction, and dots denote identity of nucleotides between the splice junction and the consensus sequence.

Such a mutation will obviously have deleterious effects and is not expected ever to become fixed in the population. However, if the newly created splice site is much weaker, then most mRNA will be of the original type, and only small quantities of the new mRNA will be made. Such a change will not obliterate the old function, and yet will create an opportunity to produce a new protein, possibly with a new useful function.

Gene sharing

From the viewpoint of creating a new function, an extremely intriguing situation arises when a gene product is recruited to serve an additional function without any changes in its amino acid sequence. This phenomenon has been termed "**gene sharing**" (Piatigorsky et al. 1988). Gene sharing means that a gene acquires and maintains a second function without duplication and without loss of the primary function. Gene sharing may, however, require a change in the regulation system of tissue specificity or developmental timing.

Gene sharing was first discovered in crystallins, which are used in the eye lens to maintain transparency and proper light diffraction. The first finding was that the ε crystallin from birds and crocodiles is identical in its amino acid sequence with lactate dehydrogenase B (LDH-B_4; see page 145), and possesses identical LDH activity (Wistow et al. 1987). Subsequent work has shown that the "two" proteins are in fact one and the same and are encoded by the same gene (Hendriks et al. 1988). A second crystallin, δ, which exists in all birds and reptiles, has also been shown to be identical in sequence with another enzyme, argininosuccinate lyase, which catalyzes the conversion of argininosuccinate into the amino acid arginine, and these proteins are likewise encoded by the same gene (Piatigorsky et al. 1988). Similarly, τ-crystallin in lampreys, bony fishes, reptiles, and birds has been shown to be identical to and encoded by the same gene as α-enolase, a glycolytic enzyme converting 2-phosphoglycerate into phosphoenolpyruvate (Piatigorsky and Wistow 1989). Thus, δ-, ϵ-, and τ-crystallins illustrate instances of gene sharing, whereby a gene acquires additional roles without being duplicated. The α, β and γ crystallins, on the other hand, are classical examples of proteins that evolved by means of gene duplication and subsequent sequence divergence from ancestral genes specifying different proteins (e.g., heat-shock genes, which encode proteins expressed following exposure to excessive heat).

Gene sharing might be a fairly common phenomenon. In fact, in the above examples, the enzymes and crystallins themselves may have more than two functions. For instance, τ-crystallin/α-enolase also serves as a heat-shock protein. Gene sharing clearly adds to the compactness of the genome, even though compactness does not seem to have a high priority in eukaryotes (Chapter 8). Also note that, in the case of crystallin gene sharing, the same polypeptide serves both as an enzyme and as a structural protein, thus blurring the traditional distinction between enzymes and nonenzymatic or structural proteins.

CONCERTED EVOLUTION OF MULTIGENE FAMILIES

From the mid-1960's to the mid-1970's a large number of DNA reannealing and hybridization studies were conducted to explore the structure and organization of eukaryotic genomes. These studies revealed that the genome of higher organisms is composed of highly and moderately repeated sequences as well as single-copy sequences (Chapter 8). They also revealed an intriguing evolutionary phenomenon, namely that the members of a repeated-sequence family are generally very similar to each other within one species, although members of the family from even fairly closely related species may differ greatly from each other. This phenomenon was first noted by Brown et al. (1972) in a comparison of the ribosomal DNAs from the African toads, *Xenopus laevis* and *X. borealis*, the latter being misidentified at the time as *X. mulleri*.

In *Xenopus* and most other vertebrates, the genes specifying the 18S and 28S ribosomal RNA are present in hundreds of copies and are arranged in one or a few tandem arrays. Each repeated unit consists of a transcribed and a nontranscribed segment (Figure 16). The transcribed segment produces a 45S RNA precursor from which the functional 18S and 28S ribosomal RNAs are derived by means of enzymatic cleavage. The transcribed repeats are separated from each other by a nontranscribed spacer (NTS).

In a comparison of the ribosomal RNA genes between *X. laevis* and *X. borealis*, Brown et al. (1972) found that, while the 18S and 28S genes of the two species were highly similar, the NTS regions differed greatly between the two species. In contrast, the NTS regions are very similar within each individual and among individuals in a species. Thus, it appears that the NTS regions in each species have evolved together, although they have diverged rapidly between species. Brown et al. (1972) concluded that a "correction" mechanism must have operated to spread a mutation from one spacer sequence to the neighboring spacers faster than new changes arise in these sequences. They called this phenomenon, which manifests itself within a single individual, **horizontal evolution,** in contrast to vertical evolution, which refers to the spread of a mutation through a breeding population. Later the terms **coincidental evolution** or **concerted evolution** were suggested. The latter, coined by Zimmer et al. (1980), is now most commonly used in the literature.

With the advent of restriction-enzyme analysis and DNA sequencing techniques, a large body of data has attested to the generality of concerted evolution in multigene families (see reviews by Ohta 1980; Dover 1982; Arnheim 1983). Figure 17 shows an example from a restriction-enzyme analysis of human and chimpanzee ribosomal genes. In humans, each repeated unit has a *Hpa* I site in the NTS region 3′ to the 28S gene, whereas the chimpanzees and the other great apes lack this site. The *Hpa* I site most probably originated in the human lineage after the human–chimpanzee split and was eventually fixed in every human repeat. Other restriction sites in the NTS region exhibit species-specific homogeneity as well.

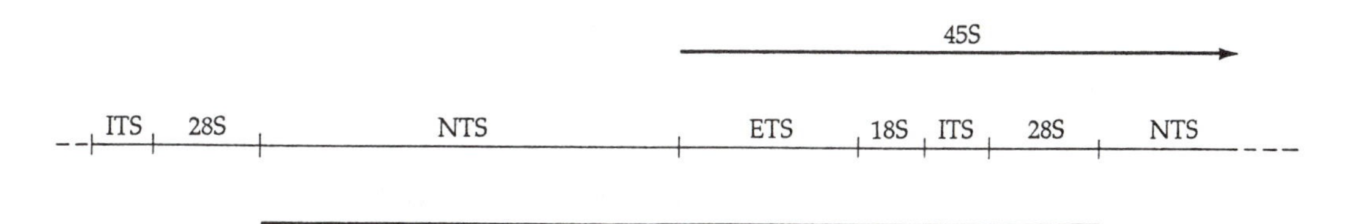

Figure 16. Diagramatic representation of a typical repeated unit of rRNA genes in vertebrates. The thick bar designates the repeat unit, and the arrow indicates the transcribed unit. ETS, external transcribed spacer; ITS, internal transcribed spacer; NTS, nontranscribed spacer. From Arnheim (1983).

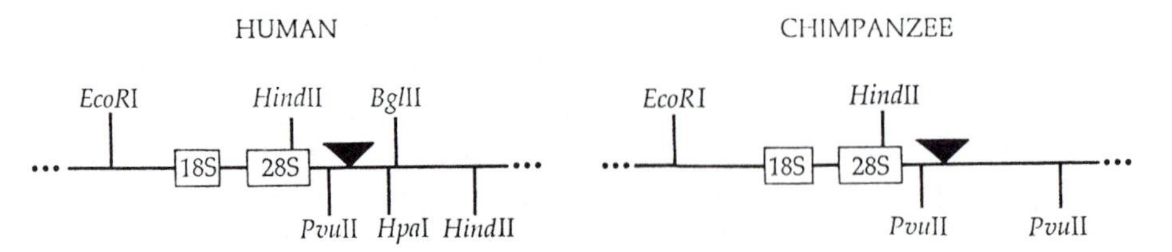

Figure 17. Restriction sites in human and chimpanzee 18S and 28S ribosomal genes. The restriction enzymes used were *EcoR*I, *Hind*II, *Pvu*II, *Bgl*II, and *Hpa*I. The restriction sites above the genes are polymorphic within the species. Those below the genes are monomorphic. The inverted triangles denote length polymorphism in the NTS. Modified from Arnheim (1983).

Concerted evolution essentially means that an individual member of a gene family does not evolve independently of the other members of the family. Through genetic interactions among its members, a multigene family evolves together in a concerted fashion, as a unit.

Mechanisms of concerted evolution

Unequal crossing-over and **gene conversion** (Figure 18) are currently considered to be the two most important mechanisms responsible for concerted evolution. Unequal crossing-over may occur either between the two sister

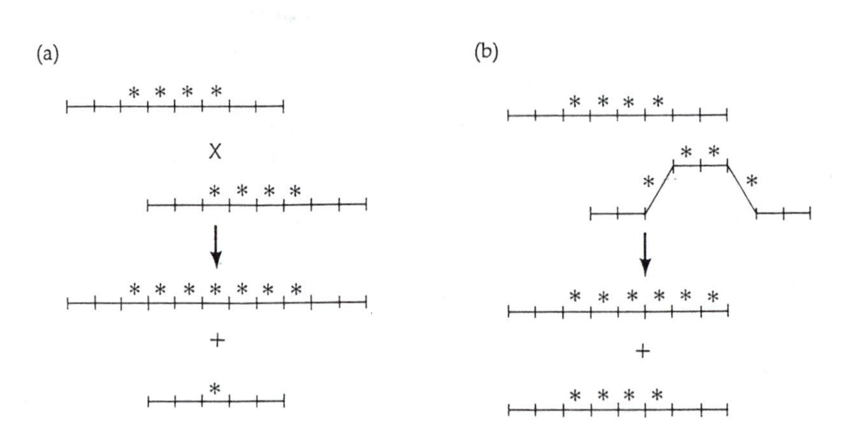

Figure 18. Models of (a) unequal crossing-over and (b) gene conversion. As a result of unequal crossing-over, both daughter chromosomes have an altered number of repeats and an altered frequency of the two repeat types (one of which is marked by an asterisk) when compared to the parental frequencies (50%). Gene conversion, on the other hand, changes the frequencies of the two types of repeats in only one of the daughter chromosomes and does not alter the total number of repeats in either chromosome. Modified from Arnheim (1983).

chromatids of a chromosome during mitosis of a germ-line cell or between two homologous chromosomes at meiosis. It is a reciprocal recombination process that creates a sequence duplication in one chromatid or chromosome and a corresponding deletion in the other. Figure 18a shows an example in which an unequal crossover event has led to duplication of three repeats in one daughter chromosome and deletion of three repeats in the other. As a result of this unequal exchange, both daughter chromosomes have become more homogeneous than the parental chromosomes. If this process is repeated, the numbers of each variant repeat on a chromosome will fluctuate with time, and eventually one type will become dominant in the family. Figure 19 gives a hypothetical example in which the type-4 gene spreads throughout a gene family due to repeated rounds of unequal crossing-over.

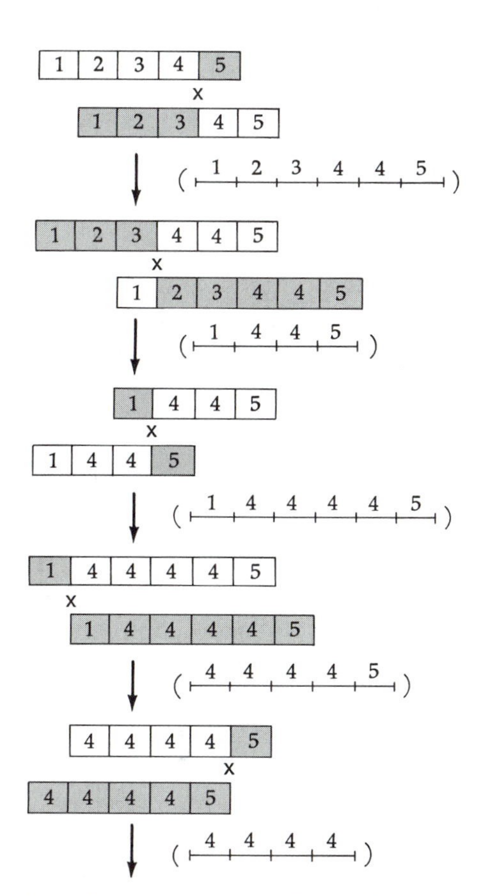

Figure 19. Concerted evolution by unequal crossing-over. Repeated cycles of unequal crossover events cause the duplicated genes on each chromosome to become progressively more homogenized. The process also affects the number of repeated sequences on each chromosome. From Ohta (1980).

Unequal crossing-over has been investigated mathematically in detail and has received considerable experimental support (see reviews by Ohta 1980; Dover 1982; Li et al. 1985a).

Gene conversion is a nonreciprocal recombination process in which two sequences interact in such a way that one is converted by the other (see Lewin 1990). From the point of view of the concerted evolutionary process, the most important type of gene conversion is nonallelic gene conversion (i.e., conversion between genes located at different loci and not between allelic forms). Figure 18b shows an example of a nonallelic gene conversion in which two of the wild-type repeats are converted to the mutant type. As a consequence, the first daughter chromosome has become more homogenous than the parental chromosome, with no change in the second daughter chromosome. Theoretical studies have shown that, like unequal crossing-over, gene conversion can produce concerted evolution (Ohta 1984; Nagylaki 1984). Gene conversion has been suggested as a mechanism of homogenization in the γ-globin genes (Jeffreys 1979; Scott et al. 1984) and many other genes (see Dover 1982).

As a mechanism for concerted evolution, gene conversion appears to have several advantages over unequal crossing-over. First, unequal crossing-over changes the number of repeated genes in a family and may sometimes cause a severe dosage imbalance. Gene conversion, on the other hand, causes no change in gene number. Second, gene conversion can act as a correction mechanism not only on tandem repeats but also on dispersed repeats. In contrast, unequal crossing-over is restricted when repeats dispersed on nonhomologous chromosomes are involved. It probably can act effectively on nonhomologous chromosomes if the repeated genes are located on the telomeric parts of the chromosome (the ends of the chromosome arms), as in the case of rRNA genes in humans and apes, but will be greatly restricted if the dispersed repeats are located in the middle of chromosomes, as in the case of rRNA genes in mice (Arnheim 1983). If the repeats are dispersed on a chromosome, unequal crossing-over can result in the deletion or duplication of the genes between the repeats. For example, Figure 20 shows a hypothetical case of unequal crossing-over between two repeat clusters, resulting in the deletion of a unique gene in one chromosome and a corresponding duplication in the other chromosome. Either one or both chromosomes could have a deleterious effect on their carriers. Third, gene conversion can have a preferred direction. Experimental data from fungi have shown that bias in the direction of gene conversion is common and often strong (Lamb and Helmi 1982), and theoretical studies have shown that a small bias can have a large effect on the probability of fixation of repeated mutants (Nagylaki and Petes 1982; Walsh 1985).

In large families of tandemly repeated sequences, unequal crossing-over may be as acceptable a process as gene conversion. First, in such families,

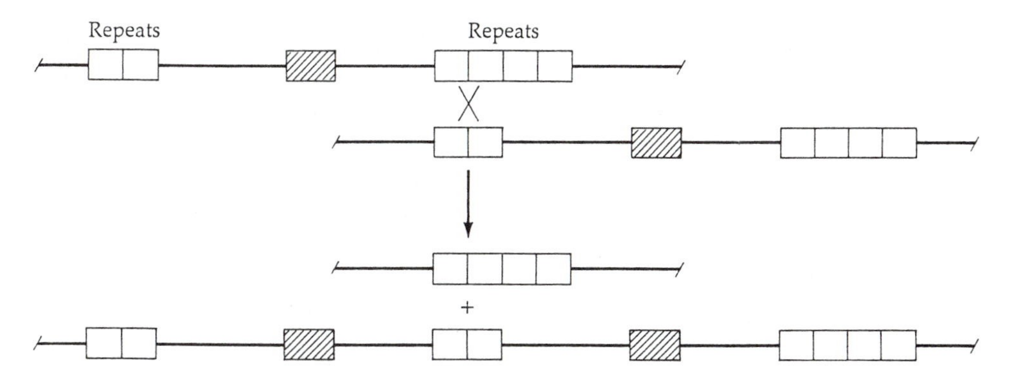

Figure 20. Crossing-over involving dispersed repeats (empty boxes). The hatched box denotes a unique gene. In the crossover event, this gene is deleted in one chromosome but duplicated in the other chromosome.

the number of repeats apparently can fluctuate greatly without causing significant adverse effects. This is suggested by the observations that the number of RNA-specifying genes in *Drosophila* varies widely among individuals of the same species and among species (Ritossa et al. 1966; Brown and Sugimoto 1973). In humans, several families of tandem repeats that exhibit extraordinary degrees of variation in copy number have been found (Nakamura et al. 1987). Second, in a gene-conversion event, usually only a small region (the heteroduplex region) is involved, whereas in unequal crossing-over the number of repeats that are exchanged between the chromosomes can be very large. Obviously, the larger the number of repeats exchanged, the higher the rate of concerted evolution will be (Ohta 1983). In some cases, this advantage of unequal crossing-over may be large enough to offset the advantages of gene conversion.

In addition to unequal crossing-over and gene conversion, there are other mechanisms, such as transposition and replication slippage (Chapters 1 and 7), that can cause gains or losses of variant genes in a family (Dover 1982). Finally, it should be noted that concerted evolution requires not only the horizontal transfer of mutations among the members of the family (homogenization), but also the spread of mutations to all individuals in the population (fixation). Thus, we need also to consider the effects of random genetic drift. Dover (1982, 1986) has proposed the term **molecular drive** for the process of concerted evolution of multigene families under the joint effect of the various mechanisms of DNA transfer and random genetic drift.

Evolutionary implications of concerted evolution

Concerted evolution allows the spreading of a variant repeat to all gene-family members. This capability of horizontal spreading has profound evolutionary consequences, because a beneficial mutant repeat can replace all other repeats and become fixed in the family. We note that the selective advantage that a single variant can confer to an organism is usually very limited. The advantage would, however, be greatly amplified if the mutation were to spread to many or all members. Thus, through concerted evolution, a small selective advantage can become a great advantage. In this respect, concerted evolution surpasses independent evolution of individual gene-family members (see Arnheim 1983; Walsh 1985).

Arnheim (1983) has compared the evolution of RNA polymerase I transcriptional control signals with that of RNA polymerase II transcriptional control signals. RNA polymerase I transcribes only rRNA genes, whereas RNA polymerase II transcribes all protein-coding genes (Chapter 1). RNA polymerase I transcriptional control signals appear to have evolved much faster than the signals for RNA polymerase II. For example, in cell-free transcription systems, a mouse rDNA clone does not work in a human cell extract, but clones of protein-coding genes from astonishingly diverse species can be transcribed in heterologous systems (e.g., silk worm genes in human cell extracts and mammalian genes in yeast). Arnheim (1983) argues that in the case of transcription units for RNA polymerase I, mutations that favorably affect transcription initiation could be propagated throughout the rDNA multigene family as a consequence of concerted evolution. On the other hand, in the case of transcription units for RNA polymerase II, advantageous mutations affecting transcription initiation that occur in any one gene would not be expected to be propagated throughout all genes, for they belong to many different families.

The traditional view concerning the creation of a new gene is that a gene-duplication event occurs and one of the two resultant genes gradually diverges and becomes a new gene. It is now clear that the process may not be as simple as previously assumed. As long as the degree of divergence between the two genes is not large, the divergent copy may be deleted by unequal crossing-over or converted to the conserved copy by gene conversion. In the former case it requires another duplication to create a new redundant copy, while in the latter case divergence must start again from scratch. Thus, divergence of duplicate genes may proceed much more slowly than traditionally thought, and for this reason the chance of creation of a new gene from a redundant copy is reduced. On the other hand, gene conversion may also prevent a redundant copy from becoming nonfunctional for long periods of time or, alternatively, may enable a "dead gene" (pseudogene) to be "resurrected" (Walsh 1987).

It has been customary to assume that, following a gene duplication, the two resultant genes will diverge monotonically with time. Under this assumption, we have previously shown that it is rather simple to infer the time of the duplication event. For example, the protein sequences of human β- and δ-globins are more similar to each other than to rabbit β1, or to mouse β major and minor sequences (Dayhoff 1972). It has therefore been inferred that the two human genes were derived from a duplication event about 40 million years ago, long after the mammalian radiation (about 80 million years ago). This conclusion may be erroneous in view of the fact that duplicate genes can correct each other. In fact, it has recently been suggested that the β and δ genes originated from a duplication that occurred before the mammalian radiation (Hardison and Margot 1984). This suggestion is based on the observation that the large intron and the 3′ untranslated region of rabbit pseudogene ψβ2 are more similar to human δ than to rabbit β1 and that the pseudogene β*h*3 in mouse is similar to δ at its 3′ end. If this hypothesis turns out to be true, it will provide an excellent example of how gene-correction events may partly or completely erase the evolutionary history of divergence between duplicate genes. In large multigene families, gene-correction events are expected to occur frequently, and in such cases it will be even more difficult to trace the evolutionary relationships among family members.

From the evolutionary point of view, there is an analogy between the evolution of multigene families and the evolution of subdivided populations. We may regard each repeat in a multigene family as a deme in a subdivided population. Then, transfer of information between repeats is equivalent to migration of genes or individuals between demes. It is well known that migration reduces the amount of genetic difference between demes but increases the amount of genic variation (e.g., the number of alleles) in a deme. Similarly, transfer of information between repeats will reduce the genetic difference between repeats but will increase the amount of genic variation at a locus (Ohta 1983, 1984; Nagylaki 1984). Some loci in the mouse major histocompatibility complex are highly polymorphic; in fact, as many as 50 alleles have been observed at a locus. Thus, it has been suggested that the high polymorphism is due to gene conversion (e.g., Weiss et al. 1983; but see Hughes and Nei 1989).

PROBLEMS

1. What are the advantages of exon duplication compared to the duplication of a randomly chosen DNA segment?

2. In overlapping genes, the number of degenerate sites can be greatly reduced. (a) If the following sequence is translated only in the first reading frame, how many nondegenerate sites are there? How many fourfold degenerate sites? (b) If, in addition to the first reading frame, the sequence is also translated in the second reading frame, how many nondegenerate and fourfold degenerate sites are there?

(c) How many nondegenerate and fourfold degenerate sites will be in the sequence if it is translated in all three reading frames? (The startpoint of each of the three reading frames is marked by an arrow.)

CATTCGTCTTTATTCGAAATCGCGTGGACAGCGGTGGATCTCTTTGCGCTGTGCAAAGCAGCGCTGGCGGTT

1 2 3

3. Many multimeric proteins are composed of subunits encoded by duplicated genes. There are two possible situations. (a) The subunits can be either all from a single locus or from different loci; the protein is said to be "homomeric" in the former case and "heteromeric" in the latter case. Suppose that the protein is a tetrameric enzyme like lactate dehydrogenase (LDH; see page 145) and that the subunits are encoded by two loci. How many different isozymes can be produced? (b) The protein is always heteromeric (i.e., always composed of subunits from different loci). For example, the adult hemoglobin in mammals is a tetramer consisting of two α chains and two β chains. Suppose that in the genome of a certain mammalian species there are three α-like loci and two β-like loci. How many different heteromeric tetramers can be produced if each tetramer consists of two subunits from an α-like locus and two subunits from a β-like locus?

4. In the genome of rats and mice there are two genes coding for insulin (preproinsulin I and II), while in the genome of mammals other than rodents there is only one gene for insulin. The preproinsulin I gene is thought to have arisen by a process called "retroposition" (Chapter 7). The preproinsulin I and II genes both contain a small intron (118 nucleotides long) in the 5′ untranslated region. The numbers of nucleotide differences between pairs of introns in rats and mice are as follows:

	Intron		
Intron	Mouse I	Mouse II	Rat I
Mouse II	21		
Rat I	15	25	
Rat II	16	24	18

(a) Explain why the numbers in this matrix indicate that the intron in the mouse preproinsulin II (i.e., mouse II) gene has evolved faster than the corresponding introns in the other genes. (b) Estimate the divergence time between the preproinsulin I and II genes by using mouse I, rat I, and rat II but excluding mouse II, assuming a constant rate of nucleotide substitution, and assuming that mice and rats diverged 15 million years ago.

5. In the intron (dashed line) of the following sequence insert a 0-0 symmetrical exon. What happens to the reading frame? What happens if you insert a 2-2 symmetrical exon? What happens if you insert an asymmetrical exon?

5′—CAT TCG TCT TTA TTC GAA ATC GCG---TGG ACA GCG GTG AAT CTC TTT GAC GCT GTG—3′

6. Explain why a family of tandem repeats can undergo concerted evolution more readily than a family of dispersed repeats.

FURTHER READINGS

Cold Spring Harbor Symposium on Quantitative Biology. 1987. *Evolution of Catalytic Function*. Vol. 52. Cold Spring Harbor Laboratory, Cold Spring Harbor, NY.

Dayhoff, M. O. 1972. *Atlas of Protein Sequence and Structure*, Vol. 5. National Biomedical Research Foundation, Silver Spring, MD.

Dover, G. A. and R. B. Flavell. (eds.). 1982. *Genome Evolution*. Academic Press, New York.

Li, W.-H. 1983. Evolution of duplicate genes and pseudogenes. Pp. 14–37. *In* M. Nei and R. K. Koehn (eds.), *Evolution of Genes and Proteins*. Sinauer Associates, Sunderland, MA.

Ohno, S. 1970. *Evolution by Gene Duplication*. Springer-Verlag, Berlin.

Ohta, T. 1980. *Evolution and Variation of Multigene Families*. Springer-Verlag, Berlin.

7

EVOLUTION BY TRANSPOSITION

Genomes used to be thought of as rather static entities, in which genes could be assigned to well-defined loci. Accordingly, genes were supposed to retain their precise chromosomal position over long periods of evolution. The static picture of the genome started crumbling in the 1940's when Barbara McClintock discovered that certain genetic elements in maize can "jump" from one genomic location to another, sometimes altering the expression of structural genes. The rigid static picture, however, was so ingrained in scientific thought that it took nearly 40 years for the significance of McClintock's seminal discovery to be appreciated. Today we recognize that the structural organization of genomes is much more fluid and prone to evolutionary changes than was previously thought. In this chapter we describe a myriad of transposable elements which facilitate the movement of genetic material from one genomic location to another, and discuss possible impacts such elements may have on the evolutionary process.

TRANSPOSITION AND RETROPOSITION

Transposition is defined as the movement of genetic material from one chromosomal location to another. DNA sequences that possess an intrinsic capability to change their genomic location are called **mobile elements** or

transposable elements. There are two types of transposition, distinguished by whether the transposable element is replicated or not. In the **conservative** type of transposition, the element moves from one site to another (Figure 1a). What happens to the donor site is unclear. One model proposes that the ends of the donor DNA are not joined to one another and that the remnant molecule is destroyed. However, loss of the donor DNA species from the cell lineage is avoided if the cell contains a duplicate of the donor sequence. In this case, although one copy is consumed, the other survives, and the resulting lineage will contain one element at the original site and a second at the new site (Berg et al. 1984). Another model proposes that the double-strand break is repaired by the host repair system.

In **replicative** (or **duplicative**) transposition, the transposable element is copied, and one copy remains at the original site, while the other inserts at a new site (Figure 1b). Thus, replicative transposition is characterized by an increase in the number of copies of the transposable element. Some transposable elements use only one type of transposition; others use both the conservative and the replicative pathways.

In the above types of transposition, the genetic information is carried by DNA. It is known that genetic information can also be transposed through RNA. In this mode, the DNA is transcribed into RNA, which is then reverse-transcribed into cDNA (Figure 1c). In order to distinguish between the two modes, the RNA-mediated mode has been termed **retroposition**. Both transposition and retroposition are found in eukaryotic and prokaryotic organisms (see Weiner et al. 1986; Temin 1989). In contrast with DNA-mediated transposition, retroposition is always of the duplicative type, because it is a reverse-transcribed copy of the element, not the element itself, that is transposed.

When a transposable element is inserted into a host genome, a small segment of the host DNA at the insertion site (usually 4–12 bp) is duplicated (Figure 1). The duplicated repeats are in the same orientation and are called **direct repeats**. This is a hallmark of transposition and retroposition.

Some transposable elements can transpose themselves in all cells; others are highly specific. For example, *P* elements in *Drosophila melanogaster* are usually mobile only in germ cells. The genomic locations of the recipient sites for transposition also show variation among different transposable elements. Some elements show an exclusive preference for a specific genomic location. For example, *IS4* incorporates itself exactly and always at the same point in the galactosidase operon of *Escherichia coli*, and thus each bacterium can contain only one copy of *IS4* (Klaer et al. 1981). Others, such as bacteriophage *Mu*, can transpose themselves at random to almost any genomic location. Many transposable elements show intermediate degrees of genomic preference. For example, 40% of all *Tn10* transposons in *E. coli* are found in the

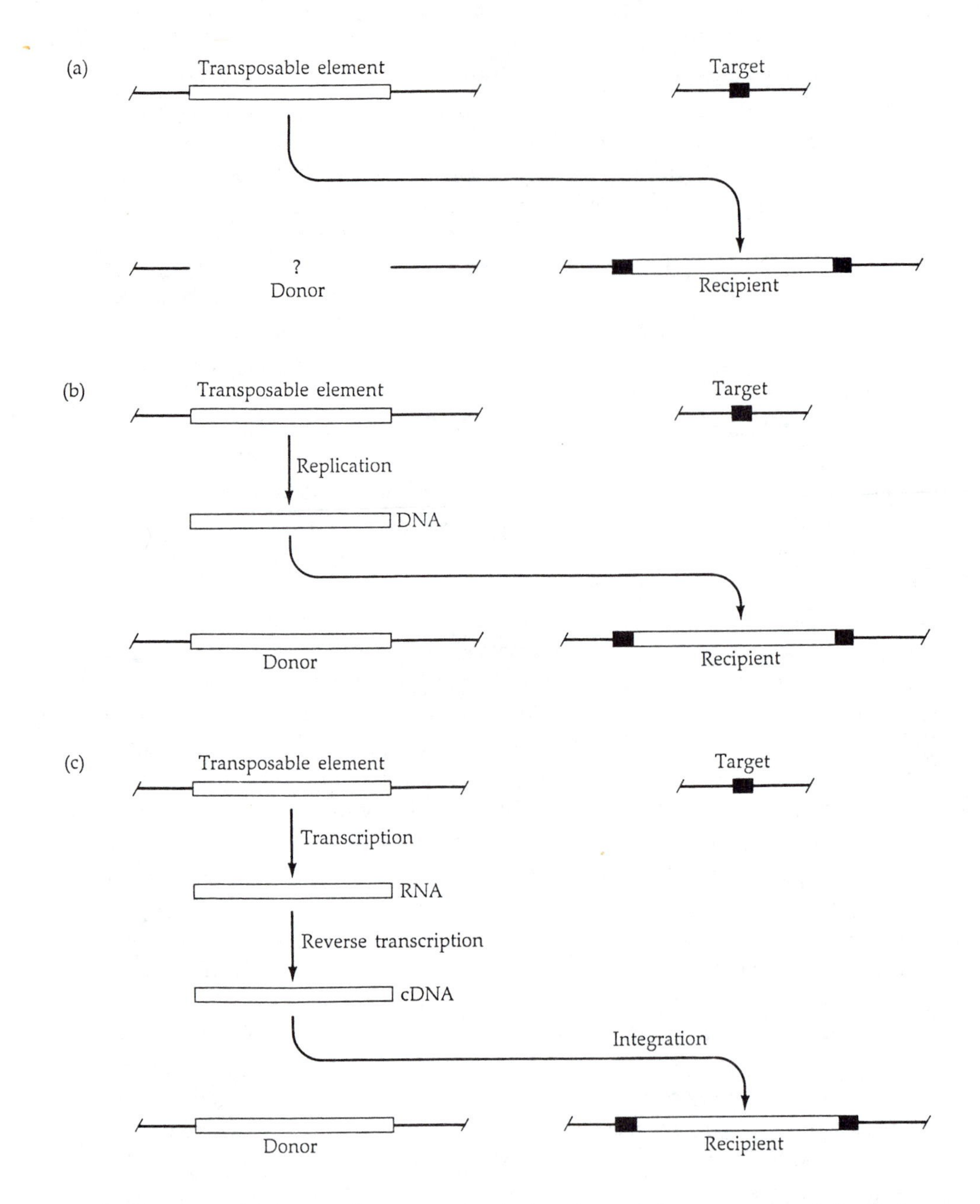

lacZ gene, which constitutes a minute fraction of the genome, while *P* elements have an affinity for the X chromosome and also prefer to insert themselves into sequences 5′ to the coding region of genes rather than into the coding region itself. Some transposable elements exhibit higher affinities for a particular type of nucleotide composition. For example *IS1* favors AT-rich insertion sites (Devos et al. 1979).

◄Figure 1. (a) Conservative transposition. The element is transposed from the donor site to a target site. What will happen to the donor site is unclear. The donor molecule may be destroyed, which can be tolerated by bacteria that possess more than one copy of the chromosome. Another possibility is that the double-strand break is repaired by the host repair system. (b) Replicative transposition. The element is replicated, and a copy is inserted at a target site, while the other copy remains at the donor site. For a more detailed explanation of conservative and replicative transposition, see Lewin (1990). (c) Retroposition. The element is transcribed into RNA, which is reverse-transcribed into DNA. The DNA copy is inserted into the host genome. For an example of retroposition, see Figure 4. Note that both transposition and retroposition create a short repeat (black box) at each end of the newly inserted element.

TRANSPOSABLE ELEMENTS

According to their mode of transposition and the number and kinds of genes they contain, transposable elements can be classified into three types: insertion sequences, transposons, and retroelements.

Insertion sequences

Insertion sequences are the simplest transposable elements. They carry no genetic information except that which is necessary for transposition. Insertion sequences are usually 700–2,500 bp in length and have been found in bacteria, bacteriophages, plasmids, and maize. Bacterial insertion sequences are denoted by the prefix ***IS*** followed by the type number. The structure of an insertion sequence, *IS1* from the intestinal bacteria *Escherichia coli* and *Shigella dysinteria*, is shown schematically in Figure 2a. *IS1* is approximately 770 nucleotides in length, including two inverted non-identical terminal repeats, 23 bp each. It contains two reading frames, *InsA* and *InsB*, which encode one or two forms of **transposase**, an enzyme that catalyzes the insertion of transposable elements into insertion sites. There are dozens of different types of insertion sequences in *Escherichia coli*, and the genomes of most strains isolated from the wild contain variable numbers of each (Sawyer et al. 1987).

Transposons

Transposons are mobile elements, usually about 2,500–7,000 bp long, that exist mostly as families of dispersed repetitive sequences in the genome. They are distinguished from insertion sequences by also carrying so-called **exogenous genes**, that is, genes that encode functions other than those related to transposition. (Note that the nomenclature is muddled in the

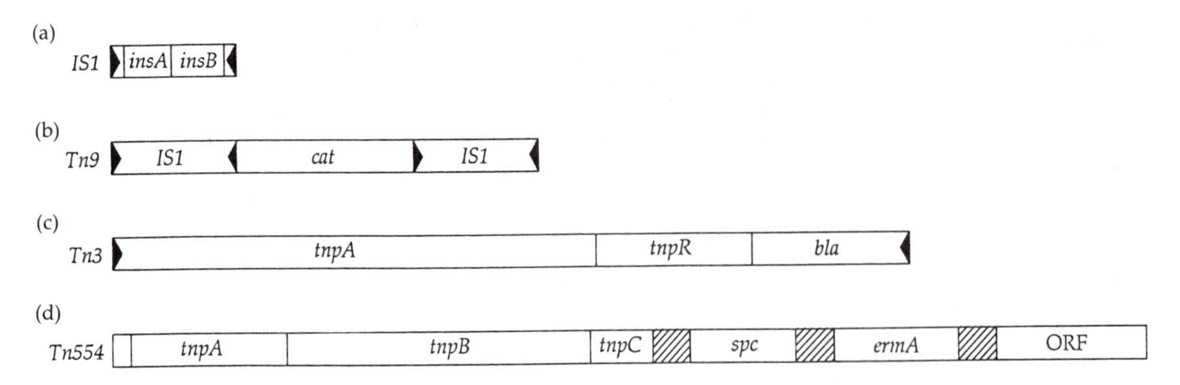

Figure 2. Schematic representation of four transposable elements in bacteria. Black triangles denote inverted repeats. (a) Insertion sequence *IS1* from *Escherichia coli* and *Shigella dysenteria* is flanked by imperfect inverted repeats 23 bp long. (b) Composite transposon *Tn9* from *E. coli* contains two copies of *IS1* flanking the *cat* gene, which encodes a protein conferring chloramphenicol resistance. (c) Transposon *Tn3* from *E. coli*, which confers streptomycin resistance, contains three genes, two of which (*tnpR* and *bla*) are transcribed on one strand, and the third (*tnpA*) on the other. *Tn3* is flanked by two perfect inverted repeats, 38 bp long. (d) Transposon *Tn554* from *Staphilococcus aureus* lacks terminal repeats and contains five genes and an open reading frame (ORF). Three of the genes encode transposases (*tnpA*, *tnpB*, and *tnpC*) and are transcribed as a unit. The *spc* and *ermA* genes confer spectinomycin and erythromycin resistance, respectively. The *spc* gene, which encodes an S-adenosylmethionine-dependent methylase, is transcribed on a different strand from the other genes. The ORF is abundantly transcribed, but it is not known whether or not it is translated. The hatched boxes contain no open reading frames.

literature, and the term transposon is sometimes used to denote any transposable element, including insertion sequences, retrotransposons, etc.) In bacteria, transposons are denoted by the prefix *Tn* followed by the type number. Some bacterial transposons are **complex transposons** or **composite transposons**, so named because two complete, independently transposable insertion sequences in either orientation flank one or more exogenous genes (Figure 2b). Interestingly, in complex transposons, not only can the entire transposon transpose as a unit but also one or both of the flanking insertion sequences can transpose independently. Since the transposition functions are encoded by the insertion sequences, complex transposons do not usually contain an independent transposase gene.

Other bacterial transposons, as well as many eukaryotic transposons, are flanked only by short repeated sequences in various orientations (Figure 2c) and contain no insertion sequences. Not all transposons, however, are symmetrical in structure. Some have asymmetrical ends, lacking either inverted or direct terminal repeats (Figure 2d). The coding regions of some transpo-

Figure 3. Schematic structure of a complete *P* element in *Drosophila melanogaster*. The element is flanked by short inverted repeats, 31 bp long, and its coding region contains four exons (white boxes) interrupted by three introns (black boxes). The element is about 2,900 bp long.

sons in animals (e.g., *P* elements in *Drosophila*) are interrupted by introns (Figure 3).

Transposons in bacteria often carry genes that confer antibiotic resistance (e.g., *Tn554*), heavy-metal resistance (e.g., *Tn21*), or heat resistance (e.g., *Tn1681*) to their carriers. Plasmids can carry such transposons from cell to cell and, as a consequence, resistance can quickly spread throughout populations of bacteria exposed to such environmental factors.

Several bacteriophages in bacteria are in fact transposons or **transposing bacteriophages**. For example, bacteriophage *Mu* is a very large transposon (~38,000 bp) that encodes not only the enzymes that regulate its transposition, but also a large number of structural proteins necessary to construct the packaging of its DNA.

Transposons of many types are quite widespread in the genomes of animals, plants, and fungi. *Drosophila melanogaster*, for instance, contains multiple copies of 50–100 different kinds of transposons (Rubin 1983).

Retroelements

Retroelements are DNA or RNA sequences that contain a gene for the enzyme reverse transcriptase, which catalyzes the synthesis of a DNA molecule from an RNA template. The resulting DNA molecule is called **complementary DNA (cDNA)**. Those retroelements that do transpose do so by the process of retroposition. There are different classes of retroelements, and in Table 1 we adopt the classification proposed by Temin (1989).

Retroviruses are RNA viruses that are similar in structure to transposons. Although they are the most complex of all retroelements, we discuss them first because the concept of retroposition originated with the discovery of the life cycle of retroviruses (Figure 4). After the retroviral particle, called the **virion**, invades a host cell, its genomic RNA is reverse-transcribed into viral DNA. This DNA can integrate into the host genome and become a **provirus**. Next, the proviral DNA is transcribed into RNAs, which can serve both as mRNAs for synthesizing viral proteins and as viral genomes that can be packaged into infectious virion particles. Once a virion is formed, the cycle can start again.

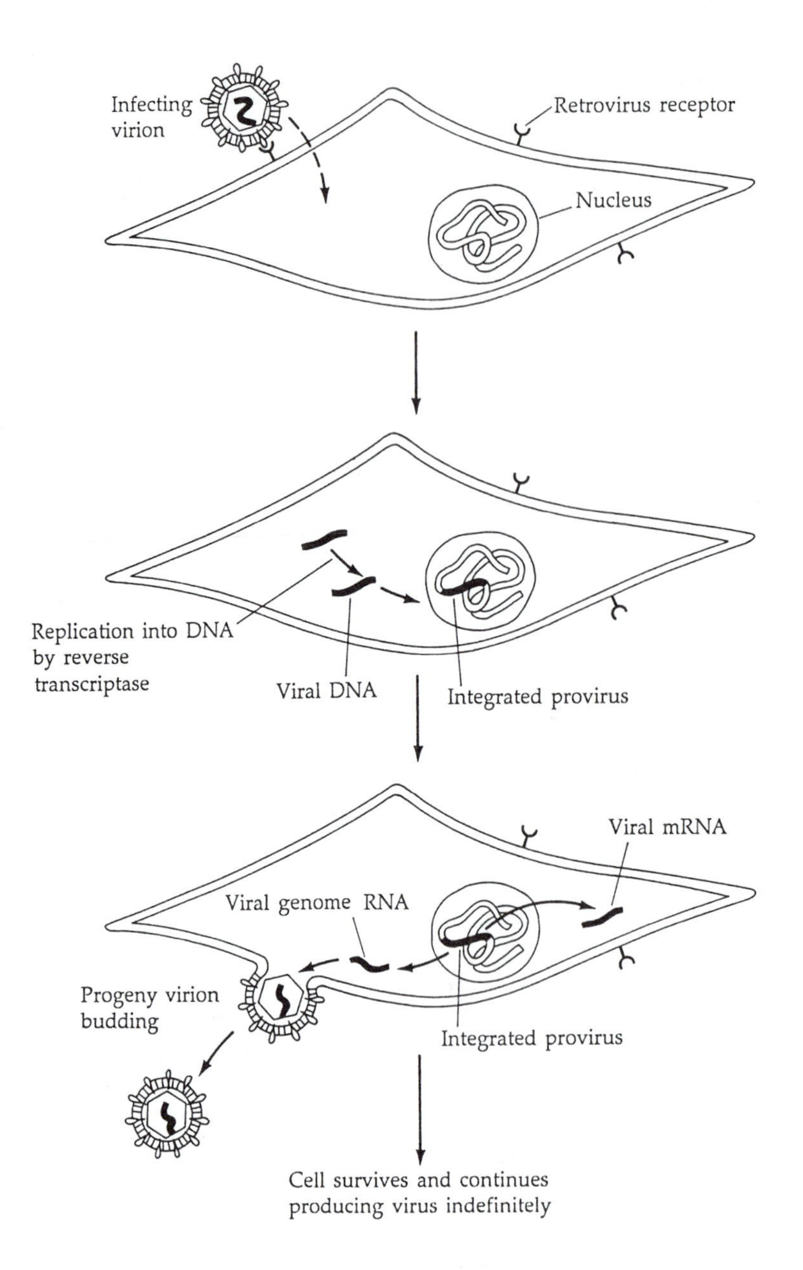

Figure 4. An overview of the life cycle of a retrovirus. The retroviral virion attaches to a receptor on the surface of the cell. The genomic RNA is injected into the cytoplasm where it is reverse-transcribed by the enzyme reverse transcriptase. The cDNA penetrates the nucleus and may become integrated within the genome of the host cell. The integrated provirus is transcribed into (1) mRNAs that are used to synthesize viral proteins, and (2) genomic RNA. The genomic RNA and the structural and enzymatic proteins assemble into infectious virion particles that bud out of the cell membrane. From Watson et al. (1987).

Table 1. **Classification of retroelements and retrosequences.**

Element	Reverse transcriptase	Transposition	Presence of LTRs[a]	Virion particles
Retron	yes	no	no	no
Retrosposon	yes	yes	no	no
Retrotransposon	yes	yes	yes	no
Retrovirus	yes	yes	yes	yes
Pararetrovirus	yes	no	yes	yes
Retrosequence	no	no	no	no

From Temin (1989).
[a] LTR, long terminal repeat.

Retroviruses possess at least three genes: *gag*, *pol*, and *env* (Figure 5). These genes encode several internal proteins, several enzymes (including a reverse transcriptase), and an envelope protein, respectively. Many retroviruses possess additional genes; for example, the AIDS virus possesses at least six additional genes. The coding region of a retrovirus is flanked by **long terminal repeats (LTRs).** The LTRs contain promoters for transcription (in the proviral stage) and reverse transcription (in the viral stage).

Retroposons and **retrotransposons** are transposable elements that do not construct virion particles, and so, unlike retroviruses, cannot independently transport themselves across cells. They are distinguished from each other by the absence or presence of terminally repeated sequences (LTRs) (Table 1). Note that some authors use retroposons and retrotransposons synonymously. The *copia* element in *Drosophila* represents a typical retrotransposon; it contains LTRs at both ends and a single long open reading frame with regions of similarity to the *pol* gene of retroviruses. Figure 5b shows another example of a retrotransposon, the *DIRS-1* element in the slime mold *Dictyostelium discoideum*. *D. discoideum* contains, on the average, about 40 intact copies of *DIRS-1* and about 200–300 *DIRS-1* fragments. Interestingly, *DIRS-1* has the propensity to insert itself into other *DIRS-1* sequences, thus obliterating them, which may be the reason for the existence of many defective *DIRS-1* fragments within the slime mold genome (Cappello et al. 1984). Transcription of the *DIRS-1* genes is induced by developmental stage as well as by heat shock.

Unlike retrotransposons, retroposons contain no LTRs. The *G3A* element in *Drosophila melanogaster* is a retroposon (Figure 5c). This retroposon contains two ORFs. ORF-1 contains a region with similarity to the *pol* gene of retro-

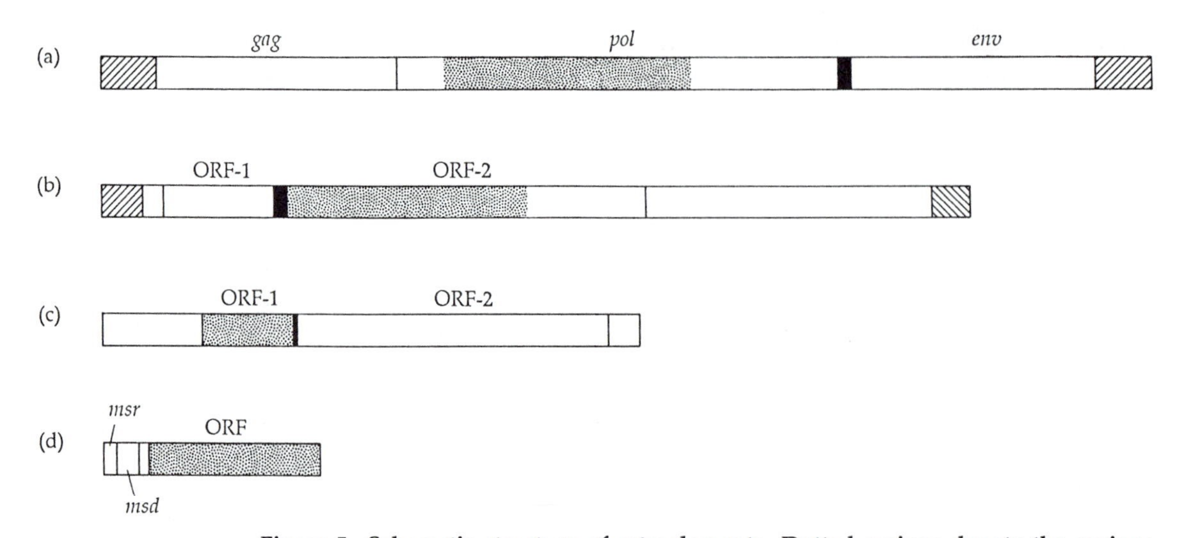

Figure 5. Schematic structure of retroelements. Dotted regions denote the regions coding for reverse transcriptase; overlapping regions between adjacent genes are shown as black boxes; and hatched regions represent long terminal repeats (LTRs). ORF, open reading frame. (a) Feline leukemia retrovirus. The coding region is flanked by two nonidentical LTRs, 482 and 472 bp long. The coding region encodes two precursor polyproteins. The precursor polyprotein encoded by the *gag–pol* region is cleaved into two polyproteins, corresponding to *gag* and *pol*. The *gag* polyprotein gives rise to four internal viral proteins, denoted p15, p12, p27, and p10. The *pol* polyprotein is cleaved into three enzymes: a protease, a reverse transcriptase, and an endonuclease/integrase. The precursor polyprotein encoded by *env* is cleaved into two proteins of the envelope denoted as p70 and p15. (b) Slime mold (*Dictyostelium discoideum*) retrotransposon *DIRS-1*. The inverted LTRs are 200–350 bp long. ORF-2 contains a region that has sequence similarity with the *pol* gene of retroviruses. (c) *Drosophila melanogaster* retroposon *G3A*. ORF-1 contains a region that has sequence similarity with the *pol* gene of retroviruses. Note the diagnostic absence of LTRs. (d) A retron from the myxobacterium, *Myxococcus xanthus*. The *msr* gene is transcribed into RNA. The *msd* gene is transcribed from the complementary strand and then reverse-transcribed into DNA by the reverse transcriptase encoded by the ORF of the retron. The two molecules are subsequently attached to each other via a 2′, 5′-phosphodiester bond to form a branched molecule called multicopy single-stranded DNA (msDNA). Note the absence of LTRs.

viruses while ORF-2 consists of seven exons separated by very short intervening sequences.

Retrons are the simplest retroelements (Figure 5d). They have been found in some bacterial genomes (Inouye et al. 1989; Lampson et al. 1989) as well as in the mitochondrial genome of the plant *Oenothera berteriana* (Schuster and Brennicke 1987). Their open reading frame has sequence similarity to the genes for other reverse transcriptases. However, retrons do not excise,

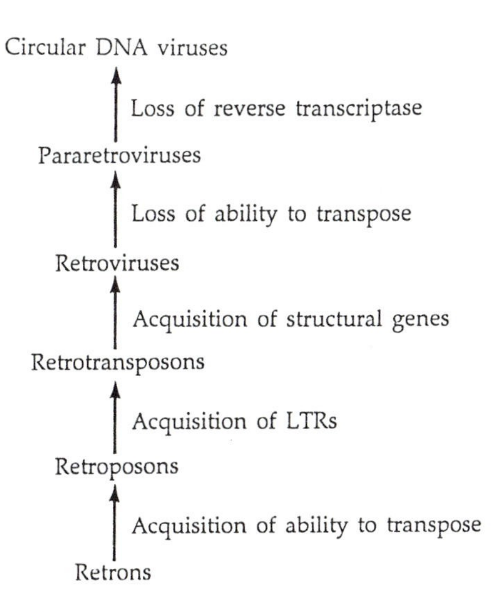

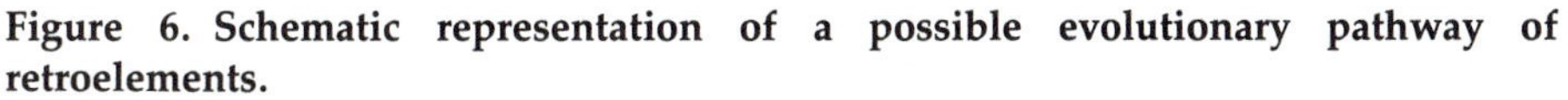

Figure 6. Schematic representation of a possible evolutionary pathway of retroelements.

and therefore are integral parts of the genome. Unlike proviruses, retrons have no LTRs and cannot construct virion particles.

Pararetroviruses, such as the hepatitis B virus, are structurally similar to retroviruses but have lost their ability to insert themselves into the host genome. This disqualifies them as transposable elements, although they clearly have a common evolutionary origin with the retroviruses.

The fact that the reverse transcriptases of all retroelements have some amino acid identity suggests a common evolutionary origin. Because of the simplicity of retrons as opposed to the complex structure of retroviruses, and because of the antiquity of bacteria, Temin (1989) has suggested that the path of evolution went from retrons to retroposons to retrotransposons to retroviruses to pararetroviruses (Figure 6). Of course, it is possible that some of the present-day retrotransposons have been derived from retroviruses, rather than the other way around.

RETROSEQUENCES

Retrosequences, or **retrotranscripts**, are genomic sequences that have been derived through the reverse transcription of RNA and subsequent integration into the genome but lack the ability to produce reverse transcriptase (Table 1). The template from which a retrosequence has been derived is usually the RNA transcript of a gene. Some authors call retrosequences "nonviral retro-

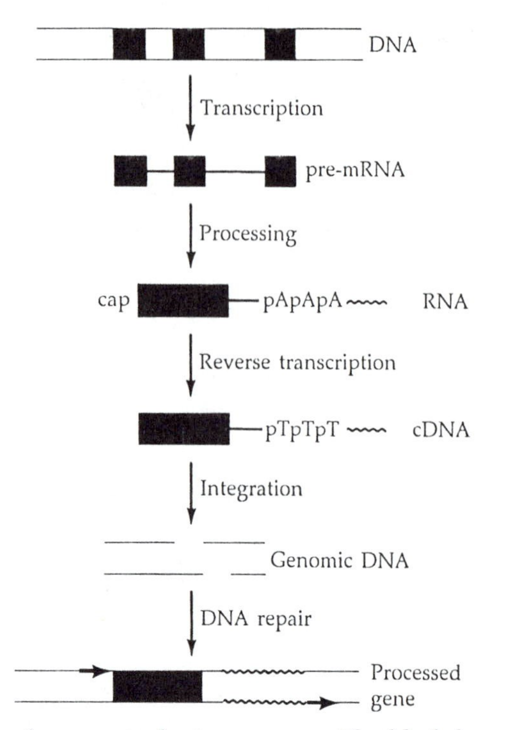

Figure 7. Creation of a processed retrosequence. The black boxes represent exons. The wavy lines indicate a poly-A tail in mRNA and the complementary poly-T in cDNA. The DNA is transcribed into pre-mRNA, then processed into mRNA. The mRNA is reverse-transcribed into cDNA, which becomes integrated into the genomic DNA. The gaps are repaired, and so two direct short repeats flanking the inserted retrosequence are created (black horizontal arrows). If the cDNA is inserted into the genome of a different cell from the one in which the RNA has been synthesized, the process of creating a retrosequence requires the mRNA to become incorporated into a retroviral particle and transported to the target cell. Such a process is called retrofection.

posons" (e.g., Weiner et al. 1986). A process of producing retrosequences is shown in Figure 7. If a gene is not transcribed within any germ-line cells, the creation of a retrosequence requires the RNA to cross cell barriers. This can happen when an RNA molecule becomes encapsulated within the virion particle of a retrovirus and is then transported to a germ-line cell where it is reverse-transcribed (Linial 1987). This process is referred to as **retrofection**.

Since retrosequences originated from RNA sequences, they bear marks of RNA processing and are, hence, also referred to as **processed sequences**. The diagnostic features of retrosequences include: (1) lack of introns, (2) precise boundaries coinciding with the transcribed regions of genes, (3) stretches of poly-A at the 3′ end, (4) short direct repeats at both ends, indicating that transposition may have been involved, (5) various posttranscriptional modifications, such as the addition or removal of short stretches of nucleotides, and (6) chromosomal position different from the locus of the original gene from which the RNA was transcribed.

There are two types of retrosequences: processed genes (retrogenes) and processed pseudogenes (retropseudogenes).

Retrogenes

A **processed gene** or **retrogene** is a functional retrosequence producing a protein that is identical or nearly identical to that produced by the gene from which the retrogene has been derived. There are several reasons why it is highly unlikely that a reverse-transcribed gene will retain its functionality. First, the process of reverse transcription is very inaccurate, and many differences (mutations) between the RNA template and the cDNA can occur. Second, unless a processed gene has been derived from a gene transcribed by RNA polymerase III, it usually does not contain the necessary regulatory sequences that reside in the untranscribed regions. Third, a processed gene may be inserted at a genomic location that may not be adequate for its proper expression. Indeed, in the vast majority of cases a processed gene is "dead on arrival."

Surprisingly, processed functional genes have been found, although they seem to be very rare. The human phosphoglycerate kinase (*PGK*) multifamily consists of an active X-linked gene, a processed X-linked pseudogene, and an additional autosomal gene. The X-linked gene contains 11 exons and 10 introns. Its autosomal homologue, on the other hand, is unusual in that it has no introns and is flanked on its 3′ end by remnants of a poly-A tail, strongly suggestive of a reverse transcription process involving mRNA. Interestingly, the autosomal *PGK* gene is expressed almost exclusively in the testes. Thus, the reverse-transcribed *PGK* gene has not only maintained an intact reading frame and the ability to transcribe and produce a functional polypeptide, but also has acquired a novel tissue-specificity (McCarrey and Thomas 1987). The muscle-specific calmodulin gene in chicken is also intronless and apparently was produced by a reverse-transcriptase-mediated event (Gruskin et al. 1987).

The rat and mouse preproinsulin I gene may represent an instance of a **semiprocessed retrogene**. The gene contains a single 119-bp intron in the 5′ untranslated region. In comparison, its homologue, preproinsulin II, contains the same small intron, as well as an additional larger (499-bp) intron within the coding region for the C peptide. All preproinsulin genes from other mammals, including other rodents, contain two introns as well. Moreover, the preproinsulin I gene is flanked by short repeats, and the polyadenylation signal is followed by a short poly-A tract (Soares et al. 1985). These features suggest that the preproinsulin I gene might be a semiprocessed retrogene derived from a partially processed preproinsulin II pre-mRNA. Indeed, based on comparisons between the two preproinsulin genes, preproinsulin I appears to have been derived from an aberrant pre-mRNA transcript that initiated 500 base pairs upstream of the normal cap site and from which only the first intron has been excised. It is precisely because the aberrant transcript contained 5′ regulatory sequences not normally transcribed that the retrogene has maintained its function following its integration into a new genomic location.

Processed pseudogenes

A **processed pseudogene** or **retropseudogene** is a retrosequence that has lost its function. It bears all the hallmarks of a functional retrosequence but has molecular defects that prevent it from being expressed. A comparison

```
         M   A   T   K   A   V   C   V   L   K   G   D   G   P   V
SOD-1    ATG GCG ACG AAG GCC GTG TGC GTG CTG AAG GGC GAC GGC CCA GTG
           • ••   •       •  •    •  •   •           • • •      •
ψ69.1    ATA ATG ATG AAG GTC ATG TAC ATG TTG AAG GGC CAG AGC CCG GTG
         I   M   M   K   V   M   Y   M   L   K   G   Q   S   P   V

SOD-1    Q   G   I   I   N   F   E   Q   K                E  S   N   G
         CAG GGC ATC ATC AAT TTC GAC CAG AAG G  intron    AA AGT AAT GGA
              ••  −       •    −    •                         •  ---  •
ψ69.1    CAG GCG A C ATC CAT TT  GAG CAG AAG G            AA AAT     GAA
         Q   V     T   S   I   *  **

         P   V   K   V   W   G   S    I   K   G   L   T   E   G   L
SOD-1    CCA GTG AAG GTG TGG GGA A GC ATT AAA GGA CTG ACT GAA GGC CTG
             • •  •       − • •   +       •     •           •    •
ψ69.1    CCA TTT ATG GTG T C AGA ATGC ATT ACA GGA TTG ACT GAA CGC CAG

         H   G   F   H   V   H   E   F   G   D   N   T   A
SOD-1    CAT GGA TTC CAT GTT CAT GAG TTT GGA GAT AAT ACA GCA intron
           • •                                −   −   •   •
ψ69.1    CAC AGA TTC CAT GTT CAT CAG TTT GGA G T A T AAC ACA
```

Figure 8. Comparison between the first two exons of the human Cu/Zn superoxide dismutase gene (*SOD-1*) and the homologous parts of a processed pseudogene (*ψ69.1*). Dots denote substitutions, "minus" symbols denote deletions, and "plus" symbols denote insertion. Note the absence of the intron and the premature termination codon (indicated by asterisks). See Table 1 in Chapter 1 for the one-letter amino acid abbreviations. Data from Danciger et al. (1986).

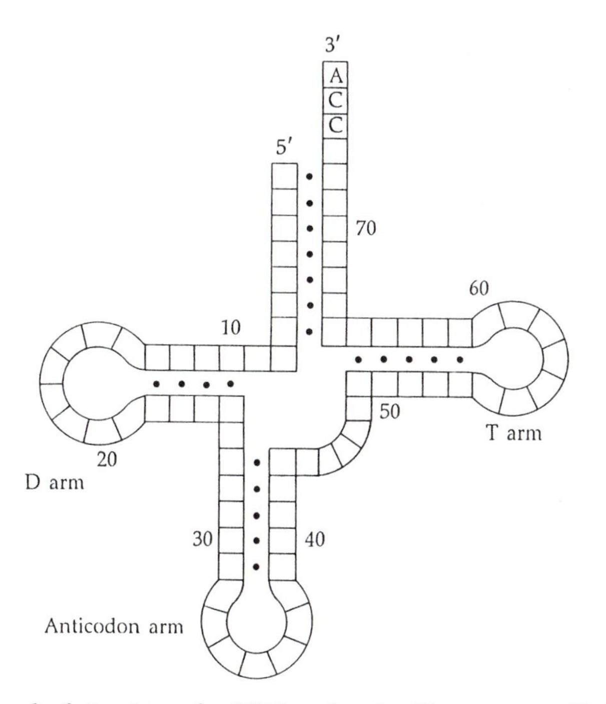

Figure 9. Cloverleaf structure of a tRNA molecule. The sequence CAA at its 3′ end is added posttranscriptionally in the functional molecule, but it often appears in the genomic sequence of tRNA pseudogenes.

between a functional gene and a processed pseudogene is shown in Figure 8. Many processed pseudogenes are truncated during retrofection; 5′ truncation of processed mRNA is particularly common, but 3′ truncation is also known. Truncation of the 5′ end can occur during (1) transcription (e.g., initiation downstream of the normal site), (2) RNA processing (e.g., faulty splicing), or (3) reverse transcription (e.g., failure of the enzyme to complete the reverse transcription to the 3′ end of the RNA molecule, which corresponds to the 5′ end of the cDNA).

Processed pseudogenes derived from all types of RNA (e.g., mRNA, tRNA, rRNA, snRNA, and 7SL RNA) are known. Transfer RNA pseudogenes are particularly interesting since they provide one of the most compelling pieces of evidence that processed pseudogenes are indeed derived through the reverse transcription of RNA. All nuclear tRNA molecules possess a CCA sequence at the 3′ end (Figure 9). This sequence is not encoded by the tRNA-specifying gene, but is added enzymatically after transcription. In contrast, genomic processed tRNA pseudogenes often contain the CCA sequence at the 3′ terminal.

Processed pseudogenes have been found in animals, plants, and even

Table 2. **The number of retropseudogenes and the number of parental functional genes.**

Species	Gene	Number of genes	Number of retropseudogenes
Human	argininosuccinate synthetase	1	14
	β-actin	1	~20
	β-tubulin	2	15–20
	Cu/Zn superoxide dismutase	1	≥4
	cytochrome *c*	2	20–30
	dihydrofolate reductase	1	~5
	nonmuscle tropomyosin	1	≥3
	glyceraldehyde-3-phosphate dehydrogenase	1	~25
	phosphoglycerate kinase	2[a]	1
	ribosomal protein L32	1	~20
	triosephosphate isomerase	1	5–6
Mouse	α-globin	2	1
	cytokeratin endo A	1	1
	glyceraldehyde-3-phosphate dehydrogenase	1	~200
	myosin light chain	1	1
	proopiomelanocortin	1	1
	ribosomal protein L7	1–2	≥20
	ribosomal protein L30	1	≥15
	ribosomal protein L32	1	16–20
	tumor antigen p53	1	1
Rat	α-tubulin	2	10–20
	cytochrome *c*	1	20–30

From Weiner et al. (1986).
[a] One of which is a retrogene.

bacteria. However, although processed pseudogenes are abundant in mammals, they are relatively rare in other organisms, such as chickens, amphibia, and *Drosophila*. Table 2 shows a list of processed pseudogenes in humans and rodents for which both the number of functional genes and the number of processed pseudogenes are known or have been estimated. On average, there seem to be more processed pseudogenes than functional genes in these species. In fact, the number of processed pseudogenes may even be underestimated in many cases because old processed pseudogenes may have diverged in sequence from their parental gene to such an extent that they are no longer detectable by molecular probes derived from the functional homologue.

In some cases, the number of processed pseudogenes can exceed the number of their functional counterparts by orders of magnitude. One such example is the *Alu* family, so named because the sequence contains a characteristic restriction site for the *Alu1* endonuclease. *Alu* sequences are approximately 300 bp long and they belong to a family of repeated sequences that appear more than 500,000 times in the human genome, constituting a remarkable 5–6% of the genome.

Ullu and Tschudi (1984) found that *Alu* sequences are in fact processed pseudogenes of the gene specifying 7SL RNA. 7SL RNA is essential in the cutting of signal sequences of secreted proteins. The active gene is highly constrained, and its sequence is conserved in such diverged organisms as humans, *Xenopus*, and *Drosophila*. Human *Alu* sequences have been derived from a functional 7SL sequence by a series of steps, involving a duplication, two deletions, and many nucleotide substitutions (Figure 10a). Most human *Alu* sequences have a dimeric structure. The human genome also contains a number of tetrameric *Alu* sequences, but only a few monomeric *Alu* elements have ever been found in humans. In contrast, the rodent *Alu* equivalent, the *B1* family, is almost exclusively monomeric (Figure 10b). Britten et al. (1988) dated the emergence of the first monomer to a time before the mammalian radiation and dated the duplication that produced the dimer to a time after the primate lineage had been established.

If a retropseudogene retains its capability to be transcribed, a cascade process may ensue, whereby new retropseudogenes are created out of the RNA transcripts of existing pseudogenes. This was suggested to be the case in the *Alu* family (Bains 1986). We note that the 7SL gene is transcribed by RNA polymerase III, which does not require promoters outside of the transcribed region. Therefore, it is conceivable that some *Alu* sequences have retained intact promoters and are continuously transcribed. However, Willard et al. (1987) and Britten et al. (1988) have recognized only a few subfamilies of *Alu* sequences. Britten et al. (1988) proposed that these subfamilies have been sequentially derived from four source genes (Figure 11). The more recently derived subfamilies have sequences that are more divergent from the progenitor 7SL sequence than the older subfamilies. The conclusion is, thus, that *Alu* sequences have been derived not directly from the 7SL functional gene, but from a small number of source sequences, which had originally been derived from 7SL RNA sequences through many steps of changes. Each of these four source genes served at one time or another as the predominant source of *Alu* sequences and was superseded by a descendant line. The successive waves of fixation did not occur in sudden bursts, but consecutive subfamilies continued to coexist within the genome for long periods of time (see also Quentin 1988).

Due to the ubiquity of reverse transcription, the genomes of mammals are

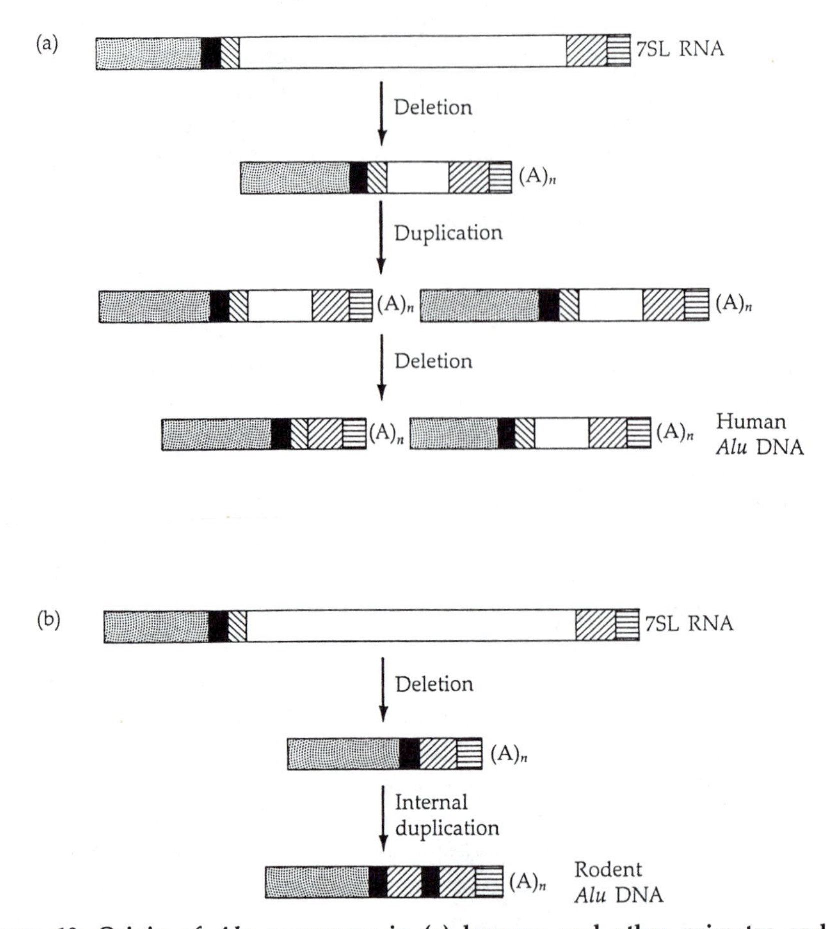

Figure 10. Origin of *Alu* sequences in (a) humans and other primates and (b) rodents. Different regions in the 7SL RNA genes are shaded differently to emphasize the deletions and rearrangements in the *Alu* sequences. $(A)_n$ means that A is repeated *n* times. Note the dimeric structure in (a) and the monomeric structure in (b).

literally bombarded with copies of reverse-transcribed sequences. The vast majority of these copies are nonfunctional from the moment they are integrated into the genome. Moreover, such sequences cannot be easily rescued by gene conversion, since they are mostly located at great chromosomal distances from the parental functional gene (Chapter 6). The phenomenon of a functional locus pumping out defective copies of itself and dispersing them all over the genome has been likened to a volcano generating lava, and the process has been termed the **Vesuvian mode of evolution** (P. Leder, cited in Lewin 1981).

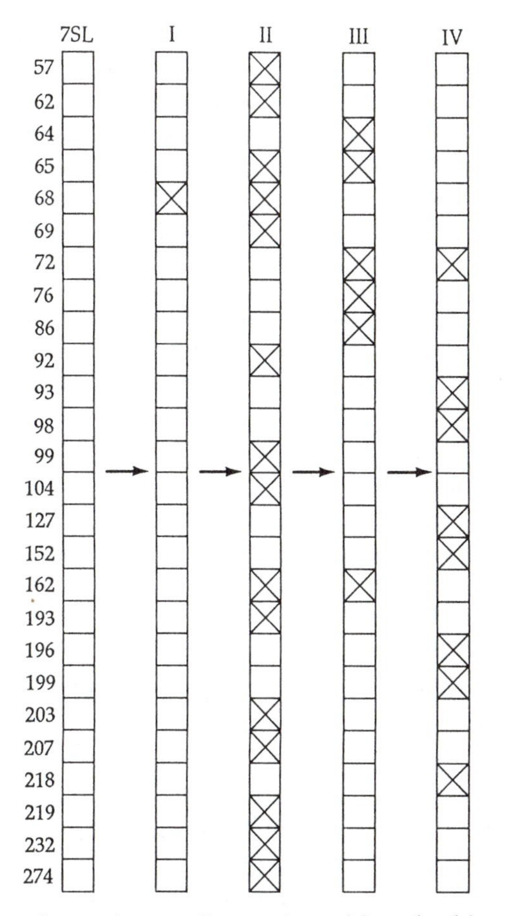

Figure 11. Sequence of mutations at diagnostic positions (arabic numerals) between different subfamilies of *Alu* sequences. Roman numerals indicate successive source genes that served at various periods as the predominant source of *Alu* sequences. Substitutions distinguishing each subfamily from the preceding one are denoted by X. Data from Britten et al. (1988).

Evolution of processed pseudogenes

As mentioned previously, as soon as a processed pseudogene is established as a chromosomal sequence within the genome, it is nonfunctional and free from all selective constraints. Because of the lack of function, pseudogenes are affected by two evolutionary processes (Graur et al. 1989b). The first involves the very rapid accumulation of point mutations. This accumulation eventually obliterates the sequence similarity between the pseudogene and its functional homologue, which evolves much more slowly. The nucleotide composition of the pseudogene will come to resemble more and more its nonfunctional surroundings, and it will eventually "blend" into it. This process has been called **compositional assimilation**.

The second evolutionary process is characterized by pseudogenes becoming increasingly shorter compared to the functional gene. This **length abridgment** is caused by the excess of deletions over insertions. It has been estimated that a mammalian processed pseudogene loses about half of its DNA in about 400 million years. This process is so slow that the human genome, for instance, still contains major chunks of pseudogenic DNA that were found in very distant ancestors. Obviously, these ancient pseudogenes have by now lost almost all similarity to the functional genes from which they have been derived.

In summary, it seems that processed genes are created at a much faster rate than the rate by which they are obliterated by deletions. It has, thus, been concluded that abridgment is too slow a process to offset the increase in genome size following the continuous Vesuvian bombardment (Chapter 8; Graur et al. 1989b).

EFFECTS OF TRANSPOSITION ON THE HOST GENOME

Transposition and retroposition can have very profound effects on the size and structure of genomes. In particular, transposable elements have been considered the best example of "**selfish DNA**," which may not confer any advantage to the host but can spread in the genome because it multiplies faster than genomic sequences (Doolittle and Sapienza 1980; Orgel and Crick 1980). For this reason, transposition can greatly increase the genome size. This effect will be dealt with in Chapter 8. Here, we shall concern ourselves with the ways in which transposable elements affect gene evolution and expression.

First, as mentioned above, transposons in bacteria often carry genes that confer antibiotic or other resistances to their carriers. Thus, transposons may enable the host species to survive in an adverse environment.

Second, the expression of a gene may be altered by the presence of a transposable element either within the gene or adjacent to it. In the simplest case, the insertion of a transposable element into the coding region of a protein-coding gene will most probably alter the reading frame and may have drastic phenotypic effects. Similarly, excision of transposable elements may be imprecise, resulting in the addition or deletion of bases. There are, however, unexpected effects. For example, a transposable element may contain regulatory elements, such as promoters, which would affect the rate of transcription of a nearby gene. Indeed, the LTRs of retroviruses often contain strong enhancers, which greatly influence the expression of nearby genes. Similarly, *Ty* (which stands for "transposon yeast") elements in *Saccharomyces cerevisiae* are known to increase the expression of downstream genes. This may be beneficial in some specific circumstances, although in most cases the metabolic imbalance produced by such a change is most probably detrimental. Transposable elements that contain splice donor or acceptor sites may affect the processing of the primary RNA transcript even if the transposable element has incorporated itself within a noncoding region of a gene, such as an intron.

Third, many transposable elements promote gross genomic rearrangements. Inversions, translocations, duplications, and large deletions and insertions can be mediated by transposable elements. These rearrangements can take place as a direct result of transposition (i.e., by moving pieces of DNA from one genomic location to another) and can alter the milieu in which they are expressed. A more indirect effect will ensue if, as a result of transposition, two sequences that had previously had little similarity with each other are now sharing a similar transposable element so that unequal crossover between them is possible. Figure 12 illustrates how an unequal crossover event, facilitated by the presence of multiple *Alu* sequences within the introns flanking exon 5 of the low-density-lipoprotein receptor gene, has given rise to a mutant gene lacking this exon (Hobbs et al. 1986). In the low-density-lipoprotein gene, the deletion of exon 14 by the same mechanism has been observed (Lehrman et al. 1986). Patients homozygous for these deletions

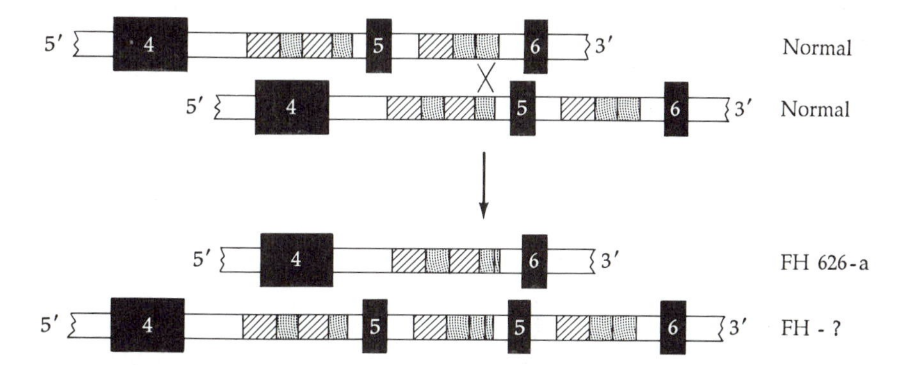

Figure 12. Unequal crossing-over in the low-density-lipoprotein receptor gene. Exons are indicated by solid bars and are numbered. *Alu* sequences in the introns are indicated by hatching (right arms) and by stippling (left arms); an Alu sequence may consist of two arms (dimeric) or one arm (monomeric). The position of the postulated crossing-over is indicated by the X. The deleted (observed) and inserted (inferred) products of the recombination are depicted as FH 626-a and FH-? below the arrow. From Hobbs et al. (1986).

have a high level of cholesterol in the blood (hypercholesterolemia). Recombination between two *Alu* elements has also been shown to be responsible for the deletion of the promoter and the first exon of the adenosine deaminase gene in patients with adenosine-deaminase deficiency (Markert et al. 1988). In general, genomic instability has been demonstrated for all regions containing *Alu* repeat sequences (Calabretta et al. 1982).

Fourth, there is evidence that some transposable elements may cause an increase in the rate of mutation. For example, strains of *Escherichia coli* that contain the transposable element *Tn10* were found to have elevated rates of insertions (Chao et al. 1983). Under most conditions this trait will be deleterious to the carrier. However, under severe environmental stress it is possible that an elevated rate of mutation might be advantageous because some mutations may be better suited to the new circumstances and their carriers will be fitter than non-carriers.

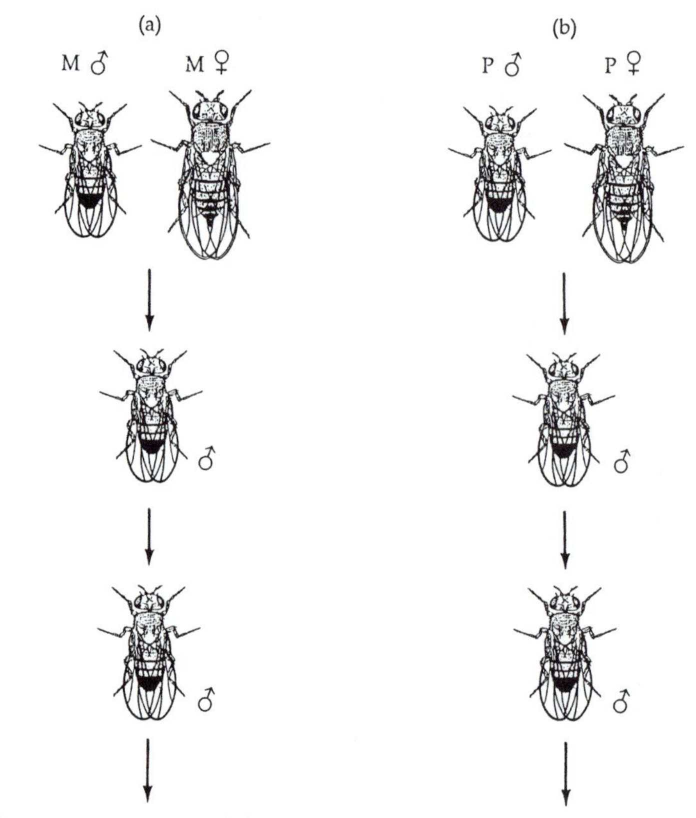

Figure 13. Hybrid dysgenesis in *Drosophila*. (a) Matings within *M* strains and (b) matings within *P* strains produce normal progeny. (c) Normal progeny are also

HYBRID DYSGENESIS

Hybrid dysgenesis in *Drosophila* is a syndrome of correlated abnormal genetic traits that is spontaneously induced in one type of hybrid between certain mutually interactive strains, but usually not in the reciprocal hybrid (Sved 1976; Kidwell and Kidwell 1976). Hybrid dysgenesis has attracted much attention from molecular and evolutionary biologists because it has been found to be caused by transposable elements and because its main manifestation is the creation of a barrier against hybridization between strains or populations, which has been speculated to be a cause of speciation (see below).

There are several dysgenic systems in *Drosophila*, and in the following we shall deal with just one of them, the *P-M* system. The asymmetry of hybrid dysgenesis is shown in Figure 13. When a male from a *P* strain mates with a female from an *M* strain, the offspring are dysgenic; in the reciprocal

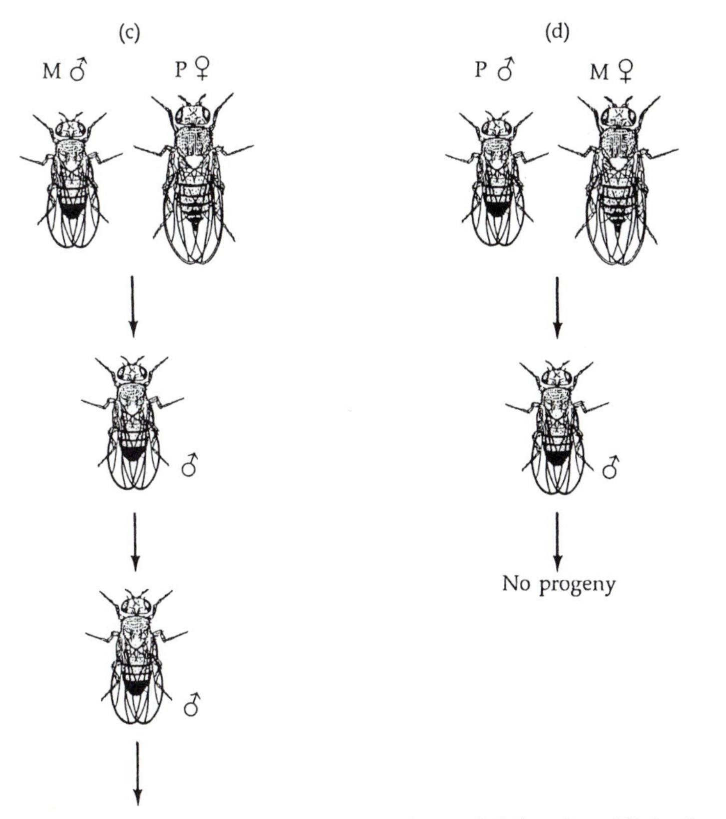

produced in crosses between *M* males and *P* females. (d) In the crosses between *P* males and *M* females, dysgenic offspring are produced, many of which are sterile.

mating, the offspring are normal. The dysgenic traits of the *P-M* system include: (1) failure of the gonads to develop in some individuals under certain conditions, (2) recombination in males (an unnatural occurrence in *Drosophila*, in which recombination is usually restricted to females), (3) chromosome breakage, (4) distortion of Mendelian transmission ratios (i.e., *M* chromosomes are preferentially transmitted to the offspring relative to *P* chromosomes), and (5) a high frequency of mutation.

The cause of *P-M* dysgenesis is a family of transposable elements called *P* elements (Figure 3). In *P* strains, there are 30–50 *P* elements in the genome, but many of them may contain deletions. They are distributed throughout all chromosomes, although in some strains transposition shows a preference for the X chromosome. *M* strains do not carry *P* elements. The asymmetry of the hybrid-dysgenesis system is thought to result from the maternal inheritance of a *P*-element-encoded repressor by F_1 progeny of *P* female × *M* male matings and the absence of such a repressor in the F_1 progeny of the reciprocal *M* female × *P* male matings. Dysgenic traits are associated with the transposition of *P* elements in germ-line cells that have maternally derived cytoplasm lacking repressors. In the context of hybrid dysgenesis, the presence or absence of repressors in the cytoplasm defines the type of reaction following the formation of the zygote, and is referred to as the **cytotype**. In the presence of the repressors, *P* element transposition may be wholly or partially inhibited. Transposition of *P* elements normally does not take place in somatic cells, because the third intron in the coding region is not excised as it is in germ-line cells.

An interesting observation has been made by Kidwell (1983) concerning the distribution of *P*-carrying strains. *P* characteristics were not found in any *Drosophila melanogaster* strains collected before 1950, and collections made subsequently showed increasing frequencies of *P* with decreasing age (Figure 14). A similar observation was made with another dysgenic system, *I–R*. Two hypotheses were suggested to explain the distribution of *P* elements. Engels (1981b) proposed that most strains in nature are of the *P* type but that they tend to lose the transposon in laboratory populations. The second hypothesis postulates a recent introduction of *P* transposons into *Drosophila melanogaster* populations followed by a rapid spread of *P* elements in formerly *M* populations (Kidwell 1979). There are several reasons why the second hypothesis seems more reasonable. First, *P*-carrying laboratory strains that have been monitored for close to 15 years did not lose their *P* characteristics. Second, there seems to be a geographical cline in the distribution of *P* strains, with North American populations showing earlier signs of carrying *P* than some European, African, and Asian strains. Lastly, there is now evidence that *P* elements have been recently acquired by *Drosophila melanogaster* from a distantly related species (see page 201).

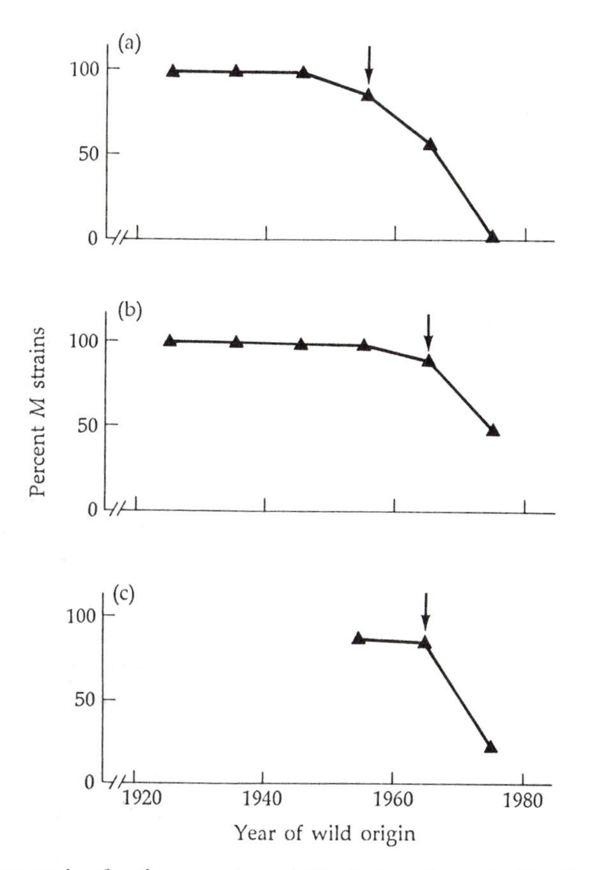

Figure 14. Changes in the frequencies of *M* strains (i.e., strains devoid of *P* elements) in natural populations in (a) North and South America, (b) Europe, Africa, and the Middle East, and (c) Australia and the Far East. Arrows denote the first appearance of *P* strains. Note that the proportion of *M* populations decreased first in the American continents and only later decreased in the other continents. Modified from Kidwell (1983).

TRANSPOSITION AND SPECIATION

Speciation or **cladogenesis** (i.e., the creation of two or more species from a parental species) is one of the most important evolutionary processes. Unfortunately, at the molecular level, it is also one of the least understood evolutionary processes. We do not know by what means new species arise from old ones. What we do know is that the process of speciation requires the creation of a reproductive barrier between two populations belonging to the same species, so that they can no longer interbreed. Hybrid dysgenesis has been thought for a while to represent an early stage in the process of speciation, by acting as a postmating reproductive isolation mechanism between different populations belonging to the same species. Indeed, the sterility of hybrids produced by crosses between the sibling species *D. melanogaster* and *D. simulans* is very similar to dysgenesis (e.g., rudimentary gonads, segregation distortion).

There are, however, problems with this view. First, although hybrids exhibit reduced fitness and are, therefore, partially isolated in terms of reproduction, the transposition of *P* elements in the germ-line virtually ensures that most of the chromosomes transmitted to the hybrids will bear *P* elements, and the cytotype will eventually change to the *P*-type too. Thus, provided the reduction in fitness of the hybrids is not too great, the *P* element will spread through the entire population. Indeed, the hybrid must be almost completely sterile for an effective reproductive isolation to last. Second, *P* elements have the ability to transpose themselves horizontally as an infectious agent from individual to individual (see page 201). Thus, an entire population may be rapidly taken over by *P*, so that hybrid dysgenesis is likely to last for very short periods of time in nature. Indeed, many species of *Drosophila* are known in which all individuals and all populations carry *P* elements or *P*-like elements, and consequently hybrid dysgenesis does not occur in any of these species. Finally, as far as we know now, hybrid dysgenesis is restricted to *Drosophila* and may not represent a universal phenomenon in nature. Even in *Drosophila* no evidence has been found for the involvement of mobile elements as barriers to gene flow between sibling species.

Since the discovery of transposable elements, numerous other mechanisms for speciation by transposition have been proposed in the literature. For example, it has been suggested that mass replicative transposition of elements containing regulatory sequences in one population may cause a so-called **genetic resetting** of the genome, whereby many genes will be subject to a novel form of regulation. Such a population will obviously become reproductively isolated from a population that retains the old form of regulation. Another suggestion invokes a mechanism of **mechanical incompatibility**, also caused by mass replicative transposition. In this case, it is assumed that in one population the transposable elements multiply to such an extent as to cause a significant increase in the size of the chromosomes. A hybrid organism that inherited big chromosomes from one parent and small chromosomes from the other would experience difficulties in chromosome pairing during meiosis, and would most probably be sterile. Unfortunately, none of the speciation models so far proposed has been substantiated by empirical data.

EVOLUTIONARY DYNAMICS OF TRANSPOSABLE-ELEMENT COPY NUMBER

The number of transposable elements within a genome is determined by three factors: (1) u, the probability that a transposable element produces a new genomic copy (i.e., the probability of replicative transposition), (2) v, the probability that the element is excised, and (3) the intensity of selection against increased numbers of transposable elements within the genome. The values of u and v have been determined experimentally for several transposable elements in populations of *Drosophila melanogaster*. Transposition rates were found to vary among the transposable elements, but on the average were on the order of 10^{-4} per element per generation. Excision rates were about one order of magnitude lower (Charlesworth and Langley 1989). Consequently, in the absence of selection against the transposable element, the number of copies in the genome is expected to increase indefinitely.

If the number of transposable elements is maintained at an equilibrium, an assumption that may not hold in nature, then selection must operate against an increase in copy number. In the simplest deterministic model, we assume that the fitness of an individual, w, decreases with copy number, n. The justification for this assumption is that the insertion of transposable elements frequently alters the expression of adjacent genes; with increasing numbers of transposable elements, the probability of a deleterious alteration of gene expression increases. It can be shown that as long as w decreases with n, then regardless of the exact relationship between n and w, the mean fitness of a population at equilibrium relative to an individual lacking transposable elements is

$$\overline{W} = e^{-n(u-v)} \tag{7.1}$$

(Charlesworth 1985).

In the case of *Drosophila melanogaster*, there are approximately 50 families of transposons, each appearing in the genome on the average 10 times (Finnegan and Fawcett 1986). Thus, $n = 500$. Since v is smaller than u by at least an order of magnitude, $u - v \approx u = 10^{-4}$. Solving Equation 7.1, we obtain $\overline{W} = 0.95$. The reduction in fitness is, thus, $s = 1 - 0.95 = 0.05$. The stability of the equilibrium given by Equation 7.1 requires that the logarithm of fitness declines more steeply than linearly with increasing n (Charlesworth 1985). For simplicity, however, if we assume linearity, then the reduction in fitness with each additional transposon is approximately $0.05/500 = 10^{-4}$. Such a small selective coefficient essentially means that the number of copies of transposable elements within an organism is mainly determined by random genetic drift. If an organism contains larger numbers of transposable elements, and the possibility exists that even in *Drosophila* the number of transposable elements greatly exceeds 500 (Rubin 1983), then a transposon's effect on fitness might be even smaller than above.

An alternative to selection against increase in copy number would be a mechanism of self-regulated transposition, i.e., a rate of transposition that decreases with copy number or a rate of excision that increases with copy number (see Charlesworth and Langley 1989).

HORIZONTAL GENE TRANSFER

Horizontal gene transfer is defined as the transfer of genetic information from one genome to another, specifically between two species. This term has been coined to distinguish this type of transfer from the usual "vertical" transfer in which the parental generation passes genetic information on to the progeny. Horizontal gene transfer requires (1) a vehicle to transport the genetic information between organisms and cells and (2) the molecular machinery for inserting the foreign piece of DNA into the host genome. Retroviruses can accomplish both tasks since they are capable of both incorporating chromosomal DNA into their genomes and crossing species boundaries (Benveniste and Todaro 1976; Bishop 1981). In the case of transposons and other types of DNA-mediated transposition, the cross-cellular transport must be provided by an infectious agent, such as a plasmid. In fact, many naturally occurring plasmids contain transposable elements, which can move from the plasmid to the bacterial chromosome and vice versa.

One may be able to detect a horizontal gene-transfer event through the discovery of an outstanding discontinuity in the phylogenetic distribution of a certain gene. For example, the bacterium *Salmonella typhimurium* contains a histone-like gene which, as far as we know, has no counterpart in other bacteria (Higgins and Hillyard 1988). Horizontal gene transfer may also be suspected when a notable discrepancy between gene phylogeny and species phylogeny is discovered, in particular when sequence similarity seems to reflect geographical proximity rather than phylogenetic affinity. Consider, for example, the phylogenetic tree in Figure 15a. Let us assume that a piece of DNA was transferred from species B to species C after the divergence of B from A. On the basis of sequence comparisons involving any gene other than the horizontally transferred one, we expect to be able to reconstruct the correct phylogenetic relationships among the species. In contrast, if we use the horizontally transferred piece of DNA we will obtain the erroneous tree in Figure 15b. We note, however, that factors other than horizontal gene transfer can contribute to a discrepancy between the species tree and the gene tree (Chapter 5).

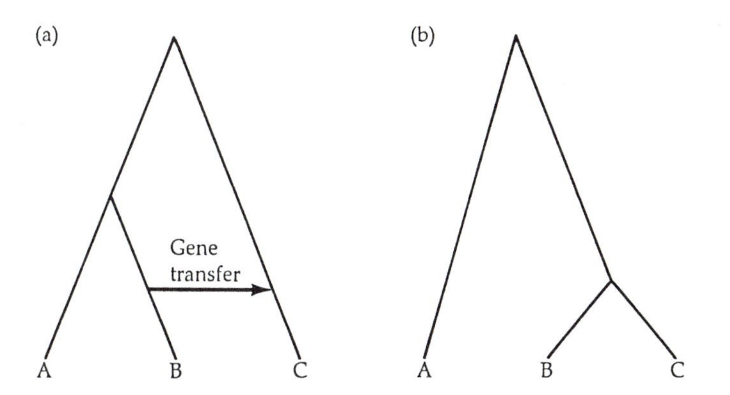

Figure 15. Phylogenetic tree reconstruction in the case of a horizontal gene transfer (a) True tree. (b) Inferred tree.

Two types of sequences can be transferred horizontally: (1) sequences derived from transposable elements, and (2) genomic sequences. There are very few cases in which horizontal gene transfer of genomic sequences has been convincingly demonstrated. Many such initial claims have subsequently been found not to be supported by molecular evidence. Moreover, we note that a horizontally transferred gene is expected to retain its functionality in the host species even less frequently than a gene transferred from one genomic location to another within the same species (see page 183).

Horizontal transfer of virogenes from baboons to cats

The vertebrate genome contains a number of sequences that are homologous to retroviruses. Such sequences, which are normal constituents of the nuclear DNA of eukaryotes, are called **endogeneous retroviral sequences** or **virogenes**. There are several examples of endogeneous retroviral sequences being transferred between vertebrate species (reviewed in Benveniste 1985). One such example involves a type-C virogene from baboons (Figure 16).

Sequences homologous to the baboon virogene have been detected in the cellular DNA of all Old World Monkeys. The sequence similarity between them is closely correlated with the taxonomic relationship among the species. Thus, the virogene has existed for at least 30 million years among primates. Interestingly, six species of cats closely related to the domestic cat (*Felix catus*) also contain this sequence, although it is present in neither more distantly related Felidae, such as lions, leopards, and bobcats, nor in any other carnivores. It is, thus, highly probable that this sequence was horizontally transferred between species some time in the past. The date and direction

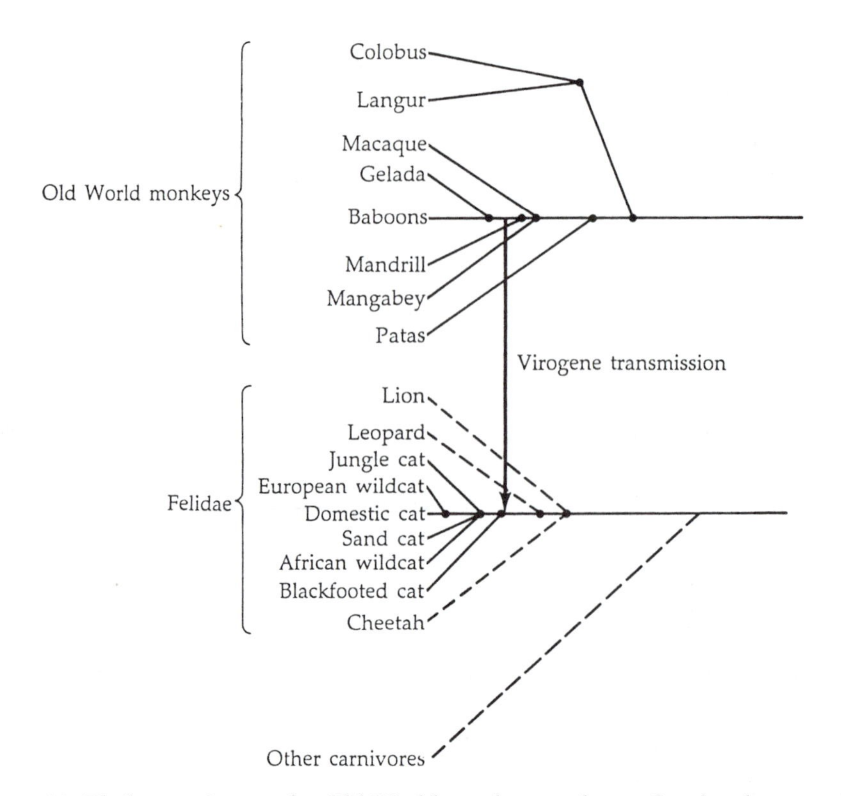

Figure 16. Phylogenetic trees for Old World monkeys and cats. Species that contain the type-C virogene are marked with solid lines. The type-C virogenes from the baboons and the gelada exhibit the highest similarity with the feline type-C virogenes. Therefore, a horizontal gene transfer has probably occurred about 10 million years ago from the ancestor of the baboons and the gelada to the ancestor of the modern cats after the divergence of lion, leopard, and cheetah lineages. Modified from Benveniste (1985).

of the horizontal transmission can be deduced from two types of data: (1) sequence similarity and (2) paleogeographical information.

All cat species that contain the baboon virogene are from the Mediterranean area, while Felidae from Southeast Asia, the New World, and Africa lack the sequence. Therefore, the transfer occurred after the major radiation of the Felidae and was limited to one geographical area. This conclusion points to a date of 5–10 million years ago for the horizontal gene-transfer event. The direction of transmission can be deduced by considering the distribution of the sequence among primates, on the one hand, and cats, on the other. Since all Old World primates possess the virogene, while only a few species of cats possess it, it is reasonable to assume that the cats acquired the sequence from the baboons and not vice versa. This conclusion is strengthened by considering that the virogene in cats is more similar to that in three species of baboons (*Papio cynocephalus*, *P. papio*, and *P. hamadryas*) and the closely related gelada baboon (*Theropithecus gelada*) than to that of any other primate sequence. Therefore, the sequence must have been transferred to the cats from the ancestor of the baboons and the gelada shortly after their divergence from the mandrill (Figure 16). The date derived from the baboons agrees well with the date derived from the cats.

Horizontal transfer of *P* elements between *Drosophila* species

Another example of horizontal gene transfer involves the *P* elements in *Drosophila melanogaster*. As mentioned previously, *P* elements have rapidly spread through natural populations of *Drosophila melanogaster* within the last four decades (see page 195). *P* elements do not exist in closely related species of the *melanogaster* group, such as *D. mauritania*, *D. séchellia*, *D. simulans*, and *D. yakuba*. Where then did these elements come from? An extensive survey of hundreds of *Drosophila* species by Daniels et al. (1990) has shown that, with the exception of *D. melanogaster*, *P* sequences are not found in any other species of the melanogaster subgroup. In contrast, all species of the distantly related *willistoni* and *saltans* groups contain *P* and *P*-like elements. In particular, the *P* element from *D. willistoni* was found to be identical to the one in *D. melanogaster* with the exception of a single base substitution, indicating that *D. willistoni* may have served as the donor species in the horizontal gene transfer of *P* elements to *D. melanogaster*.

There are several reasons to suspect that this horizontal gene transfer occurred quite recently. First, the near identity between the *P* sequences from *D. melanogaster* and *D. willistoni* suggests a very short time of divergence. Second, the near absence of genetic variability in *P* sequences from *D. melanogaster* from even very distant geographic locations indicates that the time since the introduction of *P* elements into *D. melanogaster* is too short for genetic variability to accumulate. And finally, the geographical pattern of appearance of *P* elements in *D. melanogaster*, with populations in the American continents acquiring it first, seems to indicate that a very recent invasion, probably within the last 50 years, was involved.

PROBLEMS

1. What are the differences between a retroelement and a retrosequence?

2. Given a genomic sequence, how can you tell whether or not it is a retrosequence?

Table 3. Degrees of sequence divergence between *Alu* sequences and between η-globin pseudogenes.

Species pair	Percent divergence	
	Alu sequence[a]	η Pseudogene
Human vs. chimpanzee	2.2 ± 1.4	1.7
Human vs. orangutan	3.7 ± 1.9	3.1

Data from Koop et al. (1986a) and Li et al. (1987a).

[a] Seven orthologous sequences were used to compute the means and standard deviations.

3. How is it possible to distinguish between processed and unprocessed pseudogenes?

4. Most processed pseudogenes are "dead on arrival." Explain why this condition makes them excellent materials for inferring the pattern of point mutation (see Chapter 4 and Li et al. 1984).

5. Explain why retrogenes are rare.

6. Explain why *Alu* sequences are so abundant in the genomes of humans and other primates.

7. *Alu* sequences were thought to be functional but are now commonly believed to be processed pseudogenes. Under this hypothesis, *Alu* sequences should evolve rapidly as do other pseudogenes. Are the data in Table 3 compatible with this hypothesis?

8. Enumerate the possible advantageous and disadvantageous effects of transposition on the host.

FURTHER READINGS

Berg, D. E. and M. M. Howe (eds.). 1989. *Mobile DNA*. American Society for Microbiology, Washington, DC.

Campbell, A. 1983. Transposons and their evolutionary significance. Pp. 258–279. *In* M. Nei and R. K. Koehn (eds.), *Evolution of Genes and Proteins*. Sinauer Associates, Sunderland, MA.

Charlesworth, B. and C. H. Langley. 1989. The population genetics of *Drosophila* transposable elements. Annu. Rev. Genet. 23: 251–287.

Doolittle, R. F., D.-F. Feng, M. S. Johnson and M. A. McClure. 1989. Origins and evolutionary relationships of retroviruses. Quart. Rev. Biol. 64: 1–30.

Lewin, B. 1990. *Genes IV*. Oxford University Press, Oxford.

Scaife, J., D. Leach and A. Galizzi (eds.). 1985. *Genetics of Bacteria*. Academic Press, New York.

Shapiro, J. A. (ed.). 1983. *Mobile Genetic Elements*. Academic Press, New York.

Varmus, H. 1988. Retroviruses. Science 240: 1427–1435.

Weiner, A. M., P. L. Deininger and A. Efstratiadis. 1986. Nonviral retroposons: Genes, pseudogenes, and transposable elements generated by the reverse flow of genetic information. Annu. Rev. Biochem. 55: 631–661.

8

GENOME ORGANIZATION AND EVOLUTION

A discussion of genome organization and evolution must include three different topics. The first is genome size, which varies enormously among organisms. How did this variation come into existence? What mechanisms can increase or decrease genome size to produce such variation? The second topic is the genetic information included within genomes. Do genomes consist of mostly genic DNA? Or, is the genome made of mostly nongenic sequences? Are there many repetitive sequences in the genome, and if so what is the pattern of chromosomal distribution of the repeated sequences? Does the nongenic fraction have a function, or is it merely "junk"? The third topic concerns the nucleotide composition of the genome. Is there heterogeneity in composition among different regions of the genome? What mechanisms can give rise to localized differences in nucleotide composition?

C VALUES

The amount of DNA in the haploid genomic set, such as that in the sperm nucleus, is called the **genome size** or **C value**, where C stands for "constant" or "characteristic" to denote the fact that the size of the haploid genome is fairly constant within any one species. In contrast, C values vary widely among species in both prokaryotes and eukaryotes.

Table 1. Conversion of units commonly used to measure genome sizes.

Unit	Conversion factor		
	Picograms	Daltons	Base pairs
Picogram	1	6.02×10^{11}	0.98×10^{9}
Dalton	1.66×10^{-12}	1	1.62×10^{-3}
Base pair	1.02×10^{-9}	618	1

The sizes of nuclear genomes in eukaryotes are usually measured in picograms (pg) of DNA (1 pg = 10^{-12} g). The smaller prokaryotic genomes are more commonly measured in daltons, the unit of relative atomic or molecular mass. The sizes of still smaller genomes, such as those of organelles and viruses, as well as the sizes of specific stretches of DNA, are most often expressed in base pairs (bp) or kilobase pairs (kb) of double-stranded DNA or RNA (1 kb = 1,000 bp). To avoid confusion, we shall only use bp and kb. The conversion factors are provided in Table 1.

EVOLUTION OF GENOME SIZE IN BACTERIA

Bacterial genome sizes vary over a 20-fold range, from about 6×10^5 bp in some obligatory intracellular parasites, to more than 10^7 bp in several cyanobacterial species (Table 2). The smallest free-living prokaryotes, the mycoplasmas, contain about 350 protein-coding genes including some 50 ribosomal

Table 2. Range of C values in bacteria.

Taxon	Range in genome size (kb)	Ratio (highest/lowest)
Eubacteria	650–13,200	20
Gram negative	650–7,800	12
Gram positive	1,600–11,600	7
Cyanobacteria	3,100–13,200	4
Mycoplasmas	650–1,800	3
Archaebacteria	1,600–4,100	3

Data from Cavalier-Smith (1985).

proteins, two sets of rRNA genes (5S, 16S, and 23S), and about 40 tRNA genes. Therefore, the genome of a mycoplasma consists of about 400 genes, which is assumed to be close to the minimum number sufficient to support autonomous life (Muto et al. 1986). The numbers of genes in other bacteria range very roughly from about 500 to 8,000. In other words, among characterized bacterial species, the variation in the number of genes is the same as the variation in C values. Therefore, bacteria do not seem to contain large quantities of nongenic DNA.

The genomes of bacteria can be divided into three fractions: (1) chromosomal DNA, (2) DNA that originated in plasmids, and (3) transposable elements (Hartl et al. 1986). The chromosomal fraction contains protein-coding genes required for growth and metabolic functions (70–80%), spacers and various signals (20–30%), RNA-specifying genes (~1%), and a number of short repeated sequences, usually on the order of a few dozens of base pairs in length. Bacteria carry many plasmids as extrachromosomal genetic elements. In some instances, however, genes derived from plasmids are found integrated in the bacterial chromosome. Transposable elements are common components of some bacterial genomes. For example, the chromosome of *Escherichia coli* contains 1–10 copies of at least six different types of insertion sequences (Chapter 7). The nonchromosomal fraction of the genome (including insertion sequences and plasmid-derived genes in the chromosome) seems to be one order of magnitude smaller than the chromosomal fraction.

The distribution of genome sizes in bacteria is discontinuous, showing major peaks with modal values of about 0.8×10^6, 1.6×10^6, and 4.0×10^6 bp, and several minor peaks at 7.2×10^6 and 8.0×10^6 bp (Herdman 1985). This distribution has led to the suggestion that the larger genomes evolved from smaller genomes by successive cycles of genome duplication. Since there seems to be no notable relationship between genome size and bacterial phylogeny, it has been suggested that genome duplications have occurred frequently in the evolution of bacterial lineages (Wallace and Morowitz 1973).

Using a tentative phylogeny of bacteria based on comparison of rRNA sequences, Herdman (1985) was able to relate changes in genome size to phylogenetic history. The results indicate that genome duplication occurred independently in different bacterial lineages. Interestingly, many of the duplications seem to have occurred coincidentally in several bacterial lineages at a rather specific time in evolution, i.e., soon after the appearance of oxygen in the atmosphere, approximately 1.8 billion years ago.

In summary, the distribution of genome sizes in bacteria can be explained by a combination of several processes: (1) genome duplication during the independent evolution of a respiratory metabolism in several lines, (2) subsequent genome duplications occurring independently in many lineages, (3) small-scale deletions and insertions, (4) duplicative transposition, (5) horizontal transfer of genes derived mainly from plasmids, and (6) loss of massive chunks of DNA in many parasitic lines.

GENOME SIZE OF EUKARYOTES AND THE C-VALUE PARADOX

C values in eukaryotes are usually much larger than in prokaryotes, but there are exceptions. For instance, the yeast *Saccharomyces cerevisiae* has a smaller genome than many Gram-positive bacteria and most cyanobacteria. Since the eukaryotic genome has multiple origins of replication, while most prokaryotes seem to have only one, the eukaryotes are able to replicate much larger amounts of DNA in the same amount of time as that required for a smaller prokaryotic genome.

The variation in C values in eukaryotes is much larger than that in bacteria, from 8.8×10^6 bp to 6.9×10^{11} bp, approximately a 80,000-fold range (Table 3). Unicellular protists, particularly sarcodine amoebae, show the greatest variation in C values. The three amniote classes (mammals, birds, and reptiles), on the contrary, are exceptional among eukaryotes in their small variation in genome size (up to four-fold). Other classes, for which a substantial body of C-value data exists, show variation of at least 100-fold.

Interestingly, the huge interspecific variation in genome sizes among eukaryotes seems to bear no relationship to either organismic complexity or the likely number of genes encoded by the organism. For example, several unicellular protozoans possess much more DNA than mammals (Table 4), which are presumably more complex. Moreover, organisms that seem similar in complexity (e.g., flies and locusts, onion and lily, *Paramecium aurelia* and *P. caudatum*) exhibit vastly different C values (Table 4). This lack of correspondence between C values and the presumed amount of genetic information contained within genomes has become known in the literature as the **C-value paradox**. The C-value paradox is also evident in comparisons of sibling species (i.e., species that are so similar to each other morphologically as to be indistinguishable phenotypically). In protists, bony fishes, amphibia, and flowering plants, many sibling species differ greatly in their C values, in spite of the fact that, by the definition of sibling species, no difference in organismic complexity exists. Since we cannot assume that a species possesses less DNA than the amount required for specifying its vital functions, we have to explain why so many species contain vast excesses of DNA.

The first question to be clarified is whether or not a correlation exists between genome size and the number of genes. In other words, are the differences in genome sizes attributable to genic DNA or to nongenic DNA? Eukaryotes show approximately 50-fold variation in the number of protein-coding genes, from about 3,000 in yeast to roughly 150,000 in mammals

Table 3. Range of C values in various eukaryotic groups of organisms.

Taxon	Range in genome size (kb)	Ratio (highest/lowest)
Protists	23,500–686,000,000	29,191
Euglenozoa	98,000–2,350,000	24
Ciliophora	23,500–8,620,000	367
Sarcodina	35,300–686,000,000	19,433
Fungi	8,800–1,470,000	167
Animals	49,000–139,000,000	2,837
Sponges	49,000–53,900	1
Annelids	882,000–5,190,000	6
Mollusks	421,000–5,290,000	13
Crustaceans	686,000–22,100,000	32
Insects	98,000–7,350,000	75
Echinoderms	529,000–3,230,000	6
Agnathes	637,000–2,790,000	4
Sharks and rays	1,470,000–15,800,000	11
Bony fishes	382,000–139,000,000	364
Amphibians	931,000–84,300,000	91
Reptiles	1,230,000–5,340,000	4
Birds	1,670,000–2,250,000	1
Mammals	1,420,000–5,680,000	4
Plants	50,000–307,000,000	6,140
Algae	80,000–30,000,000	375
Pteridophytes	98,000–307,000,000	3,133
Gymnosperms	4,120,000–76,900,000	17
Angiosperms	50,000–125,000,000	2,500

Data from Cavalier-Smith (1985) and other sources.

(Cavalier-Smith 1985). This 50-fold variation is obviously insufficient to explain the 80,000-fold variation in nuclear DNA content. Moreover, gene number is positively correlated with structural complexity, while genome size is not. Nor can the interspecific variation in the lengths of mRNA molecules explain the C-value paradox. While there are differences in the mean lengths of both coding and noncoding regions between different organisms, no correlation exists between gene length and the size of the genome.

A positive correlation between the degree of repetition of several RNA-specifying genes and genome size has been found (Chapter 6). Similarly, a

Table 4. C values from eukaryotic organisms ranked by genome size.

Species	C value (kb)
Navicola pelliculosa (diatom)	35,000
Drosophila melanogaster (fruitfly)	180,000
Paramecium aurelia (ciliate)	190,000
Gallus domesticus (chicken)	1,200,000
Erysiphe cichoracearum (fungus)	1,500,000
Cyprinus carpio (carp)	1,700,000
Lampreta planeri (lamprey)	1,900,000
Boa constrictor (snake)	2,100,000
Parascaris equorum (roundworm)	2,500,000
Carcarias obscurus (shark)	2,700,000
Rattus norvegicus (rat)	2,900,000
Xenopus laevis (toad)	3,100,000
***Homo sapiens* (human)**	**3,400,000**
Nicotiana tabaccum (tobacco)	3,800,000
Paramecium caudatum (ciliate)	8,600,000
Schistocerca gregaria (locust)	9,300,000
Allium cepa (onion)	18,000,000
Coscinodiscus asteromphalus (diatom)	25,000,000
Lilium formosanum (lily)	36,000,000
Amphiuma means (newt)	84,000,000
Pinus resinosa (pine)	68,000,000
Protopterus aethiopicus (lungfish)	140,000,000
Ophioglossum petiolatum (fern)	160,000,000
Amoeba proteus (amoeba)	290,000,000
Amoeba dubia (amoeba)	670,000,000

Data from Cavalier-Smith (1985), Sparrow et al. (1972), and other references.

correlation exists between genome size and the number of copies of such regulatory sequences as telomeres, centromeres, and replicator genes, which are required for chromosome replication, segregation, and recombination during meiosis and mitosis. However, all these genes constitute only a minute fraction of the genome, such that the variation in the number of RNA-specifying genes and regulatory sequences cannot explain the variation in genome sizes.

In summary, we are left with the nongenic DNA fraction as the sole culprit for the C-value paradox. In other words, a substantial portion of the eukaryotic genome consists of DNA that does not contain genetic information. It has been estimated that the amount of nongenic DNA per genome varies in eukaryotes from about 3.0×10^6 bp to over 1.0×10^{11} bp (a 100,000-fold range) and constitutes anything from less than 30% to almost 100% of the genome (Cavalier-Smith 1985).

THE REPETITIVE STRUCTURE OF THE EUKARYOTIC GENOME

The eukaryotic genome is characterized by two major features: (1) repetition of sequences and (2) compositional compartmentalization into distinct fragments characterized by specific nucleotide compositions. **Repetitive DNA** consists of nucleotide sequences of various lengths and compositions that occur several times in the genome either in tandem or in a dispersed fashion. Segments of DNA that do not repeat themselves are referred to as **single-copy DNA** or **unique DNA**. The proportion of the genome taken up by repetitive sequences varies widely among taxa. In yeast, this proportion amounts to about 20%; in mammals, up to 60% of the DNA is repetitive. In plants, the proportion can exceed 80%, and higher values than that have also been registered (Flavell 1986).

Classical studies of the kinetics of DNA reassociation by Britten and Kohne (1968) showed that the genome of higher eukaryotes can be divided roughly into four fractions: **foldback DNA**, **highly repetitive DNA**, **middle-repetitive DNA**, and single-copy DNA (Figure 1). Foldback DNA consists of palindromic DNA sequences which can form hairpin double-stranded structures as soon as the denatured DNA is allowed to renature. The highly repetitive fraction is made up of short sequences, from a few to hundreds of nucleotides long, which are repeated on the average 500,000 times. The middle-repetitive fraction consists of much longer sequences, hundreds or thousands of base pairs on the average, which appear in the genome up to hundreds of times.

On the basis of the pattern of dispersion of repeats within the genome, the repetitive fractions have been found to consist of two types of repeated families: localized and dispersed.

Localized repeated sequences

Most eukaryotic genomes contain tandemly arrayed, highly repetitive DNA sequences. In some species, these localized repetitive DNA sequences can account for the majority of the DNA in the genome. For example, in the kangaroo rat, *Dipodomys ordii*, more than 50% of the genome consists of three repeated sequences: AAG (2.4 billion times), TTAGGG (2.2 billion times), and ACACAGCGGG (1.2 billion times) (see Widegren et al. 1985). Of course,

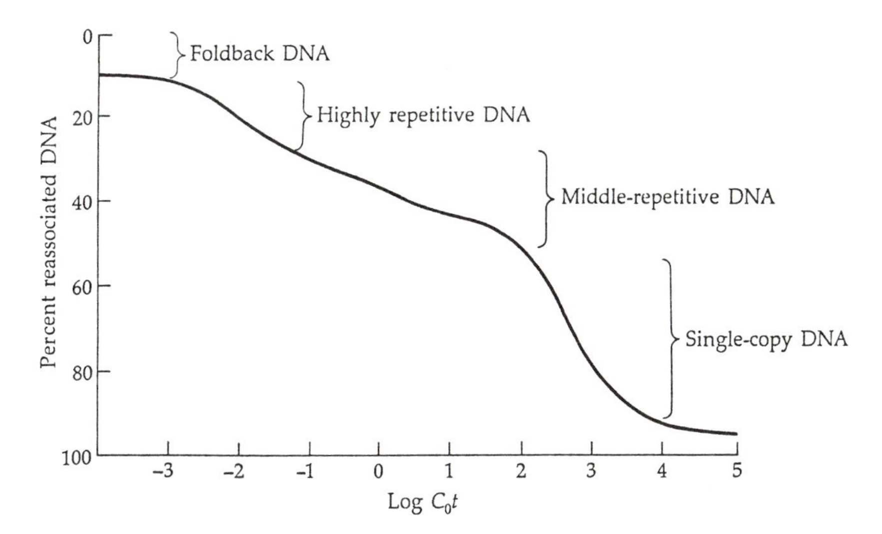

Figure 1. A reassociation profile of mammalian DNA. DNA is purified, sheared, thermally melted into single strands, and then allowed to reassociate through gradual cooling. The percentage of reassociated double-stranded DNA on the vertical axis is shown as a function of the product of DNA concentration and time (C_0t) on the horizontal axis. Modified from Schmid and Deiniger (1975).

these families are not completely homogeneous but contain many variants that differ from the consensus sequence in one or two nucleotides. For example, some sequences in the TTAGGG family are actually TTAGAG. Notwithstanding, many of the localized highly repeated sequences have such a uniform nucleotide composition that, upon fractionalization of the genomic DNA and separation by density gradient, they form one or more thick bands that are clearly distinguishable from the main band of DNA and from the smear created by the other fragments of more heterogeneous composition. These bands are called **satellite DNA**.

In some species, tandemly arrayed highly repetitive sequences are found on all chromosomes, while in others they are restricted to a particular chromosomal location. For example, 60% of the genome of *Drosophila nasutoides* consists of satellite DNA, and all of it is localized on one of the four chromosomes (Figure 2), which seems to contain little else (Miklos 1985).

Based on the evidence available at the present time, it is highly probable that localized highly repetitive sequences are devoid of function. Moreover, it is possible that localized repeated sequences neither lower nor increase the fitness of the individual. Consequently, the evolution of such sequences is not affected by natural selection but is determined mainly by gene conversion and unequal crossing-over (Chapter 6). These mechanisms will result in two

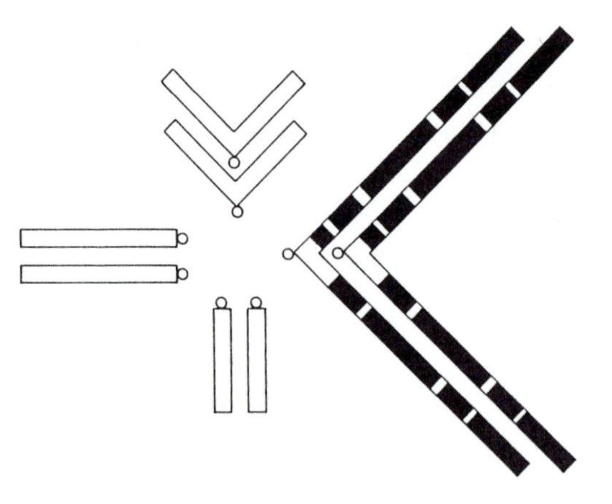

Figure 2. Highly repetitive DNA sequences (black areas) in *Drosophila nasutoides* are localized on a single chromosome. Modified from Miklos (1985).

outcomes: (1) sequence homogeneity and (2) wide fluctuations in copy-number over evolutionary times (see Charlesworth et al. 1986). It has also been suggested that the rate of turnover of localized repeated sequences is quite high; that is, existing arrays may be removed by unequal crossing-over, while new arrays may be continuously created by processes of DNA duplication (Walsh 1987).

The suggestion that most tandemly repeated sequences are merely "junk" DNA essentially implies that they have no phenotypic effects or that they do not affect the fitness of their carriers. While this may be true in the majority of cases, there is evidence pertaining to a specific array of highly repeated sequences that this is not always the case. The *Responder* (*Rsp*) locus in natural populations of *Drosophila melanogaster* consists of 20–2,500 copies of an AT-rich 120-bp-long sequence (Wu et al. 1988). In a competition experiment involving a mixed population consisting of flies containing 700 copies of the repeat and flies containing only 20 copies, it has been observed that the frequency of the flies containing 20 repeats decreased with time (Wu et al. 1989). Therefore, it has been concluded that flies containing 700 copies have a higher fitness than flies containing only 20 copies. At the present time, the function of the *Rsp* locus is not known, but it is clearly not "junk" DNA since its absence reduces the fitness. Whether these results can be applied to other tandemly repeated sequences and, in particular, to the major satellite DNA classes, is not known.

Dispersed repeated sequences

The second class of highly repetitive DNA consists of sequences that are dispersed throughout the genome. Copies of dispersed highly repetitive sequences are found in introns, flanking regions of genes, intergenic regions, and nongenic DNA. There are two major kinds of dispersed highly repeated sequences, short interspersed repeated sequences, abbreviated as SINEs, and long interspersed repeated sequences, or LINEs (Singer 1982).

SINEs are typically shorter than 500 bp and occur in 10^5 copies or more in the haploid genome. In common with tRNA-specifying genes (Chapter 1), many SINEs contain internal transcriptional signals and are transcribed by RNA polymerase III. Most SINEs are retrosequences. The most well-known SINE family in the human genome is the *Alu* family (Chapter 7).

LINEs were originally described as DNA sequences longer than 5 kb and present in 10^4 or more copies per genome (Singer 1982). The human genome apparently contains only one LINE family, L1 (Hutchison et al. 1989). The consensus L1 sequence is about 6 kb in length, has a poly-A tail at one end, and is flanked by short direct repeats generally less than 20 bp long. The copy number is about 100,000 per haploid genome. Most L1 sequences are truncated at their 5′ end. Intact L1 sequences contain two large open reading frames, ORF-1 and ORF-2, with about 375 and 1,300 codons, respectively. ORF-2 contains amino acid sequence motifs characteristic of reverse transcriptase enzymes. The above features suggest that L1 elements are produced by reverse transcription of polyadenylated mRNAs and the subsequent insertion of the reverse transcripts into the genome. Since L1 sequences do not possess long terminal repeats (LTRs), they are most probably retroposons (Chapter 7). Most L1 sequences are not transcribed. However, those few that are use RNA polymerase II for transcription.

LINEs homologous to the human L1 family were found in all mammals, including marsupials (see Hutchison et al. 1989). Nonmammalian elements related to L1 include the I, F, G, and D factors in *Drosophila* species, *Ingi* in *Trypanosoma brucei*, R2 in *Bombyx mori*, and *Cin*4 in maize. The degree of L1 sequence divergence among species is much greater than the degree of divergence among conspecific L1 copies. For example, L1 sequences from mice and humans differ on average from each other by about 30%. In comparison, the amount of sequence divergence among L1 elements in mice is only 4% (Hutchison et al. 1989).

Since most L1 sequences are truncated, they do not contain intact reading frames. As a consequence, they may not be able to transpose. Such defective elements were found to evolve much more rapidly than intact elements. Moreover, evolutionary lineages of defective L1 sequences were found to contain no branches, indicating that these elements are incapable of replicative transposition. They may, thus, be considered as pseudogenes of retroposons, on which functional constraints no longer operate. The fact that most L1 sequences are defective implies that the propagation of L1 elements within the genome depends on only a small number of source elements. As a consequence, L1 elements within the genome are highly homogeneous and the rate of sequence turnover is very high. Indeed, in rodents it has been estimated that more than half of the L1 elements are only 3 million years old or younger (Hardies et al. 1986).

It has been suggested by Hutchison et al. (1989) that SINEs and LINEs should be redefined. That is, instead of using length and copy number as diagnostic features, LINEs should include all active retroposons and retrotransposons that encode proteins likely to mediate their own retroposition, as well as their descendant sequences. SINEs, on the other hand, should include retrosequences that do not encode such functions.

The genomic location of transcribed genes

RNA–DNA hybridization experiments have shown that only a very small fraction of the transcribed genes are located within the repeated fractions of the genome. Most transcribed genes reside within the unique DNA fraction. Even in the unique fraction, most sequences are not transcribed. In fact, only about 3% of the nonrepetitive DNA sequences in humans are transcribed (see Lewin 1990). These experiments constitute further support for the view that most of the eukaryotic genome is devoid of genetic information.

MECHANISMS FOR INCREASING GENOME SIZE

In attempting to explain the existence of the vast amounts of nongenic DNA in the genome of eukaryotes, we must first deal with the processes that may bring about an increase in the size of genomes. The putative mechanisms should be able to explain not only the phenomenon of nongenic DNA per se, but also its repetitiveness and particular genomic distribution.

We distinguish between two types of genome increase: (1) global increases, in which the entire genome or a major part of it, such as a chromosome, is duplicated and (2) regional increases, in which a particular sequence is multiplied to generate repetitive DNA. In the latter case, we distinguish between mechanisms that are responsible for the creation of dispersed repetitive sequences and mechanisms that can create localized repetitive sequences.

Genome duplication

Since the genome of eukaryotes is significantly larger than the genome of eubacteria, the evolution of eukaryotes from prokaryotic ancestors must have involved an increase in the size of the genome. There are several molecular mechanisms by which an increase in genome size can be brought about. A major contributor to genome growth is **genome duplication** or **genome doubling**. Genome duplication occurs as a consequence of lack of disjunction between all the daughter chromosomes following DNA replication.

Given that the mammalian genome is about 1,000 times larger than the genome of bacteria, and assuming that genome duplication is solely responsible for genome enlargement, we can deduce that only about 10 rounds of genome duplications were required to enlarge the genome from a primordial bacterial size to its present size in mammals. To put it another way, genome duplication has occurred on average once every 300 million years. If, on the other hand, DNA content increased in a continuous fashion by the addition of small pieces of DNA, say by means of transposition or unequal crossing-over, then the rate of genome growth from bacteria to mammals has been approximately seven nucleotides per year (Nei 1969).

A polymodal distribution of genome sizes has been registered in many groups of eukaryotes (Rees and Jones 1972; Grime and Mowforth 1982), resembling the situation in bacteria (see page 206). This is particularly evident in monocotyledons, where genome sizes exhibit a polymodal distribution with peaks at 0.60, 1.18, 2.16, 4.51, and 8.53×10^9 bp (Sparrow and Nauman 1976). Similar distributions have been observed in other groups, such as echinoderms, insects, and fungi, and to a lesser extent in amphibia and bony fishes. Thus, genome duplication or **polyploidy** seems to be a major mechanism in the evolution of genome size in eukaryotes. Interestingly, each round of genome duplication involves small losses of DNA, such that the amount of DNA after each round increases by a factor slightly smaller than 2 (Sparrow and Nauman 1976). In a recently formed polyploid one cannot speak of an increase in the C value, since the value refers to the haploid size, but as the two genomes undergo mutations, translocations, chromosomal rearrangements, and changes in chromosome number, they will eventually become a single new genome. In other words, an ancient polyploid will not be distinguishable from a diploid (Cavalier-Smith 1985).

Polyploidy is a common occurrence in nature; however, during evolutionary history polyploids seem to have survived only rarely. The reason is that, in many cases, polyploidy is deleterious and will be strongly selected against. Deleterious effects of polyploidy include (1) considerable prolongation of cell division time, (2) increase in the volume of the nucleus, (3) significant increases in the number of chromosome disjunctions during segregation at meiosis, (4) genetic imbalances, and (5) interference with sexual differentiation in cases where the sex of the organism is determined by either the ratio between the number of sex chromosomes and autosomes (e.g., *Drosophila*), or ploidy (e.g., Hymenoptera). In some cases, polyploidization may be beneficial, such as in flowering plants where it reduces or eliminates hybrid infertility (see Cavalier-Smith 1985).

Chromosomal duplication

The duplication of a single chromosome, or aneuploidy, is most often deleterious. In mammals, it is frequently associated with lethality or infertility. In humans, well-known examples include such frequent abnormalities as Down's syndrome (trisomy 21), and trisomy of chromosome 18. Similar deleterious manifestations are associated with duplications of parts of chromosomes. Therefore, chromosomal duplication is not expected to contribute significantly to the increase in genome size.

In many organisms, especially in plant populations, which have been investigated most thoroughly, several cases have been observed in which extra chromosomes, called **B-chromosomes**, have become established (Jones 1985). The B-chromosomes are not precise duplicates of whole chromosomes and may be regarded as the products of partial chromosome duplication. These extra chromosomes do not seem to contain many genes, and they mostly consist of short repetitive sequences (i.e., they are largely heterochromatic). Hence, partial aneuploidy may have played a role in the process of genome size increase during evolution but probably did not contribute significantly to the evolution of gene number.

Regional increases of genome size

Regional increases in genome size can be brought about by transposition (Chapter 7) and unequal crossing-over (Chapters 1 and 6). The former mode will create dispersed repetitive sequences; the latter will create tandemly repeated sequences. Regional increases in genome size can also be brought about by the acquisition of foreign DNA (Chapter 7); however, the contribution of this process to total DNA size is probably negligible.

It has been suggested that all the middle-repetitive DNA fraction in eukaryotes has originated in transposable elements. However, most of the elements are no longer mobile, their ability to transpose being destroyed by mutations or the insertion of other elements. Much of the heterochromatin in *Drosophila*, for instance, may in fact be a graveyard of such dead elements. Inactivated mobile elements can, however, multiply locally through such processes as unequal crossing-over.

Many localized repeated sequences have been produced by unequal crossing-over. However, it seems that unequal crossing-over tends to remove tandem arrays more often than to increase their size and copy number, and so the process cannot explain the existence of all the localized repeated DNA (Walsh 1987). Furthermore, unequal crossing-over usually creates sequences consisting of relatively long repeated units. In contrast, many localized repeated sequences, such as satellite DNA, consist of very short, simple repeated motifs (see page 210). To explain the existence of such sequences, DNA amplification has been suggested. **DNA amplification** refers to any event that increases the number of copies of a gene or a DNA sequence far above the level characteristic for an organism. In particular, DNA amplification refers to events that occur within the lifespan of an organism and cause a sudden increase in the copy number of a DNA sequence.

One of the most powerful methods of amplification is the **rolling-circle mode** of DNA replication (Figure 3; Bostock 1986). This type of replication is used in the amplification of rRNA genes in amphibian oocytes. Amplification in this case involves the formation of an extrachromosomal circular copy of a DNA sequence, which can then produce many additional extrachromo-

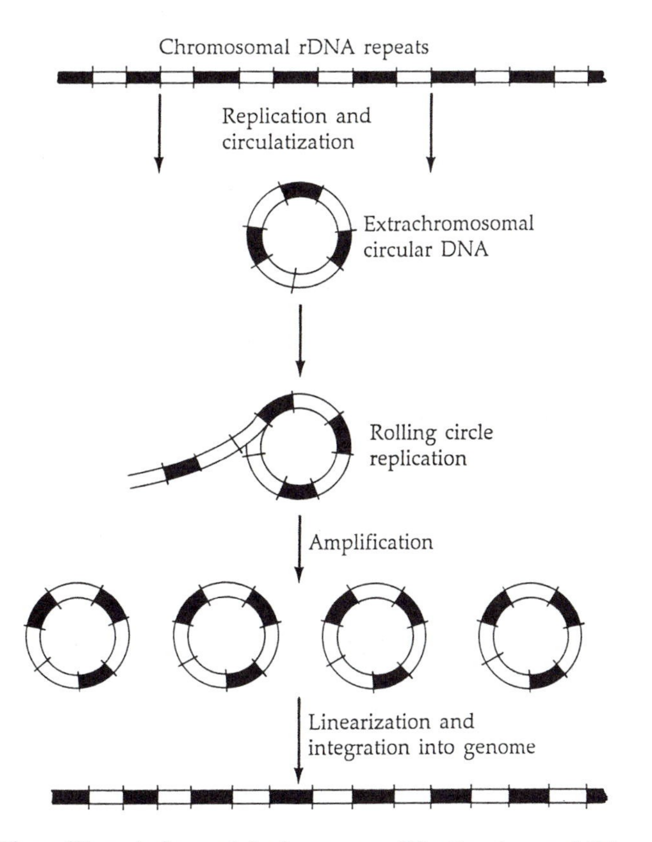

Figure 3. The rolling-circle model of gene amplification in amphibian oocytes. The chromosomal rRNA is arranged in a tandem array containing transcribed (black) and nontranscribed (white) parts. Amplification involves the formation of an extrachromosomal circular copy containing a variable number of repeats which is then amplified by multiple rounds of rolling-circle replication. Note that the periodicity of the repeats may change following rolling-circle amplification. Modified from Bostock (1986).

somal units containing tandem repeats of the original sequence. If such units become integrated back into the chromosome, there will be an addition to the genome consisting of identical repeated sequences.

Replication slippage or **slipped-strand mispairing** is a process in which the DNA polymerase turns back and uses the same template again to produce a repeat (Figure 8 in Chapter 1). Existing tandem repetitive sequences are particularly prone to replication slippage, and the process can therefore produce very long tandem arrays of short repeats. Both rolling-circle replication and replication slippage can provide mechanisms for the rapid proliferation of tandemly repeated sequences within the genome. However, the empirical evidence for these processes is limited.

MAINTENANCE OF NONGENIC DNA

Solving the C-value paradox and accounting for the structure of the eukaryotic genome requires that we provide an evolutionary mechanism for the long-term maintenance of vast quantities of nongenic, seemingly superfluous DNA. This, in turn, is intimately linked with the question of what function this DNA might have, if any. Numerous attempts have been made to provide evolutionary explanations for this phenomenon. In the following we present four such hypotheses.

1. The nongenic DNA performs essential functions, such as global regulation of gene expression (Zuckerkandl 1976). According to this hypothesis, the excess of DNA is only apparent, and the DNA is wholly functional. Consequently, deletions or removal of such DNA will have a deleterious effect on fitness.
2. The nongenic DNA is useless **junk DNA** (Ohno 1972), carried passively by the chromosome merely because of its physical linkage to functional genes. According to this view, the excess DNA does not affect the fitness of the organism and thus will be carried from generation to generation indefinitely.
3. The nongenic DNA is a functionless "parasite" (Östergren 1945) or "selfish DNA" (Orgel and Crick 1980; Doolittle and Sapienza 1980) that accumulates and is actively maintained by intragenomic selection.
4. DNA has a structural or **nucleotypic** function, i.e., a function unrelated to the task of carrying genetic information (Cavalier-Smith 1978).

There is very little evidence for the first hypothesis. In fact, most indications are that what is now considered nongenic DNA is indeed devoid of function and that most of it can be deleted without discernable phenotypic effects. It also seems that excess DNA in eukaryotes does not tax the metabolic system and that the cost in energy and nutrients of maintaining and replicating large amounts of nongenic DNA is not excessive. It is thus possible that most nongenic DNA is indeed junk or selfish DNA.

However, there may be some drawbacks in maintaining large amounts of nongenic DNA. First, large genomes have been found to exhibit greater sensitivity to mutagens than small genomes (Heddle and Athanasiou 1975). Second, maintaining and replicating large amounts of nongenic DNA may impose a certain burden on the organism, especially when the vast majority of the genome is nongenic.

Cavalier-Smith (1978, 1985) argued that there must be a "major evolutionary force" maintaining large genomes. His hypothesis is that the DNA acts as a nucleoskeleton which determines the nuclear volume. Since larger cells require larger nuclei, selection for a particular cell volume will secondarily result in selection for a particular genome size. According to this scheme, excess DNA is maintained by selection, but its nucleotide composition can change at random. An additional nucleotypic function of nongenic DNA may be mechanical. For example, a highly repetitive, tandemly arrayed satellite DNA may affect the architecture of the chromosome, in particular its curvature and the phasing of the nucleosomes. Some satellite DNAs are also known to bind specific chromosomal proteins that may influence chromosome condensation during meiosis and mitosis and may, therefore, have an effect on gene regulation (Levinger and Varshavsky 1982; James and Elgin 1986).

No single explanation is likely to solve the C-value paradox. All the above mechanisms, and many additional ones, may contribute to the maintenance of "excess" genomic size, and our task in the future will be to determine the relative contribution of each.

GC CONTENT IN BACTERIA

Among eubacteria, the mean percentage of guanine and cytosine in genomic sequences, or the **GC content,** varies from approximately 25% to 75%. In many cases, bacterial GC content seems to be related to phylogeny, with closely related bacteria having similar GC contents (Figure 4).

There are essentially two types of hypotheses to explain the variation in GC content in bacteria. The selectionist view regards the GC content as a form of adaptation to environmental conditions. For example, in thermophilic bacteria, which inhabit very hot niches, strong preferential usage of thermally stable amino acids encoded by GC-rich codons (e.g., alanine and arginine) and strong avoidance of thermally unstable amino acids encoded by GC-poor codons (e.g., serine and lysine) have been reported (e.g., Argos et al.

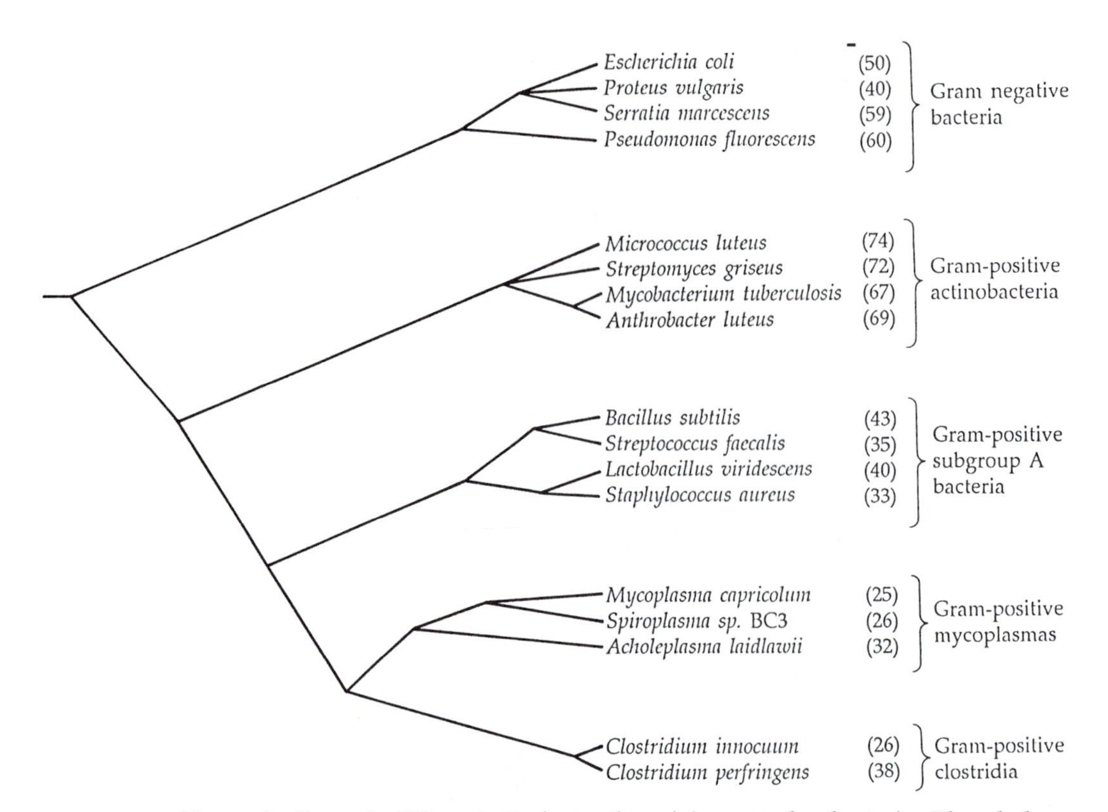

Figure 4. Genomic GC contents (parentheses) in several eubacteria. The phylogenetic tree is based on 5S rRNA sequences. The branches are unscaled. Combined data from Hori and Osawa (1986), and Muto et al. (1986, 1987).

1979; Kagawa et al. 1984; Kushiro et al. 1987). Therefore, GC content may be a trait that is determined by selection. Another selectionist scenario invokes UV radiation as the selective force. Since T—T dimers are sensitive to radiation, microorganisms in the upper layers of the soil, which are exposed to sunlight, should have a higher GC content than bacteria that are not exposed, say intestinal bacteria such as *Escherichia coli* (Singer and Ames 1970).

The mutationist view invokes biases in the mutation patterns to explain the variation in GC contents (Sueoka 1964; Muto and Osawa 1987). According to this view, the GC content of a given bacterial species is determined by the balance between (1) the rate of substitution from G or C to T or A, denoted as u, and (2) the rate of substitution from A or T to G or C, denoted as v. At equilibrium, the GC content is expected to be

$$P_{GC} = v/(v + u) \tag{8.1}$$

Therefore,

$$u/v = (1-P_{GC})/P_{GC} \quad \textbf{(8.2)}$$

The ratio u/v is also called the **GC mutational pressure**. When u/v is 3.0, the GC content at equilibrium will be 25%. Such is the situation in *Mycoplasma capricolum*. When the ratio is 1, the GC content will be 50%, as in *Escherichia coli*. When it is 0.33, the GC content will be 75%, as in *Micrococcus luteus*. However, to estimate the GC mutational pressure, it is advisable to use DNA sites at which selective constraints are absent instead of the total GC content. For example, the GC content at fourfold degenerate sites in *Mycoplasma capricolum* protein-coding genes is lower than 10% (Figure 5). Consequently, u/v must be higher than 9. Similarly, in *Micrococcus luteus*, $P_{GC} > 0.9$ at degenerate sites and, therefore, $u/v < 0.11$.

In addition to GC mutational pressure, mutational changes are also subject to selective constraint. In other words, the pattern of substitution is determined by the pattern of mutation and the pattern of purifying selection against certain mutations (Chapter 4). The weaker the selective constraint is in a particular region, the stronger the effect the GC mutational pressure will have on the GC composition. Figure 5 shows the correlation between the total GC content and the GC content at the three codon positions for 11 bacterial species covering a broad range of GC-content values. We see that the correlation at the third position resembles the expectation for the case of no selection. On the other hand, the correlations at the first and second positions, while positive, show a more moderate slope. This is easily explained by the fact that selection at the mostly degenerate third position of codons is expected to be much less stringent than at the first and second positions (Chapter 4), so that the GC level at the third position is largely determined by mutation pressure.

COMPOSITIONAL ORGANIZATION OF THE VERTEBRATE GENOME

Figure 6 shows the GC content in different groups of organisms. Interestingly, while the genome sizes of multicellular eukaryotes are generally larger than those of prokaryotes, GC content exhibits a much smaller variation in eukaryotes. In particular, vertebrate genomes show quite a uniform GC content, ranging from about 40% to 45% (Sueoka 1964). Part of the reason for the small range in GC content in vertebrates might be because vertebrates, unlike bacteria, have not diverged long enough from one another to allow for considerable differences in GC content to accumulate.

Notwithstanding the uniformity of the genomic GC content, vertebrate genomes have a much more complex compositional organization than pro-

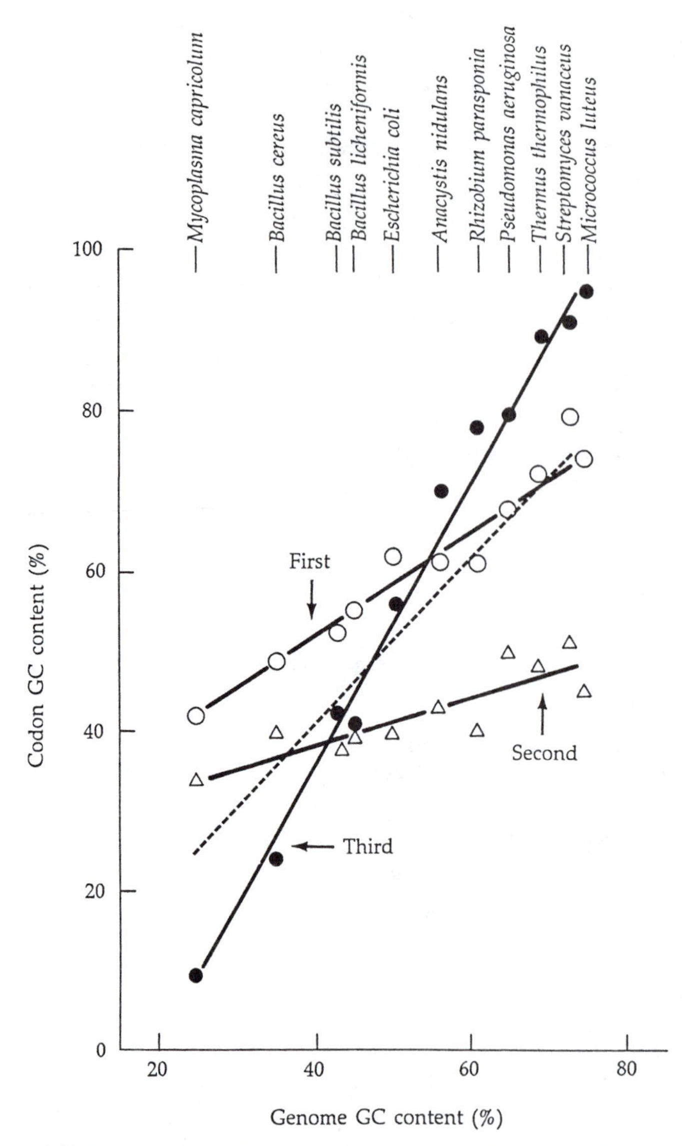

Figure 5. Correlation of the GC content between total genomic DNA and the first, second, and third codon positions. The dashed line represents the theoretical expectation of a perfect correspondence between the GC content in the genome and that in the codons. From Muto and Osawa (1987).

karyotic genomes. When vertebrate genomic DNA is randomly sheared into fragments 30–100 kb in size and the fragments are separated by their base composition, the fragments cluster into a small number of classes distinguished from each other by their GC content (GC-rich fragments being

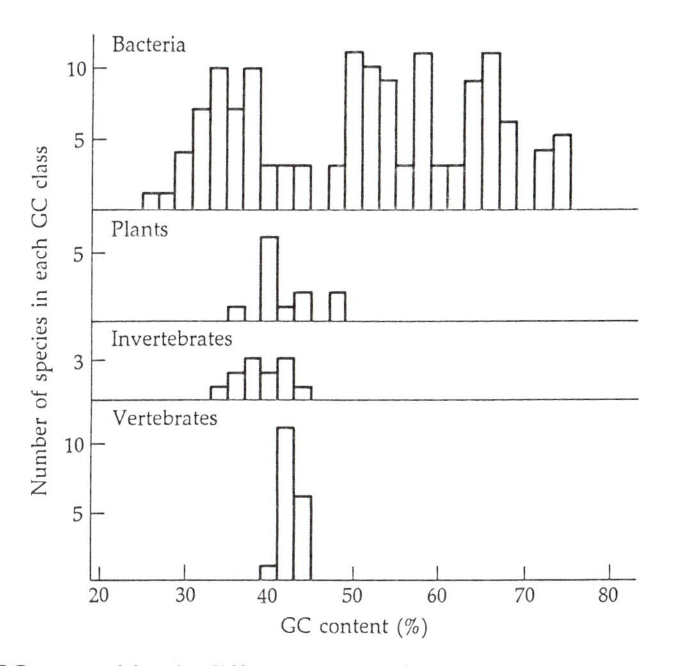

Figure 6. GC composition in different groups of organisms. From Sueoka (1964).

heavier than AT-rich fragments). Each class is characterized by bands of similar, although not identical, base compositions (Bernardi et al. 1985; Bernardi 1989).

There are conspicuous differences in compositional organization between the genomes of warm- and cold-blooded vertebrates (Bernardi et al. 1985, 1988). Figure 7a shows the relative amounts and buoyant densities of the major DNA components from the carp *Cyprinus carpio* and the amphibian *Xenopus laevis* (left panel) and from three warm-blooded vertebrates: chicken, mouse, and human (right panel). In the genome of chickens, mice, and humans, there are two light components (L_1 and L_2), representing about two-thirds of the genome, and two or three heavy components (H_1, H_2, and H_3), representing the remaining third. In contrast, genomic DNA from most cold-blooded vertebrates comprises mostly light components (Figure 7a). For example, in *Xenopus*, DNA fragments with a density higher than 1.704 g/cm^3 represent less than 10% of the genome, as compared with 30–40% for warm-blooded vertebrates.

The compositional distribution of DNA fragments is largely independent of the size of the fragments, indicating a compositional homogeneity over very long DNA stretches. Such homogeneous stretches have been termed **isochores**. Figure 7b shows the mosaic organization of nuclear DNA from

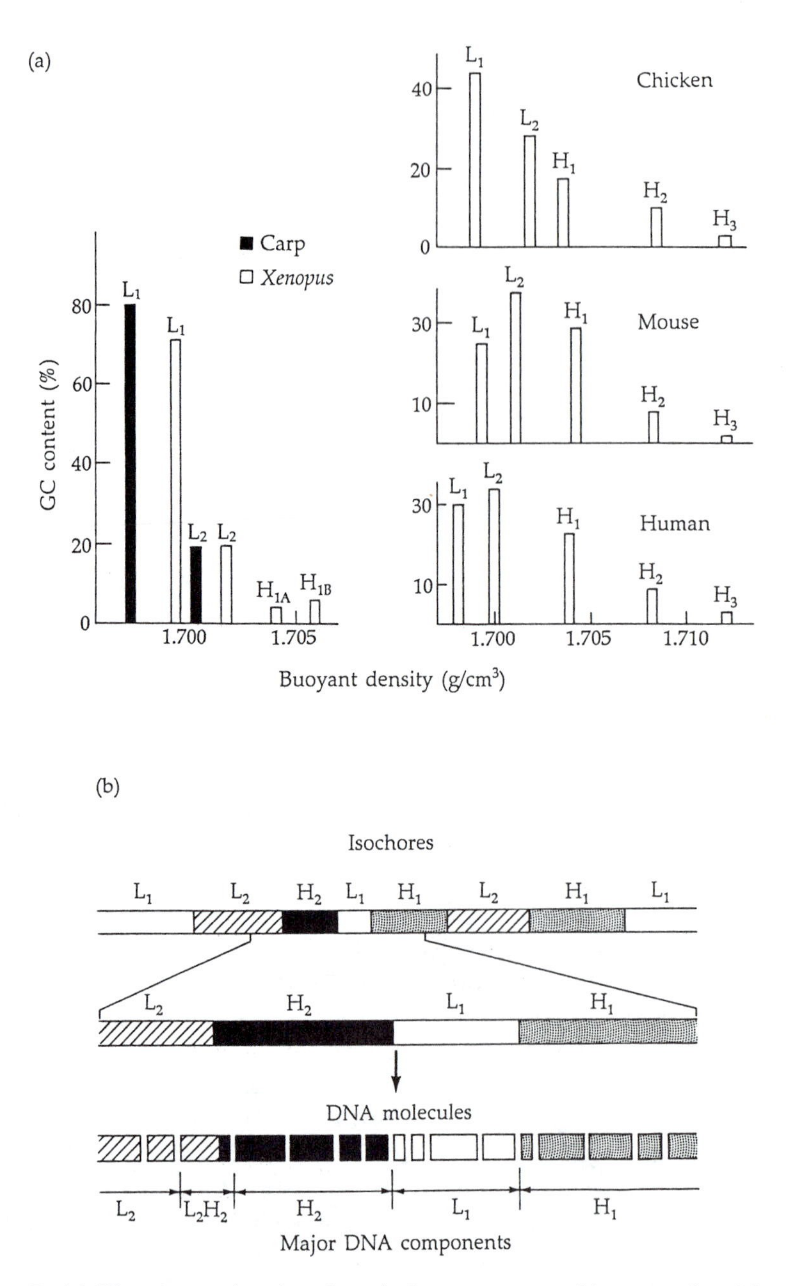

Figure 7. (a) Histograms showing the relative amounts and buoyant densities of the major DNA isochores from *Cyprinus carpio* and *Xenopus laevis* (left panel) and from chickens, mice, and humans (right panel). (b) Scheme depicting the mosaic organization of nuclear DNA from warm-blooded vertebrates. When the isochores undergo random breakage during DNA preparation, four major families of molecules with different GC contents are generated. Several minor hybrid families (e.g., L_2H_2) are also generated. From Bernardi et al. (1985).

warm-blooded vertebrates (i.e., the alternation of light and heavy isochores). When the isochores break during DNA shearing, four major families of molecules having different GC contents are generated. Bernardi et al. (1985) concluded that the GC-rich (heavy) isochores represent about one-third of the genome of warm-blooded vertebrates but are nearly absent in cold-blooded vertebrates.

The finding that the genome of warm-blooded vertebrates is mosaic is consistent with chromosome-banding studies. Metaphase chromosomes of warm-blooded vertebrates show distinct Giemsa dark bands (G-bands) and light bands (R-bands) when treated with fluorescent dyes, proteolytic digestion, or differential denaturing conditions. In contrast, metaphase chromosomes of cold-blooded vertebrates show either little banding or no banding at all. Therefore, it has been suggested that GC-poor and GC-rich isochores correspond roughly to the G- and R-bands, respectively (Comings 1978; Cuny et al. 1981). Studies on the replication timing of genes show that genes localized in GC-rich isochores (R-bands) replicate early in the cell cycle, whereas genes localized in GC-poor isochores (G-bands) replicate late (Goldman et al. 1984; Bernardi et al. 1985; Bernardi 1989).

Location of genes within isochores

A number of genes from the genomes of humans and other warm-blooded vertebrates have been localized on the compositional isochores by hybridization with appropriate probes (Bernardi et al. 1985). These studies suggest a highly nonrandom distribution of genes throughout the genome, with most of the genes residing in the heaviest component (H_3), which represents only 3–5% of the genome.

In most cases, a gene is embedded in DNA fragments that have a GC content similar to that of the gene itself. This finding provides independent evidence for the existence of isochores, as well as for the large size of the isochore. Indeed, since the fragments making up the DNA preparations were produced by random degradation, the narrow compositional range of the gene-carrying fragments indicates that they are very homogeneous in base composition over sizes roughly twice as large as the fragments themselves (i.e., up to 200 kb).

Analyses of DNA sequence data have revealed a positive correlation between the GC levels of genes, exons, and introns and the GC level in the large DNA regions in which they are embedded (Bernardi and Bernardi 1985; Bernardi et al. 1985; Ikemura 1985; Aota and Ikemura 1986). Figure 8 contrasts the α- and β-globin clusters in humans. The β and β-like globin genes are low in GC content, and are embedded in a low-GC region. The α and α-like globin genes, on the other hand, are GC-rich and are embedded in a GC-

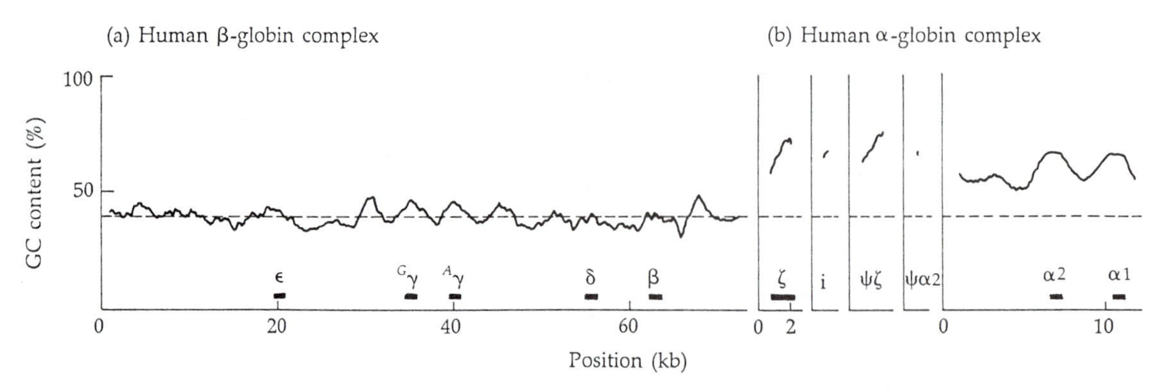

Figure 8. Distribution of GC content along human globin DNA sequences: (a) The β-globin gene cluster; (b) the α-globin gene cluster (incomplete). The genes (bars) are arranged in the same order as in Figure 7 of Chapter 6. The gene names are shown at the bottom of the figure; region i is the intergenic region between ζ and ψζ. In the β-globin cluster and the region covering the α1- and α2-globin genes each point represents the average of the GC composition of the 2,001 nucleotides surrounding the point, while in the other regions each point represents the average of 1,401 nucleotides. The horizontal broken line represents the overall GC content of the human genome (40%). Modified from Ikemura and Aota (1988).

rich region. The same situation is found in rabbits, goats, and mice. In chickens, both β- and α-globin genes are GC-rich, and both are embedded in GC-rich regions. In contrast, the α- and β-globin genes in *Xenopus* are GC-poor, and both are embedded in a GC-poor region.

In the vast majority of cases, the GC content in coding regions tends to be higher than that in the flanking regions (Figure 9). We also see that the GC level at the third codon position is on average higher than that in introns, which in turn is higher than that in the 5′ and 3′ flanking regions. The GC level in the 5′ flanking region tends to be higher than that in the 3′ flanking region, probably because the promoter and its surrounding regions tend to be GC-rich.

Origins of isochores

The origin of GC-rich isochores is as mysterious as it is controversial. Note that what is at issue is the general tendency of long DNA segments (30 kb or more) to be either GC-rich or GC-poor, not the localized variation in GC content, such as that observed among the various regions of a gene. Bernardi et al. (1985, 1988) proposed that isochores arose because of a functional (i.e., selective) advantage. Their main argument is that, in warm-blooded organisms, an increase in GC content can protect DNA, RNA, and proteins from

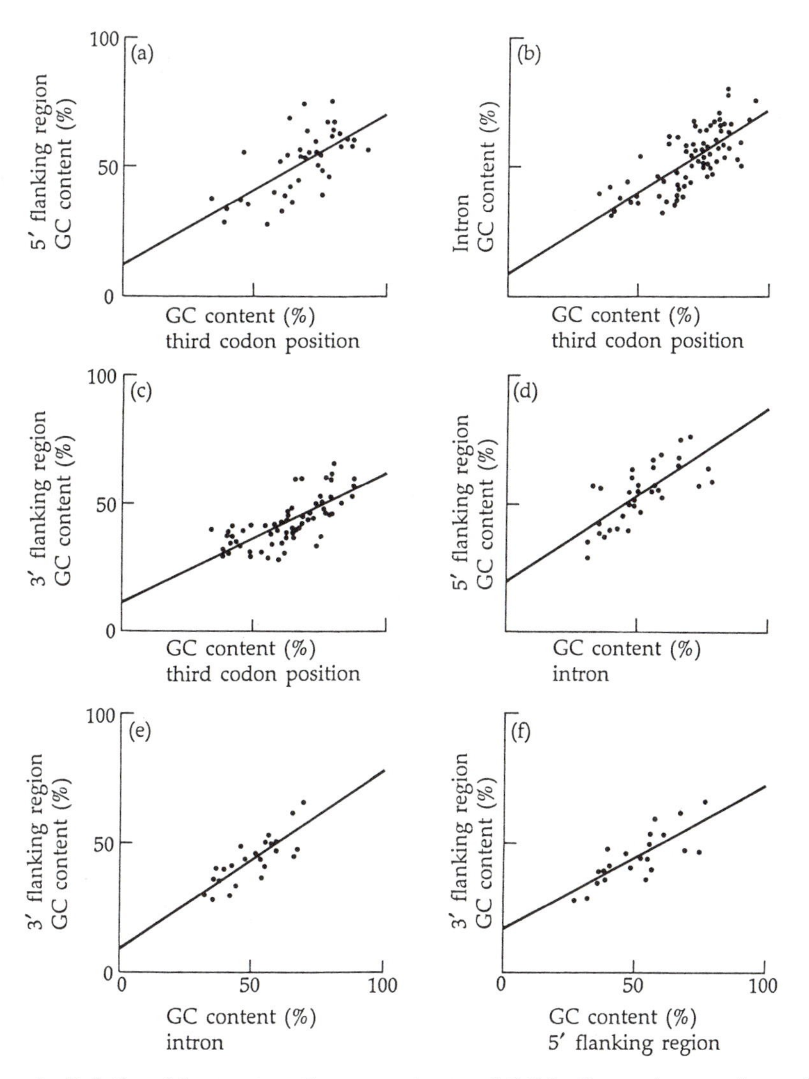

Figure 9. Relationships among the percentages of GC in the various regions of the gene. (a) The third codon position and the 5′ flanking region. (b) The third codon position and introns. (c) The third codon position and the 3′ flanking region. (d) Introns and the 5′ flanking region. (e) Introns and the 3′ flanking region. (f) 5′ and 3′ flanking regions. From Aota and Ikemura (1986).

degradation by heat (see below), because G—C bonds are stronger than A—T bonds (Chapter 1). We call this view the **selectionist hypothesis.**

Wolfe et al. (1989a) proposed that isochores arose from mutational biases due to compositional changes in the precursor nucleotide pool during the replication of germ-line DNA. The GC-rich isochores are carried on DNA regions that replicate early in the germ-line cell cycle during which the precursor pool has a high GC content and, thus, a propensity to mutate to GC. The AT-rich isochores, on the other hand, are replicated late in the cell cycle, when the precursor pool has a high AT content and a propensity to mutate to AT. We call this the **mutationist hypothesis.** This hypothesis is based on the observations that the composition of the nucleotide precursor pool changes during the cell cycle and that such changes can in fact lead to altered base ratios in the newly synthesized DNA (Leeds et al. 1985). Note that the replication of the mammalian genome is quite a lengthy process, taking eight hours or more (Holmquist 1987).

An argument for the selectionist hypothesis is that increases in GC content at the first and second codon positions may confer thermal stability to proteins, while increases in GC content in introns, third codon positions, and untranslated regions can both increase the thermal stability of primary mRNA transcripts and stabilize chromosomal structures, possibly through effects on DNA–protein interactions. Indeed, in thermophilic bacteria, strong preferential usage of GC-rich codons has been reported (see page 219). However, the body temperature of warm-blooded vertebrates is much lower than the temperatures experienced by thermophilic bacteria, and so temperature may not be a very important factor in the evolution of proteins and DNA sequences in vertebrates.

One difficulty with the selectionist hypothesis is the fact that a substantial proportion of the mammalian and avian genes are low in GC content. This hypothesis also cannot explain why some duplicate genes have opposite GC contents. For example, in mammals, the β-globin cluster is low in GC, while the α-globin cluster is GC-rich (Figure 8), although the two types of globin genes are expressed in the same cells at the same time, and serve the same function. Similarly, some immunoglobulin genes are located in GC-rich regions while others are located in AT-rich regions (see Aota and Ikemura 1986). Bernardi et al.'s (1985, 1988) explanation is that isochores represent the unit of selection and the α cluster in mammals has been translocated to a high-GC isochore, whereas the β cluster remained in a low-GC isochore. According to this argument, the α and β clusters in chickens should have both been translocated to GC-rich isochores. This argument, however, raises the issue of the functional advantage of a GC-rich isochore. If the advantage does not come from the genes it contains, where does it come from?

The mutationist hypothesis can explain the large difference in GC content between the α- and β-globin clusters in the mammalian genome by assuming that they are located in an early- and a late-replicating region, respectively. However, it also has difficulties (Bernardi et al. 1988). For example, constitutive heterochromatin, such as satellite DNAs, which are mostly GC-rich, and facultative heterochromatin, such as the inactive X chromosome, replicate at the end of the cell cycle, and in these cases, no connection between changes in nucleotide pools and DNA composition is observed.

In conclusion, the presently available data seem to be insufficient to distinguish between the two hypotheses. It is also possible that both mutation pressure and natural selection have played roles in shaping the compositional organization of the genome of warm-blooded vertebrates.

PROBLEMS

1. Table 5 shows rough estimates of gene numbers, average gene sizes, and C values of four species of protozoans. (a) Under the assumption that the genic fraction of the genome is made of protein-coding genes only, what is the proportion of nongenic DNA in each organism? (b) Is there a relationship between genome size and the proportion of nongenic DNA? (Note that C values are given in pg.)

Table 5. C values and rough estimates of the average size and the number of pre-mRNAs in four species of protozoans.

Species	Genome size (pg)	Gene size (nucleotides)	Gene number
Physarum polycephalum	0.57	1,500	20,000
Oxytricha nova	0.4	2,200	24,000
Euplotes aediculatus	0.3	1,800	40,000
Dictyostelium discoideum	0.036	1,500	6,500

Data from Cavalier-Smith (1985).

2. Table 6 lists the C values and rough estimates of the average sizes of pre-mRNA and mature mRNA in four organisms. Can the variation in C values be explained by the differences in the size of noncoding regions in genes among the organisms? (Note that C values are given in pg.)

Table 6. C values and rough estimates of average gene size and mRNA size in four types of animals.

Genus	Genome size (pg)	pre-mRNA size (nucleotides)	mRNA size (nucleotides)
Drosophila (fruit fly)	0.18	4,200	2,100
Aedes (mosquito)	0.83[a]	8,400	2,100
Strongylocentrotus (sea urchin)	0.89	8,800	2,100
Homo (human)	3.50	10,000	2,100

Data from Cavalier-Smith (1985).

[a] Mean from different species.

3. Suppose that in a certain organism the entire genome is made of protein-coding genes, the 20 amino acids are used with equal frequencies, and the choice of synonymous codons is determined by selection for the highest GC content possible. (a) What will be the GC content of this organism? (b) What will be the GC content in an organism in which the entire genome is made up of protein-coding genes, the 20 animo acids are used with equal frequency, but the choice of synonymous codons is determined by selection for the lowest possible GC content?

4. Assume that in a DNA sequence the rate of substitution from G or C to T or A is u per nucleotide site per generation and that the rate of substitution from A or T to G or C is v. Let p_t be the proportion of G and C nucleotides in the sequence at generation t. Then, the GC proportion in the next generation is

$$p_{t+1} = (1 - u)p_t + v(1 - p_t)$$

From this equation, show that the equilibrium GC proportion is given by Equation 8.1.

5. If the initial GC content in a sequence is p_0, then the GC proportion at generation t can be shown to be given by

$$p_t = \frac{v}{u + v} + \left(p_0 - \frac{v}{u + v}\right) e^{-(u+v)t}$$

Assume that $u = 3 \times 10^{-8}$, $v = 5 \times 10^{-8}$, and $p_0 = 0.20$. Compute the p_t value for $t = 10^7$. Under these conditions, is the change in GC content a slow or a fast process?

FURTHER READINGS

Bernardi, G., B. Olofsson, J. Filipski, M. Zerial, J. Salinas, G. Cuny, M. Meunier-Rotival and F. Rodier. 1985. The mosaic genome of warm-blooded vertebrates. Science 228: 953–958.

Blake, R. D. and S. Early. 1986. Distribution and evolution of sequence characteristics in the *E. coli* genome. J. Biomol. Struct. Dynamics 4: 291–307.

Britten, R. J. and D. E. Kohne. 1968. Repeated sequences in DNA. Science 161: 529–540.

Brutlag, D. L. 1980. Molecular arrangement and evolution of heterochromatic DNA. Annu. Rev. Genet. 14: 121–144.

Cavalier-Smith, T. (ed.). 1985. *The Evolution of Genome Size.* Wiley, New York.

Cold Spring Harbor Symposia on Quantitative Biology. 1986. *Molecular Biology of Homo sapiens.* Vol. 51. Cold Spring Harbor Laboratory, Cold Spring Harbor, NY.

Deininger, P. L. and G. R. Daniels. 1986. The recent evolution of mammalian repetitive DNA elements. Trends Genet. 2: 76–80.

Dover, G.A. and R.B. Flavell (eds.). 1982. *Genome Evolution.* Academic Press, New York.

Jelinek, W. R. and C. W. Schmid. 1982. Repetitive sequences in eukaryotic DNA and their expression. Annu. Rev. Biochem. 51: 813–844.

John, B. and G. Miklos. 1988. *The Eukaryotic Genome in Development and Evolution*. Allen & Unwin, London.

Singer, M. F. 1982. Highly repeated sequences in mammalian genomes. Int. Rev. Cytol. 76: 67–112.

GLOSSARY

A (1) In DNA or RNA, adenine. (2) In proteins, alanine.
acceptor site The 3′ end of an intron.
active site Part of a protein, usually an enzyme, the structural integrity of which is required for function (e.g., substrate binding).
additive tree A phylogenetic tree in which the distance between any two terminal nodes is equal to the sum of the branch lengths connecting them.
advantageous mutation A mutation that increases the fitness of the organism carrying it.
alignment The pairing of two homologous sequences for the purpose of identifying the location of insertions and deletions.
allele (allelomorph) Alternative form of a gene at a locus.
allele frequency (gene frequency) The percentage of the copies of an allele at a given locus in a population.
allozyme (allelozyme) An allelic form of an enzyme.
alternative splicing The production of two or more mRNA molecules from a single pre-mRNA sequence by using different acceptor and donor sites.
amino acid An organic molecule of the general formula R—$CH(NH_2)COOH$, possessing both basic (NH_2) and acidic (COOH) groups, as well as a side group (R) specific for each amino acid. The subunit building block of proteins.
amino terminal (N-terminal) The NH_2 end of a polypeptide.
amplification A large increase in the number of copies of a gene or a DNA sequence beyond that which is characteristic of the organism's haploid genome.
analogy Similarity by convergent evolution, but not by common evolutionary ancestry.
aneuploidy (chromosomal duplication) The presence of extra chromosomes, such that the chromosomal composition of a cell is not an exact multiple of the haploid set.
anticodon A triplet of nucleotides in a tRNA molecule that associates by complementary base pairing with a specific codon in the mRNA during translation, and specifies the placement of a particular amino acid during translation.
antiparallel The opposite orientation of the two complementary strands of a DNA duplex.
apoprotein A protein without its coenzymes, cofactors, and prosthetic groups that are required for its functionality (e.g., apolipoprotein).
archaebacteria Prokaryotes that do not incorporate muramic acid into their cell walls. A highly diverse group of methanogenic (producing methane), thermophilic, acidophilic, and hallophilic (living in high salt concentrations) bacteria that is presumed to represent one of the three primary lines of descent in the living world.
arithmetic mean The sum of n terms divided by n.

asymmetrical exon An exon flanked by introns of different phase classes.

autosome Any chromosome other than the sex chromosomes.

back (backward) mutation A mutation that reverts a nucleotide site to a previous state.

bacteriophage (abbreviated as **phage**) A virus that parasitizes bacteria.

balanced polymorphism A polymorphism that is stable over time and is maintained by balancing selection.

balancing selection A selection regime that results in the maintenance of two or more alleles at a locus in a population (e.g., overdominance).

banding Areas of light and dark staining on chromosomes.

base (see **nucleotide**)

base pair (1) A nucleotide on one strand of a nucleic acid that hydrogen-bonds with a nucleotide on the other strand according to the pairing rules between a purine and a pyrimidine. (2) A unit of measurement of double-stranded DNA length.

biased codon usage (see **unequal codon usage**)

bifurcation (dichotomy) The graphical representation in a phylogenetic tree of an evolutionary speciation event whereby an ancestral taxon splits into two.

bottleneck A severe reduction in population size.

box A short DNA sequence adjacent to or residing within a gene that performs a regulatory function (e.g., TATA box).

branch The graphical representation of an evolutionary relationship in a phylogenetic tree.

C (1) In DNA or RNA, cytosine. (2) In proteins, cysteine.

C-terminal (see **carboxy terminal**)

C value (genome size) The characteristic amount of DNA in the haploid genome of a species.

C-value paradox The apparent lack of correlation between the C value and the level of morphological complexity.

capping The modification of the 5′ end of pre-mRNA in eukaryotes whereby a GTP is added to the molecule via a 5′–5′ triphosphate bond.

cap site (transcription initiation site) In the DNA, the site where transcription starts. In the RNA, the site that is capped during the process of mRNA maturation.

carboxy terminal (C-terminal) The COOH end of a polypeptide.

carrying capacity The maximum number of individuals of a given species that can be sustained in a defined habitat.

catalyst A compound that lowers the energy necessary to activate a chemical reaction without being consumed or altered by it.

cDNA (see **complementary DNA**)

census population size (see **population size**)

central dogma The path of information flow in DNA organisms (DNA→RNA→proteins).

chimeric protein (see **mosaic protein**)

chloroplast A chlorophyll-containing, membrane-enclosed organelle which is the site of photosynthesis in the cells of plants and some protists.

chromatid (1) Each of the two copies produced by chromosome replication. (2) Each of the two DNA strands that comprise the chromosome.

chromosomal duplication (see **aneuploidy**)

chromosome In prokaryotes, the DNA molecule containing the genome. In eukaryotes, a linear DNA molecule complexed with proteins forming a thread-like structure containing the genetic information.

cis The arrangement of two sequences or genes on the same chromosome.

clade (1) According to the rigorous definition, a taxon consisting of a single species and all its descendants representing a monophyletic branch on an evolutionary tree. (2) In looser usage, as above, except that some descendants are not represented. (3) In reference to extant organisms, a subgroup of organisms from among a larger group under consideration sharing a common ancestor not shared by the other organisms in the group.

cladogenesis (see **speciation**)

cladogram A graphic representation that portrays or attempts to portray the evolutionary relationships among a number of populations, species, or higher taxa.

classification (see **taxonomy**)

coding region All exon parts of a protein-coding gene that are ultimately translated.

codominance (genic selection) The equal contribution to fitness made by the two alleles at a locus in a diploid organism.

codon A triplet of adjacent nucleotides in mRNA that either codes for an amino acid carried by a specific tRNA or specifies the termination of the translation process.

codon family All the codons that code for the same amino acid and differ from each other only at the third position (e.g., among the six codons for leucine, UUA and UUG form one family, and CUU, CUC, CUA and CUG form another).

codon usage The frequency with which members of a codon family are used in protein-coding genes.

coenzyme An organic nonprotein molecule that does not bind the enzyme but is required for the function of the enzyme by acting as an intermediate carrier of electrons, atoms, or groups of atoms.

cofactor An inorganic molecule required by an enzyme in order to function.

coincidental evolution (see **concerted evolution**)

coincidental substitution The occurrence of two substitutions at the same nucleotide site in two homologous sequences.

colinearity The exact correspondence between the DNA sequence of intronless genes and the amino acid sequence of the encoded protein.

complementarity The antiparallel pairing of nucleotides in double-stranded DNA, double-stranded RNA, or DNA-RNA duplexes.

complementary DNA (cDNA) DNA synthesized from an RNA template by the enzyme reverse transcriptase.

complex (composite) transposon A transposon flanked by two complete, independently transposable insertion sequences.

compositional assimilation The accumulation of point mutations in a pseudogene that eventually obliterates its sequence similarity to the functional gene from which it has been derived and makes its nucleotide composition similar to neighboring DNA sequences.

concerted evolution (horizontal evolution, coincidental evolution) Maintenance of homogeneity of nucleotide sequences among members of a gene family in a species, although the nucleotide sequences change over time.

conditional fixation time The time until fixation of a mutant allele that will eventually become fixed in the population.

consensus sequence A sequence that represents the most prevalent nucleotide or amino acid at each site in a number of homologous sequences.

conservative substitution The substitution of an amino acid by another with similar chemical properties.

conservative transposition The movement of a transposable element from one genomic position to another without replication of the element.

constant site or **constant region** A site or region within the DNA that is occupied by the same nucleotide in all homologous sequences under comparison.

convergence The independent evolution of similar genetic or phenotypic traits.

convergent substitution The substitution of two different nucleotides by the same nucleotide at the same nucleotide site in two homologous sequences.

crossing-over The process of exchange of genetic material between two homologous chromosomes leading to recombination of linked genes. Presumed to occur through breakage of both chromosomes at homologous sites followed by reunion after exchange.

cyanobacteria A type of photosynthetic eubacteria possessing the ability to photosynthesize. Formerly called blue-green algae.

D Aspartic acid.

Darwinian fitness (see **fitness**)

decoding (see **translation**)

degenerate code A genetic code in which the number of sense codons is larger than the total number of amino acids and, consequently, some amino acids are specified by more than one codon. All known genetic codes are degenerate.

degenerate site A nucleotide site in a codon that can be occupied by more than one nucleotide and still code for the same amino acid.

degree of divergence The extent to which two homologous sequences differ from each other.

deleterious mutation A mutation that lowers the fitness of its carriers.

deletion The removal of one or more bases from a DNA sequence.

denaturation The loss of a protein's tertiary structure. Sometimes used as a synonym of DNA **melting**.

deoxyribonucleic acid (DNA) A macromolecular polymer of linked nucleotides in which the sugar residue is deoxyribose. Usually, double-stranded. The carrier of genetic information in all eukaryotes and prokaryotes, and in many viruses.

deterministic process A process, the outcome of which can be predicted exactly from knowledge of initial conditions.

diagnostic position (see **informative site**)

dichotomy (see **bifurcation**)

digestion The cutting of a double-stranded DNA by a restriction endonuclease.

diploid A chromosomal complement that contains two copies of each chromosome.

directional selection A selective regime that changes the frequency of an allele in a specific direction, either toward fixation or toward elimination.

disjunction The separation of homologous chromosomes during meiosis or the separation of complementary chromatids during mitosis.

distance (see **genetic distance**)

distance matrix A matrix of genetic distances between taxa in a group under study.

divergence The splitting of a taxonomic unit into two. (See also **sequence divergence**)

DNA (see **deoxyribonucleic acid**)

DNA–DNA hybridization The formation of heteroduplex DNA.

domain (see **functional domain**)

dominance The property of an allele to manifest its entire phenotypic effect in the heterozygote.

donor site The 5′ end of an intron.

dose repetition The presence of multiple copies of a DNA sequence which can be shown to produce increased quantities of a gene product relative to a single copy sequence.

dot matrix A method of sequence alignment in which two sequences are written as column and row headings of a matrix and dots are put in those matrix elements that have identical column and row headings.

downstream In the direction 3′ of a reference point on a nucleic acid. In the direction of transcription.

drift (see **random genetic drift**)

duplex A double-stranded DNA or RNA, or a double helix formed by the complementary pairing of a single-stranded DNA with an RNA molecule.

duplication The presence or the creation of two copies of a DNA segment in the genome.

duplicative transposition (see **replicative transposition**)

E Glutamic acid.

effective population size The population size that is relevant for random genetic drift. The actual number of individuals in a population that are reproducing.

electromorph A variant protein (isozyme or allozyme) detected by its distinct electrophoretic mobility.

electrophoresis A technique that separates dissolved or colloidal particles subjected to an electrical field according to their mobilities. Electrophoretic mobility depends on the size, three-dimensional geometry, and electrical charge of the particle.

endosymbiosis A mutually beneficial relationship between two organisms in which one, the **endosymbiont**, lives within the tissues or the cells of the other, the **host**.

endosymbiotic theory The proposal that self-replicating cellular organelles, such as the mitochondria and the chloroplasts, were originally free-living organisms that entered into a symbiotic relationship with nucleated cells and subsequently lost their ability to survive independently.

enzyme A protein or complex of proteins that catalyzes a specific chemical reaction.

eubacteria Prokaryotes that incorporate muramic acid into their cell walls. All bacteria exclusive of the archaebacteria. One of the three primary lines of descent in the living world.

eukaryote An organism having a true nucleus and membraneous organelles. One of the three primary lines of descent in the living world.

exon A DNA segment of a gene, the transcript of which appears in the mature RNA molecule.

exon duplication The creation of duplicate copies of an exon within a single gene.

exon insertion The incorporation of one or more exons from one gene into another.

exon shuffling Strictly, exon duplication and exon insertion. Often used synonymously with exon insertion.

expected heterozygosity (see **heterozygosity, gene diversity**)

extinction The termination of an evolutionary lineage.

F Phenylalanine.

fecundity A fitness component. The number of births or eggs per individual of a given genotype.

fertility A fitness component. The number of live offspring per individual of a given genotype.

fitness (Darwinian fitness) A measure of the relative survival and reproductive success of an individual or a genotype. The relative contribution of an individual or a genotype to future generations.

fixation The situation achieved when an allele reaches a frequency of 100% in a population. A condition in which all members of a diploid population are homozygous for the same allele.

fixation probability The probability that a particular allele will become fixed in a population.

fixation time The time it takes for a mutant allele to become fixed in a population.

flanking sequence Untranscribed sequences at the 5′ or 3′ terminal of transcribed genes.

foldback DNA DNA that contains a perfect or nearly perfect palindrome and is able to form a hairpin-like structure by folding back on itself when single-stranded.

fourfold degenerate site A nucleotide site within a codon at which all possible substitutions are synonymous.

frameshift mutation A mutation disturbing the reading frame of a protein-coding gene. A deletion or an insertion of a DNA segment that is not three nucleotides or a multiple of three.

frameshifted protein A protein encoded entirely or in part by a reading frame different from the original or main reading frame of a gene.

functional constraint (selective constraint) The degree of intolerance characteristic of a site or a locus toward nucleotide substitutions.

functional domain (domain) A well-defined region within a protein that can perform a specific function. May not consist of a continuous stretch of amino acids, although it almost always consists of amino acids that are adjacent to each other as far as the tertiary structure of the protein is concerned.

G (1) In DNA or RNA, guanine. (2) In proteins, glycine.

gamete A reproductive cell with a haploid number of chromosomes.

gap An insertion or a deletion. In sequence alignment, a pair containing a null base.

gap penalty The assessment of how frequent a gap event occurs in evolution in comparison with the frequency of occurrence of point substitutions. In alignment algorithms, either a factor multiplied by the total length of gaps or a function multiplied by the number of gaps of a given length, which is used to compare the likelihood of gaps and substitutions.

gene A sequence of genomic DNA or RNA that is essential for a specific function.

gene conversion A nonreciprocal recombination process resulting in a sequence becoming identical with another.

gene diversity A measure of genetic variability in a population. The mean expected heterozygosity per locus in a population.

gene duplication Generally, the production of two copies of a DNA sequence. Specifically, the duplication of an entire gene sequence.

gene family (see **multigene family**)

gene frequency (see **allele frequency**)

gene pool All the genes in a sexually reproducing population.

gene sharing The situation whereby a gene acquires and maintains a second function without duplication and without loss of its primary function.

gene substitution The process whereby a new mutant allele reaches fixation in a population.

gene tree A phylogenetic tree that has been constructed from one or a few genes from each species.

generation time The average time span between two successive generations. Sometimes defined as the mean age of the parents at which they give birth to their middle child.

genetic code A set of rules for the translation of codons into amino acids.

genetic distance (distance) Broadly, any of several measures of the degree of genetic difference between individuals, populations, or species. In reference to molecular evolution, a measure of the number of nucleotide substitutions per nucleotide site between two homologous DNA sequences that have accumulated since the divergence between the sequences.

genetic drift (see **random genetic drift**)

genetic polymorphism (see **polymorphism**)

genic DNA The portion of the genome that contains genes.

genic selection (see **codominanace**)

genome The entire complement of genetic material carried by a cell or an individual.

genome doubling (see **polyploidy**)

genome duplication (see **polyploidy**)

genome size (see **C value**)

genomic compartmentalization The existence of independently replicated genomes within a cell. Usually, in reference to the genomes of organelles.

genotype The specific allelic constitution of an organism including alleles not expressed at the phenotypic level. Often, the allelic composition of one or few genes under investigation.

geometric mean The *n*th root of the product of *n* terms.

germ-line cell A sperm or egg cell, or one of their precursor cells.

H Histidine.

haploid A cell or organism having a single set of unpaired chromosomes.

haploid set The chromosomes in a haploid cell or individual.

haplotype The specific allelic constitution of a chromosome. Often, the allelic composition of one or a few linked genes under investigation.

Hardy-Weinberg equilibrium A condition under which the genotypic frequencies in a diploid population are equal to the products of the allele frequencies involved.

heteroduplex A double-stranded nucleic acid molecule in which each strand has been derived from a different individual.

heterogenous nuclear RNA (heterogeneous RNA, heteronuclear RNA, hnRNA) RNA transcripts in the nucleus, representing precursors and processing intermediates of rRNA, mRNA, and tRNA, as well as mature RNA transcripts not yet transported into the cytoplasm.

heterosis (see **overdominance**)

heterozygosity A measure of genetic variation in a population calculated either as the mean frequency of heterozygotes over all loci (**observed heterozygosity**), or as the mean frequency of heterozygotes expected in a population in Hardy-Weinberg equilibrium (**expected heterozygosity** or **gene diversity**).

heterozygote A diploid individual with different alleles at the locus in question.

heterozygote advantage (see **overdominance**)

higher taxon A taxon above the species level.

highly repetitive DNA The fraction of genomic DNA consisting of sequences repeated on the average hundreds of thousands of times.

highly repetitive genes Functional genes appearing in numerous copies in the haploid genome.

homoduplex A double-stranded DNA, the complementary strands of which are derived from the same individual.

homology Similarity by common ancestry or genetic relatedness.

homozygote A diploid individual with identical alleles at one or more loci.

horizontal evolution (see **concerted evolution**)

horizontal gene transfer The transfer of genetic information from one genome to another, specifically between different species.

hotspot of mutation A segment of genomic DNA that shows a high propensity to mutate either spontaneously or under the action of a particular mutagen.

hybrid dysgenesis A syndrome of correlated abnormalities that is spontaneously induced in one type of hybrid between certain mutually interactive strains of *Drosophila*, but not in the reciprocal hybrid.

hybrid vigor (see **overdominance**)

hydrogen bond A weak, noncovalent bond between a hydrogen atom and an electronegative atom such as oxygen.

hypervariable site or **hypervariable region** A DNA or protein region that exhibits excessive intraspecific variability. Maintenance of so much variability usually requires the locus to be subject to a form of balancing selection, such as overdominance.

I Isoleucine.

independent assortment (**Mendel's second law**) The principle that in unlinked loci, the alleles of one locus segregate independently of the alleles of the other.

inferred tree A phylogenetic tree based on empirical data pertaining to extant taxa.

informative site (diagnostic position) A site that is used to choose the most-parsimonious tree from among all the possible phylogenetic trees. In molecular evolution, a site where there are at least two different kinds of nucleotides or amino acids, and each of them is represented in at least two sequences.

initiation codon The first codon in the reading frame of a protein-coding gene; usually, ATG encoding methionine in eukaryotes and formylmethionine in prokaryotes.

in-phase overlapping The condition in which two or more proteins are translated in the same reading frame.

insertion A mutation in which one or more nucleotides are inserted into a DNA sequence.

insertion sequence A transposable element carrying no genetic information except that which is necessary for transposition.

internal gene duplication (partial gene duplication) Repeated sequences within a gene that have been derived from duplications involving less than the entire gene sequence.

internal node The graphical representation of an ancestral organism or gene in a phylogenetic tree.

intron (intervening sequence) A DNA segment of a transcribed gene, the transcript of which is removed in the process of RNA maturation and, therefore, does not appear in the mature RNA molecule. Resides between exons.

invariant repetition The existence of repeated DNA segments that are identical or nearly identical in sequence to one another.

inversion A mutation that causes a DNA segment to assume a reverse polarity.

isoaccepting tRNA A tRNA molecule that recognizes more than one codon.

isochore A genomic DNA segment that is homogeneous in base composition.

isozyme (isoenzyme) Any of the distinct forms of an enzyme that have identical or nearly identical chemical properties but are encoded by different loci.

junk DNA The fraction of genomic DNA that is devoid of function.

K Lysine.

L Leucine.

lagging strand The DNA strand synthesized in a discontinuous fashion, 5′ to 3′ away from the replication fork.

leading strand The DNA strand synthesized in a continuous fashion, 5′ to 3′ towards the replication fork.

length abridgment The gradual shortening of pseudogenes during evolution that is caused by an excess of deletions over insertions.

lethal mutation A mutation that causes either death or sterility of its homozygous carriers.

ligation The formation of a phosphodiester bond to link two adjacent bases separated by a nick in double-stranded DNA. Used to link sticky ends produced by restriction-endonuclease digestion. The process is catalyzed by the enzyme **ligase**.

LINE (Acronym of **L**ong **IN**terspersed **E**lement) An interspersed repetitive sequence typically longer than 5,000 base pairs found in 10^4 or more copies in the genomes of multicellular eukaryotes.

linkage A condition in which two or more nonallelic genes reside in close proximity to each other and tend to be inherited together.

localized repeated sequences Tandemly arrayed, repetitive sequences, usually made up of short simple repeated motifs (e.g., satellite DNA)

locus (plural **loci**) The site on a chromosome where a particular gene or DNA segment is located.

lowly repetitive genes Genes appearing in only a few copies in the haploid genome.

M Methionine.

match In sequence alignment, the existence of the same base in a homologous position in both sequences.

maturation The formation of mRNA from pre-mRNA.

maximum parsimony (parsimony) The selection of the phylogenetic tree requiring the least number of substitutions from among all possible phylogenetic trees as the most likely to be the true phylogenetic tree.

meiosis (reduction division) The eukaryotic cell division process used in producing haploid gametes from diploid cells. Meiosis is characterized by a reduction division which ensures that each gamete contains one representative of each pair of autosomes and half the sex chromosomes.

meiotic drive (see **segregation distortion**)

melting Denaturation of duplexes into single-stranded nucleic acids.

Mendelian segregation (Mendel's first law, segregation) The Mendelian principle that the two different alleles of a gene pair in a heterozygote separate from each other during meiosis to produce two kinds of gametes in equal ratios, each bearing a different allele.

Mendel's second law (see **independent assortment**)

—mer Suffix denoting the number of subunits in a protein. The prefix denotes the number of subunits (e.g., **monomer, dimer, tetramer, multimer**), or the similarity between the units (e.g., **homomer, heteromer**), or a combination of number and type of units (e.g., **homotrimer, heteromultimer**).

messenger RNA (mRNA): An RNA molecule that is processed from a primary RNA transcript and used for translation into the amino acid of a polypeptide.

middle-repetitive DNA The fraction of genomic DNA consisting of relatively long sequences repeated on the average from tens to hundreds of times.

migration In population genetics, the movement of individuals or genes among populations.

mismatch In sequence alignment, the existence of different bases in a homologous position in the two sequences.

missense mutation (see **nonsynonymous mutation**)

mitochondrion (plural, **mitochondria**) A DNA-containing organelle in eukaryotic cells that uses an oxygen-requiring electron-transport system to transfer chemical energy derived from the breakdown of food molecules to ATP.

mitosis The mode of eukaryotic cell division that produces two daughter cells possessing the same chromosomal complement as the parent cell.

mobile element (see **transposable element**)

moderately repetitive genes Genes with a moderate number of copies in the haploid genome.

module (structural domain) In globular proteins, a structurally independent, stable, and compact spatial unit that can be distinguished from all other parts. Usually, consisting of a contiguous stretch of amino acids.

molecular clock (1) The rate at which mutations accumulate in a given genomic segment. (2) The hypothesis that, in any given gene or DNA sequence, mutations accumulate at an approximately constant rate in all evolutionary lineages as long as the gene or the DNA sequence retains its original function. The extent to which the clock applies to all genes and all organisms is controversial.

monomorphic A population in which virtually all individuals have the same allele at a locus.

monophyletic Sharing a common ancestor.

mortality A fitness component. The average probability of an individual of a given genotype to die before reaching a certain age (e.g., mortality to mean reproductive age).

mosaic protein (chimeric protein) A protein encoded by a gene that contains regions from different genes. Also, an artificial protein derived through genetic engineering.

mRNA (see **messenger RNA**)

multifurcation A graphic representation of an unknown branching order in a phylogenetic tree involving three or more taxa. Rarely, a graphic representation of a speciation event resulting in the simultaneous production of more than two species (e.g., **trichotomy**).

multigene family (gene family) A set of genes derived by duplication of an ancestral gene that display more than 50% similarity among them. Frequently in close linkage with each other, and possessing similar or overlapping functions.

multiple substitutions The successive occurrence of two or more substitutions at the same nucleotide site in a DNA sequence.

mutagen A physical or chemical entity that increases the mutation rate.

mutant A new variant form of a gene.

mutation The alteration of a DNA sequence to produce a different form than the original.

mutation rate The number of mutations arising in an individual per nucleotide site or per gene per unit time.

mutational bias A pattern of mutation in which the four nucleotides show different propensities to mutate or in which mutations result in a certain nucleotide more often than others. Often results in the uneven accumulation of certain nucleotides.

N (1) In DNA or RNA, an unknown nucleotide. (2) In proteins, asparagine.

natural selection (selection) Differential reproduction of different members of a species due to the variability in fitness among individuals or genotypes, leading to changes in allele frequencies over time.

negative selection (see **purifying selection**)
neighboring taxa (see **sister taxa**)
neutral mutation A mutation that does not change the fitness of the organism.
neutral theory (neutral-mutation theory or **neutral-mutation hypothesis)** The proposal that evolution at the molecular level is primarily determined by mutational input and random genetic drift, rather than by natural selection.
node The graphical representation in a phylogenetic tree of an extant or ancestral operational taxonomic unit.
nondegenerate site A nucleotide site in the coding region at which all substitutions are nonsynonymous.
nondisjunction The failure of homologous chromosomes to separate during meiosis.
nonfunctionalization (silencing) The turning of a functional gene into a pseudogene following the occurrence of an incapacitating mutation.
nongenic DNA The portion of the genome that does not contain genes.
nonsense codon (see **termination codon**)
nonsense mutation A mutation that alters a sense codon into a termination codon.
nonsense strand The transcribed strand of a gene, the sequence of which is complementary to the RNA transcript.
nonsynonymous substitution (missense substitution) A substitution that alters a codon to that for another amino acid.
N-terminal (see **amino terminal**)
nucleic acid DNA or RNA.
nucleotide (base) A molecule composed of a nitrogen base, a sugar, and a phosphate group. Any of the basic building blocks of nucleic acids.
nucleotide diversity A measure of polymorphism applied to nucleic acid sequences. The mean number of nucleotide differences per site between any two randomly chosen sequences from a population.
nucleotide substitution A mutation in which one nucleotide is substituted for another. In evolution, the substitution of a nucleotide by another nucleotide that becomes fixed in a population.
nucleotypic Referring to a function of a DNA sequence other than as a carrier of genetic information (e.g., serving as a skeleton for the nucleus).
nucleus (plural, **nuclei)** A membrane-enclosed organelle containing the chromosomes in eukaryotes.
observed heterozygosity (see **heterozygosity**)
ontogeny The sequence of events in the development of an organism from zygote to adult.
open reading frame (ORF) A DNA sequence that is potentially translatable into a protein.
operational taxonomic unit (OTU) Any of the extant taxonomic units under study.
operon A genetic unit or cluster that consists of one or more genes that are transcribed as a unit and expressed in a coordinated manner.
ORF (see **open reading frame**)
organelle Strictly, functional membrane-enclosed structures within eukaryotic cells (e.g., nuclei, mitochondria, chloroplasts). Usually, excluding the nucleus.
orthology Sequence similarity as a consequence of a speciation event.
OTU (see **operational taxonomic unit)**
outgroup A species or a set of species that is the least related to the others in a group of species. The taxon that diverged from a group of taxa before the others diverged from each other.

out-of-phase overlapping The encoding of two or more proteins in different frames of the same DNA sequence.

overdominance (heterosis, heterozygote advantage, hybrid vigor) A selection regime resulting from the heterozygote having a higher fitness than either homozygote.

P proline.

palindromic sequence A DNA or RNA sequence that reads the same on the complementary strand (e.g., AATGCATT). A DNA or RNA sequence that shows symmetry about a central axis point.

parallel substitutions The independent occurrence of the same mutation at the same nucleotide site in two or more lineages.

paralogy Sequence similarity between the descendants of a duplicated ancestral gene.

pararetrovirus A virus that contains a gene for reverse transcriptase but cannot insert itself into the host chromosome.

parsimony Literally, the use of a minimum number of means to achieve an end. (see **maximum parsimony**)

partial gene duplication (see **internal gene duplication**)

pattern of mutation The relative frequency with which a nucleotide tends to mutate into another.

pattern of substitution (substitution scheme) The relative frequency with which a nucleotide or an amino acid changes into another during evolution.

PCR (see **polymerase chain reaction**)

phage (see **bacteriophage**)

phase class The position of an intron relative to the reading frame of the two adjacent protein-coding exons.

phase-0 intron An intron that lies between two codons.

phase-1 intron An intron that lies between the first and second nucleotides of a codon

phase-2 intron An intron that lies between the second and third nucleotides of a codon.

phenogram A graphic representation that portrays or attempts to portray the taxonomic relationships among a number of individuals, species, or higher taxa on the basis of overall similarities between them.

phenotype The observable characteristics of a genetically controlled trait.

phylogenetic tree A graphic representation of the phylogeny of a group of taxa or genes.

phylogenetics The reconstruction of the evolutionary history of a group of taxa or genes.

phylogeny The evolutionary history of a group of taxa or genes and their ancestors.

plasmid An autonomous, self-replicating extrachromosomal circular DNA.

point mutation A mutation affecting only one nucleotide site. Usually, in reference to a nucleotide substitution.

polarity The property of nucleic acids to be read one way from 5′ to 3′ and differently in the opposite direction.

polyadenylation signal A box on most eukaryotic mRNA molecules that specifies the location of the polyadenylation site.

polyadenylation site (poly(A)-addition site) The 3′ end of most mRNA molecules in eukaryotes. The site at which a poly-A tail is added.

polygamy A mating system in which a male mates with more than one female (**polygyny**) or a female mates with more than one male (**polyandry**).

polymerase chain reaction (PCR) A method of amplification of a chosen DNA sequence from unpurified mixtures.

polymorphism (genetic polymorphism) The coexistence of two or more alleles at a locus.

polypeptide A molecule made of amino acids covalently linked to each other by peptide bonds. Often, a term used to denote the amino acid chain of a protein before it assumes a functional three-dimensional configuration.

polyphyletic Descended from different ancestors.

polyploidy (genome doubling, genome duplication) The presence in a cell or an individual of more than two haploid sets of chromosomes (e.g., **tetraploidy**, **hexaploidy**).

polyprotein A polypeptide that is cleaved after translation to give rise to two or more functional proteins.

population A group of individuals in a species that share a common gene pool.

population size (census population size) The number of individuals in a population.

positive selection Selection for an advantageous mutant allele.

pre-messenger RNA (pre-mRNA) The primary transcript of a protein-coding gene before maturation.

preproprotein The primary product of translation before any posttranslational changes have been made.

pretermination codon A codon that requires only one mutation to become a termination codon.

primary amino acid One of the 20 amino acids specified by the universal genetic code.

primary structure The sequence of amino acids in a polypeptide chain. The sequence of nucleotides in a DNA or RNA molecule.

processed gene (see **retrogene)**

processed pseudogene (see **retropseudogene**)

processed sequence (see **retrosequence)**

prokaryote (bacterium) An organism lacking nuclear membranes, histone-bound DNA, and cellular organelles. Eubacteria and archaebacteria.

proprotein A product of translation after the signal peptide has been removed and before additional posttranslational modifications have been made.

prosthetic group A nonprotein molecule attached to an apoprotein that is required for functionality (e.g., heme in hemoglobin).

protein A molecule composed of one or more polypeptide chains. May or may not contain prosthetic groups.

protein-coding gene A gene that contains a reading frame, the mRNA of which is translated.

provirus A viral genome integrated into the genome of the host cell.

pseudogene A functionless segment of DNA exhibiting sequence homology to a functional gene. A nonfunctional member of a gene family.

purifying selection (negative selection) A selection regime resulting in the removal of an allele from the population.

purine A type of nitrogen base present in nucleotides and composed of two joined ring structures, one five-membered and one six-membered. The purine bases in DNA and RNA are adenine and guanine.

pyrimidine A type of nitrogen base present in nucleotides and composed of a single six-membered ring. The pyrimidine bases in DNA are cytosine and thymine. The pyrimidine bases in RNA are cytosine and uracil.

Q Glutamine.

quaternary structure Types and modes of interaction between two or more polypeptide chains within a protein molecule with two or more subunits.

R Arginine.

radical substitution The substitution of an amino acid by another with markedly different chemical properties.

random genetic drift (drift, genetic drift) The fluctuation in allele frequencies from generation to generation caused by chance events, such as gamete sampling.

rate of gene substitution The number of gene substitutions per locus per unit time.

rate of mutation The number of mutations per locus or nucleotide site per unit time, usually per generation time.

rate of nucleotide substitution The number of nucleotide substitutions per nucleotide site per unit time.

reading frame Linear sequence of codons in a protein-coding gene starting with the initiation codon and ending in the termination codon.

recessiveness Lack of expression of an allele in the heterozygote.

recognition sequence The sequence recognized by a restriction endonuclease. In many cases, a short palindrome.

recombination The situation arising following a crossover event, in which new combinations of alleles are found in cis.

recombinator gene A regulatory gene providing a recognition site for the recombination enzymes.

reduction division (see **meiosis**)

regional duplication A duplication involving less than the entire genome.

regulatory gene A nontranscribed gene. Sometimes used to denote a structural gene engaged in the regulation of gene expression.

relative-rate test A calibration-free test for checking the constancy of the rate of nucleotide substitutions in different lineages during their evolution, thus determining whether or not the molecular clock operates at the same rate among different lineages.

repetitive DNA DNA sequences present in many copies in the haploid genome.

replacement The result of a nonsynonymous substitution at the protein level.

replication The process of DNA synthesis on a DNA template.

replication slippage A process in which a certain sequence of DNA is used more than once in a row as a template during DNA replication, thus creating a tandemly repeated sequence in the newly synthesized DNA.

replicative transposition (duplicative transposition) The insertion of a copy of a transposable element into a new chromosomal position while the element itself remains in the original position.

replicator gene A regulatory gene specifying the sites for initiation and termination of DNA replication.

replicon A chromosomal region that contains the DNA sequences necessary for the initiation of DNA replication, and that is replicated as a unit.

reproductive barrier (reproductive isolation) Any of several biological or environmental mechanisms that prevent gene exchange between populations.

restriction endonuclease (restriction enzyme) An enzyme that hydrolyzes internal phosphodiester bonds in the DNA.

restriction-fragment pattern The number and sizes of the restriction fragments resulting from a DNA sequence being digested by a restriction endonuclease.

restriction site The point at which a restriction endonuclease hydrolyzes (cuts) an internal phosphodiester bond in the DNA. May or may not reside in proximity to the recognition sequence.

restriction-site map The schematic representation of a DNA sequence showing the location of the restriction sites.

retroelement A DNA or RNA sequence that possesses the ability to produce reverse transcriptase.

retrofection The process of transfer of an RNA molecule from one cell to another, in particular to a germ-line cell, by means of a retroviral particle, into which the RNA is encapsulated; the RNA is then reverse-transcribed and incorporated into the host genome. Transduction by means of a retrovirus.

retrogene (processed gene) A functional retrosequence producing a protein that is identical or nearly identical to that produced by the gene from which the mRNA was derived.

retron A genomic sequence encoding reverse transcriptase but lacking the ability to transpose.

retroposition An RNA-mediated mode of transposition.

retroposon A transposable retroelement that neither constructs virion particles nor is flanked by terminally redundant sequences.

retropseudogene (processed pseudogene) A pseudogene derived from the reverse transcription of an RNA molecule and subsequent incorporation of the cDNA into the genome. Diagnostic signs include lack of introns, polyadenine tails, flanking repeats, truncation, evidence of posttranscriptional modifications, and lack of physical linkage with either functional or nonprocessed functionless members of the gene family.

retrosequence (retrotranscript, processed sequence) A genomic sequence that has been derived through the reverse transcription of RNA but by itself lacks the ability to produce reverse transcriptase. Retrogenes and retropseudogenes.

retrotransposon A transposable retroelement that does not construct virion particles and that is flanked by terminally redundant sequences.

retrovirus Any of a group of small single-stranded RNA viruses that encode reverse transcriptase.

reverse transcriptase The enzyme that catalyzes reverse transcription.

reverse transcription The synthesis of a single-stranded DNA molecule on an RNA template.

ribonucleic acid (RNA) A macromolecular polymer of linked nucleotides in which the sugar residue is ribose. Usually, single-stranded.

ribosomal RNA (rRNA) The RNA molecules that constitute the structural components of a ribosome.

ribosome An intracellular particle composed of rRNA and proteins that furnishes the site at which mRNA is translated.

RNA (see **ribonucleic acid**)

RNA-specifying gene A gene that is transcribed, but the RNA of which is not translated. A gene, the RNA transcript of which is functional.

rolling-circle replication A mode of amplification in which a circular extrachromosomal copy of a DNA sequence is created and replicated in a continuous fashion.

root In rooted trees, the common ancestor of all the taxa under study.

rooted tree A phylogenetic tree that specifies ancestral and descendant species, thus indicating the direction of the evolutionary path.

rRNA (see **ribosomal RNA**)

S Serine.

secondary structure In proteins and nucleic acids, the structure of the molecule brought about by the formation of hydrogen bonds between amino acids or nucleotides, respectively. In proteins, the localized structures in the polypeptide chain (e.g., α-helix, β-sheet, turn). In single-stranded DNA or RNA, the localized double-stranded structures (e.g., hairpins).

segregation (see **Mendelian segregation**)

segregation distortion (meiotic drive) Aberrant segregation ratios among the gametes of heterozygotes.

segregator gene A regulatory gene providing chromosomal attachment sites for the segregation machinery during meiosis and mitosis.

selection (see **natural selection**)

selection coefficient A quantitative measure of the reduction in fitness of a genotype in comparison with the fittest genotype in the population. A measure of selective disadvantage.

selection intensity (stringency of selection) The difference in the fitness values between the various genotypes in a population.

selective constraint (see **functional constraint**)

selfish DNA A DNA segment that may not confer any advantage to its carrier (or host) but is concerned only with its own propagation. Transposable elements are thought to be selfish DNA.

self-splicing intron An intron that can be cleaved out of the pre-mRNA without the aid of an external catalyst.

sense codon A codon specifying an amino acid.

sense strand The nontranscribed strand of a gene, the DNA sequence of which is identical to the RNA transcript.

sequence divergence (divergence) The differences between two homologous sequences due to the independent accumulation of genetic changes in each lineage.

sex-linkage The situation in which a gene is located on the sex chromosomes. Often the use of the term is restricted to genes on the X-chromosome.

sibling species Species that are indistinguishable morphologically but are reproductively isolated.

signal peptide A leader peptide that is cleaved out after the synthesis of the polypeptide and before the protein assumes its correct tertiary structure.

silencing (see **nonfunctionalization**)

silent substitution A substitution that does not alter the phenotype of its carrier. Includes substitutions in nongenic DNA and synonymous substitutions.

simple transposon A transposon that does not contain insertion sequences.

SINE (Acronym of **S**hort **IN**terspersed **E**lement) Any interspersed repeated sequence shorter than 500 base pairs found in 10^5 or more copies in the genomes of multicellular eukaryotes.

single-copy DNA (see **unique DNA**)

sister taxa (neighboring taxa) In general use, the pair of species among a group of species under study that are evolutionarily the closest to each other. In a phylogenetic tree, two taxa connected through a single internal node.

somatic cell A cell that is not destined to become a gamete.

somatic mutation A mutation occurring in a somatic cell.

spacer DNA The DNA found between two genes. Can be either transcribed or nontranscribed.

speciation (cladogenesis) The splitting of one population into two or more populations that are reproductively isolated. The process by which new species arise.

species A basic taxonomic category for which there are various definitions, among them: (1) a group of actually or potentially interbreeding individuals that is reproductively isolated from other such groups (the biological species concept); (2) a lineage evolving separately from others (the evolutionary species concept); and (3) a group of organisms resembling each other more than they resemble any other organism outside the group (the taxonomic species concept).

species tree A phylogenetic tree that represents the evolutionary relationships of a group of species.

splicing The removal of introns in the process of RNA maturation.

splicing site or **junction** The border between exons and introns.

split gene A gene containing introns.

standard nucleotide Adenine, cytosine, guanine, thymine, or uracil.

sticky ends Single strands of DNA that protrude from opposite ends of a duplex. Usually, generated by staggered cuts of a double-stranded DNA by a restriction enzyme.

stochastic process A process, the outcome of which cannot be predicted exactly from knowledge of initial conditions. However, given the initial conditions, each of the possible outcomes of the process can be assigned a certain probability.

stop (see **termination codon**)

stringency of selection (see **selection intensity**)

strong bond In reference to double-stranded nucleic acids or segments (e.g., codon–anticodon interactions), the three hydrogen bonds between C and G. Confers increased stability and high melting temperatures.

structural domain (see **module**)

structural gene A DNA nucleotide sequence that codes for a protein or specifies an RNA molecule. Some authors refer only to protein-coding genes as structural.

subspecies A geographically or morphologically distinct population in a species.

substitution (see **gene substitution** and **nucleotide substitution**)

substitution matrix The representation of the pattern of substitution in the form of a matrix, the elements of which denote the relative rate of substitution between any two nucleotides.

substitution scheme (see **pattern of substitution**)

superfamily A collection of genes, all products of gene duplication, that have diverged from each other to a considerable extent (in protein-coding genes, usually a similarity of less than 50% at the amino acid level).

symbiosis The coexistence of two or more organisms in a mutually beneficial relationship.

symmetrical exon An exon residing between two same-phase introns.

synonymous substitution (silent substitution) A nucleotide substitution resulting in a codon specifying the same amino acid as before.

systematics Taxonomy and phylogenetics.

tandem duplication A duplication, the products of which reside in close proximity to each other on the chromosome.

taxon (plural, **taxa**) A taxonomic group of any rank (e.g., species, genus, kingdom) to which individual organisms are assigned.

taxonomy (classification) The principles and procedures according to which species are named and assigned to taxonomic groups.

terminal node The graphical representation of an extant taxon in a phylogenetic tree.

termination codon (nonsense codon, stop) A codon for which no normal tRNA exists, the presence of which terminates the process of translation. The three termination codons in the universal genetic code are UAG, UAA, and UGA.

tertiary structure In proteins and nucleic acids, the three-dimensional structure of the molecule brought about by its folding upon itself.

thermal stability The degree of resistence against denaturation or melting.

T (1) In DNA, thymine. (2) In proteins, threonine.

T_m The median melting temperature of duplex DNA.

ΔT_m The difference between the median melting temperature of homoduplex DNA and heteroduplex DNA.

topology The branching pattern of a phylogenetic tree.

trans The arrangement of two sequences or genes on different chromosomes.

transcription The synthesis of an RNA molecule on a DNA template.

transcription-initiation site (see **cap site**)

transcription-termination site The point at which the transcription of RNA is terminated. In RNAs that are polyadenylated, it may or may not be identical with the polyadenylation site.

transduction The transfer of host genetic information from one cell to another by means of a virus.

transfer RNA (tRNA) A small ribonucleic acid molecule that contains an anticodon, a binding site for a specific amino acid, and recognition sites for interaction with the ribosome and the enzyme that links it to its specific amino acid.

transition The substitution of a purine for a purine or a pyrimidine for a pyrimidine.

translation (decoding) The derivation of the amino acid sequence of a polypeptide from the sequence of an mRNA molecule via tRNA intermediates which occurs in association with a ribosome.

transposable element (mobile element) A sequence that can move about in the genome of an organism.

transposition The movement of genetic material from one genomic location to another.

transposon A transposable element carrying additional genes beside those encoding the transposition functions.

transversion The substitution of a purine for a pyrimidine or vice versa.

trisomy The presence of three copies of a chromosome in an otherwise diploid cell.

tRNA (see **transfer RNA**)

true tree A phylogenetic tree that represents the true evolutionary history of a group of taxa.

twofold degenerate site A nucleotide site in the coding region at which one of the three possible nucleotide changes is synonymous and the other two are nonsynonymous.

U Uracil.

unequal codon usage (biased codon usage) In protein-coding genes, the disproportionate use of one or more of the codons in a codon family.

unequal crossing-over A crossing-over between two homologous sequences at different chromosomal positions producing an unequal number of copies of the regions involved in the resulting chromosomes.

unidentified reading frame (URF) An ORF, the product of which is not known.

unique DNA (single-copy DNA) DNA sequences present in only one copy in the haploid genome.

universal genetic code The genetic code regulating the translation of the vast majority of genomes.

unprocessed pseudogene A pseudogene that has been derived through gene duplication and the subsequent nonfunctionalization or silencing of one of the copies.

unrooted tree An evolutionary tree that specifies neither the root nor the direction of the evolutionary path.

untranscribed sequence (see **flanking sequence**)

upstream In the direction 5′ of a reference point on a nucleic acid (opposite to the direction of transcription).

URF (see **unidentified reading frame**)

V Valine.

variable site or region (1) A site or region within the DNA that is occupied by different nucleotides among the sequences under comparison. (2) Strictly, a site or region within the DNA that can vary with time.

variant repetition The existence of repeated products of gene duplication that have diverged from each other since the duplication event.

viability A fitness component. The probability that an individual of a given genotype survives from conception to reproductive age.

virion A virus particle.

virus A small parasite that depends on the host cell to replicate its genetic material and synthesize its proteins. Its genome can be either RNA or DNA and can be either single- or double-stranded.

W Tryptophan.

weak bond In reference to double-stranded nucleic acids (e.g., codon–anticodon interactions), the two hydrogen bonds between A and T or between A and U. Confers reduced thermal stability and low melting temperatures.

wild type The most prevalent allele at a locus in a population, if such an allele exists.

wobble pairing The ability of some tRNAs to recognize more than one codon by nonstandard pairing between the first anticodon position and the third codon position (e.g., U in the anticodon with G in the codon).

X-linkage (see **sex-linkage**)

xenology Sequence similarity as a consequence of a horizontal gene transfer event.

Y Tyrosine.

zygote The diploid cell produced by the fusion of haploid gametic nuclei.

ANSWERS TO PROBLEMS

Chapter 2

2. $\Delta q = \dfrac{p^2 q(w_{22} - w_{11})}{p^2(w_{11} - w_{22}) + w_{22}}$

4. (a) $P_0 = 9.776 \times 10^{-4}$. (b) $P_5 = 0.246$. (c) $P_{10} = 9.776 \times 10^{-4}$.

5. $N_e/N = 8/9$.

6. $N_e = 59.701$.

7. $P = 3.769 \times 10^{-4}$.

8. (a) The coding regions of 1-*S* and 2-*S* are identical. Therefore, $x_i = 2/3$ and $x_j = 1/3$, and $\pi = 1.729 \times 10^{-3}$. (b) $\pi = 3.891 \times 10^{-3}$.

Chapter 3

6. (a) $P_S = 0.615$. (b) $P_A = 0.244$.

9. When $r = 4$, (a) $K = 0.061$, and (b) $K = 0.056$. When $r = 6$, (a) $K = 0.041$, and (b) $K = 0.037$.

Chapter 4

2. $K_I = 0.012$, $K_{II} = 0.048$, and $K_{III} = 0.295$ substitutions per site.

3. (a) $K = 0.076$ substitutions per site. (b) $K = 0.075$ substitutions per site.

4. The mouse sequence evolves 1.63 times faster than the rat sequence.

5. (a) $r_{max} = 0.057$ substitutions per site per year. (b) $r_{min} = 0.004$ substitutions per site per year.

Chapter 5

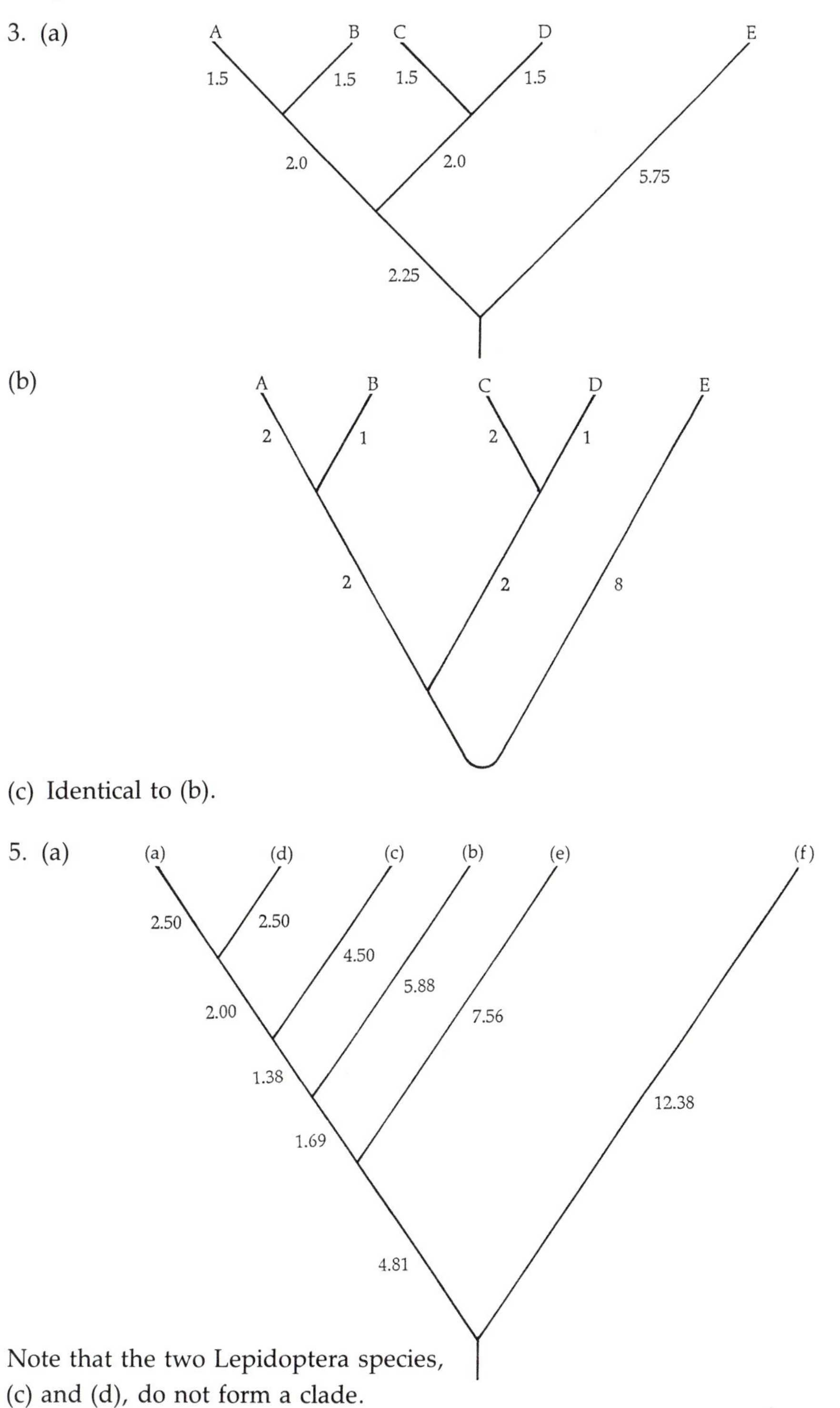

(c) Identical to (b).

Note that the two Lepidoptera species, (c) and (d), do not form a clade.

(b) Two equally parsimonious trees are obtained:

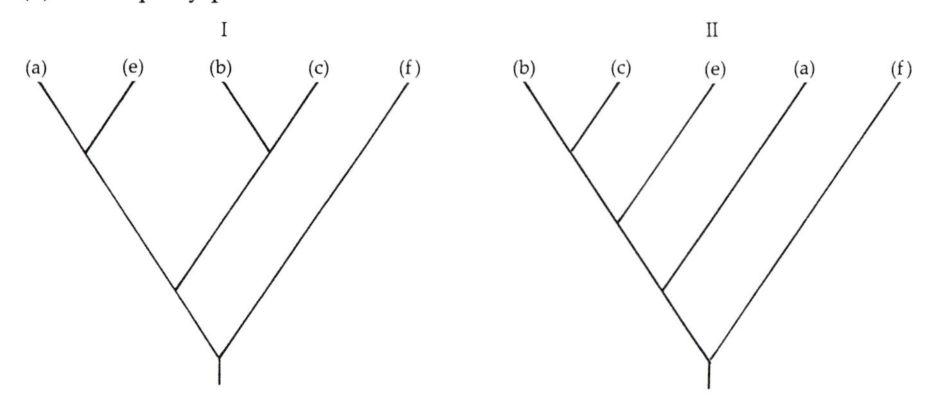

Tree II agrees with classical taxonomy, since the Holometabola, (b), (c), and (e), form a clade.

Chapter 6

2. (a) 47 nondegenerate and 13 fourfold degenerate sites. (b) 70 nondegenerate and no fourfold degenerate sites. (c) All 72 sites are nondegenerate.

3. (a) 5. (b) 6.

4. (b) 17.24×10^6 years.

Chapter 8

1. (a) *Physarum polycephalum*, 5.37%; *Oxytricha nova*, 13.46%; *Euplotes aediculatus*, 24.48%; *Dictyostelium discoideum*, 27.63%. (b) There is a negative correlation between genome size and proportion of nongenic DNA.

3. (a) 56.67%. (b) 30.00%.

5. $p_t = 0.43$.

LITERATURE CITED

Anderson, S., A. T. Bankier, B. G. Barrell, M. H. L. de Bruijn, A. R. Coulson, J. Drouin, I. C. Eperon, D. P. Nierlich, B. A. Roe, F. Sanger, P. H. Schreier, A. J. H. Smith, R. Staden and I. G. Young. 1981. Sequence and organization of the human mitochondrial genome. Nature 290: 457–464.

Aota, S.-I. and T. Ikemura. 1986. Diversity in G+C content at the third position of codons in vertebrate genes and its cause. Nucleic Acids Res. 14: 6345–6355.

Argos, P., M. G. Rossmann, U. M. Grau, A. Zuber, G. Franck and J. D. Tratschin. 1979. Thermal stability and protein structure. Biochemistry 18: 5698–5703.

Arnheim, N. 1983. Concerted evolution of multigene families. pp. 38–61. *In* M. Nei and R. K. Koehn (eds.), *Evolution of Genes and Proteins*. Sinauer Associates, Sunderland, MA.

Avers, C. J. 1989. *Process and Pattern in Evolution*. Oxford University Press, New York.

Avise, J. C. and W. S. Nelson. 1989. Molecular genetic relationships of the extinct dusky seaside sparrow. Science 243: 646–648.

Bagshaw, J. C., H. B. Skinner, T. C. Burn and B. A. Perry. 1987. Nucleotide sequence of the 5S RNA gene and flanking regions interspersed with histone genes in *Artemia*. Nucleic Acids Res. 15: 3628.

Bains, W. 1986. The multiple origins of human *Alu* sequences. J. Mol. Evol. 23: 189–199.

Baker, B. S. 1989. Sex in flies: The splice of life. Nature 340: 521–524.

Banyai, L., A. Varadi and L. Patthy. 1983. Common evolutionary origin of the fibrin-binding structures of fibronectin and tissue-type plasminogen activator. FEBS Lett. 163: 37–41.

Barker, W. C., L. K. Ketcham and M. O. Dayhoff. 1978. Duplications in protein sequences. pp. 359–362. *In* M. O. Dayhoff (ed.), *Atlas of Protein Sequence and Structure*, Vol. 5, Supplement 3. Natl. Biomed. Res. Found., Silver Spring, MD.

Beale, D. and H. Lehmann. 1965. Abnormal haemoglobin and the genetic code. Nature 207: 259–261.

Benveniste, R. E. 1985. The contributions of retroviruses to the study of mammalian evolution. pp. 359–417. *In* R. I. MacIntyre (ed.), *Molecular Evolutionary Genetics*. Plenum, New York.

Benveniste, R. E. and G. J. Todaro. 1976. Evolution of type C viral genes: Evidence for an Asian origin of man. Nature 261: 101–108.

Berg, D. E., C. M. Berg and C. Sasakawa. 1984. The bacterial transposon *Tn5*: Evolutionary inferences. Mol. Biol. Evol. 1: 411–422.

Bernardi, G. 1989. The isochore organization of the human genome. Annu. Rev. Genet. 23: 637–661.

Bernardi, G. and G. Bernardi. 1985. Codon usage and genome composition. J. Mol. Evol. 22: 363–365.

Bernardi, G., B. Olofsson, J. Filipski, M. Zerial, J. Salinas, G. Cuny, M. Meunier-Rotival and F. Rodier. 1985. The mosaic genome of warm-blooded vertebrates. Science 228: 953–958.

Bernardi, G., D. Mouchiroud, C. Gautier and G. Bernardi. 1988. Compositional patterns in vertebrate genomes: Conservation and change in evolution. J. Mol. Evol. 28: 7–18.

Betz, J. L., P. R. Brown, M. J. Smyth and P. H. Clarke. 1974. Evolution in action. Nature 247: 261–264.

Bishop, J. M. 1981. Enemies within: The genesis of retrovirus oncogenes. Cell 23: 5–6.

Blanchetot, A., V. Wilson, D. Wood and A. J. Jeffreys. 1983. The seal myoglobin gene: An unusually long globin gene. Nature 301: 732–734.

Bodmer, W. F. and L. L. Cavalli-Sforza. 1976. *Genetics, Evolution, and Man*. Freeman, San Francisco.

Bonen, L., W. F. Doolittle and G. E. Fox. 1979. Cyanobacterial evolution: Results of 16S ribosomal ribonucleic acid sequence analysis. Can. J. Biochem. 57: 879–888.

Bostock, C. J. 1986. Mechanisms of DNA sequence amplification and their evolutionary consequences. Phil. Trans. Roy. Soc. London 312B: 261–273.

Bourbon, H.-M., M. Prudhomme and F. Amalric. 1988. Sequence and structure of the nucleolin promoter in rodents: Characterization of a strikingly conserved CpG island. Gene 68: 73–84.

Britten, R. J. 1986. Rates of DNA sequence evolution differ between taxonomic groups. Science 231: 1393–1398.

Britten, R. J. and D. E. Kohne. 1968. Repeated sequences in DNA. Science 161: 529–540.

Britten, R. J., D. E. Graham and B. R. Neufeld. 1974. Analysis of repeating DNA sequences by reassociation. pp. 363–406. *In* L. Grossman and K. Moldave (eds.), *Methods in Enzymology*. Academic Press, New York.

Britten, R. J., W. F. Baron, D. B. Stout and E. H. Davidson. 1988. Sources and evolution of human *Alu* repeated sequences. Proc. Natl. Acad. Sci. USA 85: 4770–4774.

Brown, D. D. and K. Sugimoto. 1973. 5S DNAs of *Xenopus laevis* and *Xenopus mulleri*: Evolution of a gene family. J. Mol. Biol. 78: 397–415.

Brown, D. D., P. C. Wensink and E. Jordan. 1972. A comparison of the ribosomal DNAs of *Xenopus laevis* and *Xenopus mulleri*: Evolution of tandem genes. J. Mol. Biol. 63: 57–73.

Brown, W. M., M. George, Jr. and A. C. Wilson. 1979. Rapid evolution of animal mitochondrial DNA. Proc. Natl. Acad. Sci. USA 76: 1967–1971.

Brown, W. M., E. M. Prager, A. Wang and A. C. Wilson. 1982. Mitochondrial DNA sequences of primates: Tempo and mode of evolution. J. Mol. Evol. 18: 225–239.

Caccone, A. and J. R. Powell. 1989. DNA divergence among hominoids. Evolution 43: 925–942.

Cairns-Smith, A. G. 1982. *Genetic Takeover and the Mineral Origins of Life*. Cambridge University Press, Cambridge.

Calabretta, B., D. L. Robberson, H. A. Barrera-Saldana, T. P. Lambrou and G. F. Saunders. 1982. Genome instability in a region of human DNA enriched in *Alu* repeat sequences. Nature 296: 219–225.

Cappello, J., S. M. Cohen and H. F. Lodish. 1984. *Dictyostelium* transposable element *DIRS-1* preferentially inserts into *DIRS-1* sequences. Mol. Cell. Biol. 4: 2207–2213.

Cavalier-Smith, T. 1975. The origin of nuclei and of eukaryotic cells. Nature 256: 463–468.

Cavalier-Smith, T. 1978. Nuclear volume control by nucleoskeletal DNA, selection for cell volume and cell growth rate and the solution to the DNA C-value paradox. J. Cell. Sci. 34: 247–278.

Cavalier-Smith, T. (ed.). 1985. *The Evolution of Genome Size*. Wiley, New York.

Cave, M. D., H. Benes and C. Liarakos. 1987. Nucleotide sequence of two repeating units of the 5S rRNA gene from the house cricket *Acheta domesticus*. Gene 51: 287–289.

Cedergren, R., M. W. Gray, Y. Abel and D. Sankoff. 1988. The evolutionary relationships among known life forms. J. Mol. Evol. 28: 98–112.

Chao, L., C. Vargas, B. B. Spear and E. C. Cox. 1983. Transposable elements as mutator genes in evolution. Nature 303: 633–635.

Chao, S., R. Sederoff and C. S. Levings III. 1984. Nucleotide sequence and evolution of the 18S ribosomal RNA gene in maize mitochondria. Nucleic Acids Res. 12: 6629–6645.

Charlesworth, B. 1985. The population genetics of transposable elements. pp. 213–232. *In* T. Ohta and K. Aoki (eds.), *Population Genetics and Molecular Evolution*. Springer-Verlag, Berlin.

Charlesworth, B. and C. H. Langley. 1989. The population genetics of *Drosophila* transposable elements. Annu. Rev. Genet. 23: 251–287.

Charlesworth, B., C. Langley and W. Stephan. 1986. The evolution of restricted recombination and the accumulation of repeated DNA sequences. Genetics 112: 947–962.

Comings, D. E. 1978. Mechanisms of chromosome banding and implications for chromosome structure. Annu. Rev. Genet. 12: 25–46.

Coulondre, C., J. H. Miller, P. J. Farabaugh and W. Gilbert. 1978. Molecular basis of base substitution hotspots in *Escherichia coli*. Nature 274: 775–780.

Cuny, G., P. Soriano, G. Macaya and G. Bernardi. 1981. The major components of the mouse and human genomes. 1. Preparation, basic properties, and compositional heterogeneity. Eur. J. Biochem. 111: 227–233.

Curtis, S. E. and M. T. Clegg. 1984. Molecular evolution of chloroplast DNA sequences. Mol. Biol. Evol. 1: 291–301.

Czelusniak, J., M. Goodman, D. Hewett-Emmett, M. L. Weiss, P. J. Venta and R. E. Tashian. 1982. Phylogenetic origins and adaptive evolution of avian and mammalian haemoglobin genes. Nature 298: 297–300.

Danciger E., N. Dafni, Y. Bernstein, Z. Laver–Rudich, A. Neer and Y. Groner. 1986. Human Cu/Zn superoxide dismutase gene family: Molecular structure and characterization of four Cu/Zn superoxide dismutase-related pseudogenes. Proc. Natl. Acad. Sci. USA 83: 3619–3623.

Daniels, S. B., K. R. Peterson, L. D. Strausbaugh, M. G. Kidwell and A. Chovnick. 1990. Evidence for horizontal transmission of the *P* transposable element between *Drosophila* species. Genetics 124: 339–355.

Darwin, C. 1871. *The Descent of Man and Selection in Relation to Sex*. Appleton, New York.

Dayhoff, M. O. 1972. *Atlas of Protein Sequence and Structure*, Vol. 5. National Biomedical Research Foundation, Silver Spring, MD.

Dayhoff, M. O. 1978. *Atlas of Protein Sequence and Structure*, Vol. 5, Supplement 3. National Biomedical Research Foundation, Silver Spring, MD.

Devos, R., R. Contreras, J. van Emmela and W. Fiers. 1979. Identification of the translocatable element *IS1* in a molecular chimera constructed with pBR 322 into which MS2 DNA copy was inserted by poly (dA.dT) linked method. J. Mol. Biol. 128: 621–632.

Dickerson, R. E. and I. Geis. 1983. *Hemoglobin*. Benjamin/Cummings, Menlo Park, CA.

Doolittle, R. F. 1985. The genealogy of some recently evolved vertebrate proteins. Trends Biochem. Sci. 10: 233–237.

Doolittle, R. F. 1987. The evolution of the vertebrate plasma proteins. Biol. Bull. 172: 269–283.

Doolittle, W. F. and C. Sapienza. 1980. Selfish genes, the phenotype paradigm and genome evolution. Nature 284: 601–603.

Dover, G. A. 1982. Molecular drive: A cohesive mode of species evolution. Nature 299: 111–117.

Dover, G. A. 1986. Molecular drive in multigene families: How biological novelties arise, spread and are assimilated. Trends Genet. 2: 159–165.

Dyson, F. 1985. *Origins of Life*. Cambridge University Press, Cambridge.

Eck, R. V. and M. O. Dayhoff. 1966. *Atlas of Protein Sequence and Structure*. National Biomedical Research Foundation, Silver Spring, MD.

Efstratiadis, A., J. W. Posakony, T. Maniatis, R. M. Lawn, C. O'Connell, R. A. Spritz, J. De Riel, B. G. Forget, S. M. Weissman, J. L. Slightom, A. E. Blechl, O. Smithies, F. E. Baralle, C. C. Shoulders and N. J. Proudfoot. 1980. The structure and evolution of the human β-globin gene family. Cell 21: 653–668.

Engelke, D. R., P. A. Hoener and F. S. Collins. 1988. Direct sequencing of enzymatically amplified human genomic DNA. Proc. Natl. Acad. Sci. USA 85: 544–548.

Engels, W. R. 1981a. Estimating genetic divergence and genetic variability with restriction endonucleases. Proc. Natl. Acad. Sci. USA 78: 6329–6333.

Engels, W. R. 1981b. Hybrid dysgenesis in *Drosophila* and the stochastic loss hypothesis. Cold Spring Harbor Symp. Quant. Biol. 45: 561–565.

Everse, J. and N. Kaplan. 1975. Mechanisms of action and biological functions of various dehydrogenase isozymes. pp. 29–43. *In* C. L. Markert (ed.), *Isozymes II: Physiological Function*. Academic Press, New York.

Farris, J. S. 1977. On the phenetic approach to vertebrate classification. pp. 823–850. *In* M. K. Hecht, P. C. Goody and B. M. Hecht (eds.), *Major Patterns in Vertebrate Evolution*. Plenum, New York.

Felsenstein, J. 1978. The number of evolutionary trees. Syst. Zool. 27: 27–33.

Felsenstein, J. 1988. Phylogenies from molecular sequences: Inference and reliability. Annu. Rev. Genet. 22: 521–565.

Finnegan, D. J. and D. H. Fawcett. 1986. Transposable elements in *Drosophila*. Oxford Surv. Eukaryotic Genes 3: 1–62.

Fitch, W. M. 1971. Toward defining the course of evolution: Minimum change for a specific tree topology. Syst. Zool. 20: 406–416.

Fitch, W. M. 1977. On the problem of discovering the most parsimonious tree. Am. Nat. 111: 223–257.

Fitch, W. M. 1981. A non-sequential method for constructing trees and hierarchical classifications. J. Mol. Evol. 18: 30–37.

Fitch, W. M. and E. Margoliash. 1967. Construction of phylogenetic trees. A method based on mutation distances as estimated from cytochrome c sequences is of general applicability. Science 155: 279–284.

Flavel, R. B. 1986. Repetitive DNA and chromosome evolution in plants. Phil. Trans. Roy. Soc. London 312B: 227–242.

Gilbert, W. 1978. Why genes in pieces? Nature 271: 501.

Giovannoni, S. J., S. Turner, G. J. Olsen, S. Barns, D. J. Lane and N. R. Pace. 1988. Evolutionary relationships among cyanobacteria and green chloroplasts. J. Bacteriol. 170: 3584–3592.

Gō, M. 1981. Correlation of DNA exonic regions with protein structural units in haemoglobin. Nature 291: 90–92.

Gō, M. and M. Nosaka. 1987. Protein architecture and the origin of introns. Cold Spring Harbor Symp. Quant. Biol. 52: 915–924.

Gojobori, T., W.-H. Li and D. Graur. 1982. Patterns of nucleotide substitution in pseudogenes and functional genes. J. Mol. Evol. 18: 360–369.

Goldman, M. A., G. P. Holmquist, M. C. Gray, L. A. Caston and A. Nag. 1984. Replication timing of genes and middle repetitive sequences. Science 224: 686–692.

Goodman, M. 1961. The role of immunochemical differences in the phyletic development of human behavior. Hum. Biol. 33: 131–162.

Goodman, M. 1962. Immunochemistry of the primates and primate evolution. Ann. N.Y. Acad. Sci. 102: 219–234.

Goodman, M. 1963. Serological analysis of the systematics of recent hominoids. Hum. Biol. 35: 377–424.

Goodman, M. 1981. Decoding the pattern of protein evolution. Prog. Biophys. Mol. Biol. 38: 105–164.

Goodman, M., J. Barnabas, G. Matsuda and G. W. Moor. 1971. Molecular evolution in the descent of man. Nature 233: 604–613.

Goodman, M., B. F. Koop, J. Czelusniak, M. L. Weiss and J. L. Slightom. 1984. The η-globin gene: Its long evolutionary history in the β-globin gene family of mammals. J. Mol. Biol. 180: 803–823.

Grantham, R., C. Gautier, M. Gouy, R. Mercier and A. Pave. 1980. Codon catalog usage and the genome hypothesis. Nucleic Acids Res. 8: r49–r62.

Graur, D. 1985. Amino acid composition

and the evolutionary rates of protein-coding genes. J. Mol. Evol. 22: 53–63.

Graur, D., M. Bogher and A. Breiman. 1989a. Restriction endonuclease profiles of mitochondrial DNA and the origin of the B genome of bread wheat, *Triticum aestivum*. Heredity 62: 335–342.

Graur, D., Y. Shuali and W.-H. Li. 1989b. Deletions in processed pseudogenes accumulate faster in rodents than in humans. J. Mol. Evol. 28: 279–285.

Gray, M. W., R. Cedergren, Y. Abel and D. Sankoff. 1989. On the evolutionary origin of the plant mitochondrion and its genome. Proc. Natl. Acad. Sci. USA 86: 2267–2271.

Grime, J. P. and M. A. Mowforth. 1982. Variation in genome size and ecological interpretation. Nature 299: 151–153.

Gruskin, K. D., T. F. Smith and M. Goodman. 1987. Possible origin of a calmodulin gene that lacks intervening sequences. Proc. Natl. Acad. Sci. USA 84: 1605–1608.

Gu, X.-R., K. Nicoghosian and R. J. Cedergren. 1982. 5S RNA sequence from the *Philosamia* silkworm: Evidence for variable evolutionary rates in insect 5S RNA. Nucleic Acids Res. 10: 5711–5721.

Hagelberg, E., B. Sykes and R. Hedges. 1989. Ancient bone DNA amplified. Nature 342: 485.

Hahn, B. H., G. M. Shaw, M. E. Taylor, R. R. Redfield, P. D. Markham, S. Z. Salahuddin, F. Wong-Staal, R. C. Gallo, E. S. Parks and W. P. Parks. 1986. Genetic variation in HTLV-III/LAV over time in patients with AIDS or at risk for AIDS. Science 232: 1548–1553.

Haldane, J. B. S. 1932. *The Causes of Evolution*. Longmans and Green, London.

Hardies, S. C., S. L. Martin, C. F. Voliva, C. A. Hutchison III and M. H. Edgell. 1986. An analysis of replacement and synonymous changes in the rodent L1 repeat family. Mol. Biol. Evol. 3: 109–125.

Hardison, R. C. and J. B. Margot. 1984. Rabbit globin pseudogene $\psi\beta2$ is a hybrid of δ- and β-globin gene sequences. Mol. Biol. Evol. 1: 302–316.

Hartl, D. L. and A. G. Clark. 1989. *Principles of Population Genetics*, 2nd Ed. Sinauer Associates, Sunderland, MA.

Hartl, D. L., M. Medhora, L. Green and D. E. Dykhuizen. 1986. The evolution of DNA sequences in *Escherichia coli*. Phil. Trans. Roy. Soc. London 312B: 191–204.

Heddle, J. A. and K. Athanasiou. 1975. Mutation rate, genome size and their relation to the *rec.* concept. Nature 258: 359–361.

Hendriks, W., J. W. M. Mulders, M. A. Bibby, C. Slingsby, H. Bloemendal and W. W. De Jong. 1988. Duck lens ε-crystallin and lactate dehydrogenase B_4 are identical: A single-copy gene product with two distinct functions. Proc. Natl. Acad. Sci. USA 85: 7114–7118.

Herdman, M. 1985. The evolution of bacterial genomes. pp. 37–68. *In* T. Cavalier-Smith (ed.), *The Evolution of Genome Size*. Wiley, New York.

Higgins, N. P. and D. Hillyard. 1988. Primary structure and mapping of the *hupA* gene of *Salmonella typhimurium*. J. Bacteriol. 170: 5751–5758.

Hobbs, H. H., M. S. Brown, J. L. Goldstein and D. W. Russell. 1986. Deletion of exon encoding cysteine rich repeat of low density lipoprotein receptor alters its binding specificity in a subject with familial hypercholesterolemia. J. Biol. Chem. 261: 13114–13120.

Holmquist, G. P. 1987. Role of replication time in the control of tissue-specific gene expression. Am. J. Hum. Genet. 40: 151–173.

Hori, H. and S. Osawa. 1986. Evolutionary change in 5S rRNA secondary structure and a phylogenetic tree of 352 rRNA species. BioSystems 19: 163–172.

Hudson, R. R., M. Kreitman and M. Aguade. 1987. A test of neutral molecular evolution based on nucleotide data. Genetics 116: 153–159.

Hughes, A. L. and M. Nei. 1989. Nucleotide substitution at major histocompatibility complex class II loci: Evidence for overdominant selection. Proc. Natl. Acad. Sci. USA 86: 958–962.

Hunt, J. A., T. J. Hall and R. J. Britten. 1981. Evolutionary distance in Hawaiian *Drosophila* as measured by DNA reassociation. J. Mol. Evol. 17: 361–367.

Hutchison III, C. A., S. C. Hardies, D. D. Loeb, W. R. Shehee and M. H. Edgell. 1989. LINEs and related retroposons: Long interspersed repeated sequences in the eukaryotic genome. pp. 593–617. *In* D. E. Berg and M. M. Howe (eds.), *Mobile DNA*. American Society for Microbiology, Washington, DC.

Ikemura, T. 1981. Correlation between the abundance of *Escherichia coli*: Transfer RNAs and the occurrence of the respective codons in its protein genes: A proposal for a synonymous codon choice that is optimal for the *E. coli* translational system. J. Mol. Biol. 151: 389–409.

Ikemura, T. 1982. Correlation between the abundance of yeast transfer RNAs and the occurrence of the respective codons in protein genes: Differences in synonymous codon choice patterns of yeast and *Escherichia coli* with reference to the abundance of isoaccepting transfer RNAs. J. Mol. Biol. 158: 573–597.

Ikemura, T. 1985. Codon usage and tRNA content in unicellular and multicellular organisms. Mol. Biol. Evol. 2: 13–34.

Ikemura, T. and S.-I. Aota. 1988. Global variation in G+C content along vertebrate genome DNA: Possible correlation with chromosome band structures. J. Mol. Biol. 203: 1–13.

Inouye, S., M.-Y. Hsu, S. Eagle and M. Inouye. 1989. Reverse transcriptase associated with the biosynthesis of the branched RNA-linked msDNA in *Myxococcus xanthus*. Cell 56: 709–717.

Jacobs, G. H. and J. Neitz. 1986. Spectral mechanisms and color vision in the tree shrew (*Tupaia belangeri*). Vision Res. 26: 291–298.

James, T. C. and S. C. R. Elgin. 1986. Identification of a nonhistone chromosomal protein associated with heterochromatin in *Drosophila melanogaster* and its gene. Mol. Cell. Biol. 6: 3862–3872.

Jeffreys, A. 1979. DNA sequence variants in $^{G}\gamma$-, $^{A}\gamma$-, δ- and β-globin genes of man. Cell 18: 1–10.

Jones, R. N. 1985. Are B chromosome "selfish"? pp. 397–425. *In* T. Cavalier-Smith (ed.), *The Evolution of Genome Size*. Wiley, New York.

Jukes, T. H. and C. R. Cantor. 1969. Evolution of protein molecules. pp. 21–132. *In* H. N. Munro (ed.), *Mammalian Protein Metabolism*. Academic Press, New York.

Kagawa, Y., H. Nojima, N. Nukiwa, M. Ishizuka, T. Nakajima, T. Yasuhara, T. Tanaka and T. Oshima. 1984. High guanine plus cytosine content in the third letter of codons of an extreme thermophile. J. Biol. Chem. 259: 2956–2960.

Kaplan, N. 1983. Statistical analysis of restriction enzyme map data and nucleotide sequence data. pp. 75–106. *In* B. S. Weir (ed.), *Statistical Analysis of DNA Sequence Data*. Marcel Dekker, New York.

Kawata, Y. and H. Ishikawa. 1982. Nucleotide sequence and thermal property of 5S rRNA from the elder aphid, *Acyrthosiphon magnoliae*. Nucleic Acids Res. 10: 1833–1840.

Kidwell, M. G. 1979. Hybrid dysgenesis in *Drosophila melanogaster*: The relationship between the *P–M* and *I–R* interaction systems. Genet. Res. 33: 205–217.

Kidwell, M. G. 1983. Evolution of hybrid dysgenesis determinants in *Drosophila melanogaster*. Proc. Natl. Acad. Sci. USA 80: 1655–1659.

Kidwell, M. and J. F. Kidwell. 1976. Selection for male recombination in *Drosophila melanogaster*. Genetics 84: 333–351.

Kimura, M. 1962. On the probability of fixation of mutant genes in populations. Genetics 47: 713–719.

Kimura, M. 1968a. Evolutionary rate at the molecular level. Nature 217: 624–626.

Kimura, M. 1968b. Genetic variability maintained in a finite population due to mutational production of neutral and nearly neutral isoalleles. Genet. Res. 11: 247–269.

Kimura, M. 1980. A simple method for estimating evolutionary rate of base substitution through comparative studies of nucleotide sequences. J. Mol. Evol. 16: 111–120.

Kimura, M. 1983. *The Neutral Theory of Molecular Evolution*. Cambridge University Press, Cambridge.

Kimura, M. and T. Ohta. 1969. The average number of generations until fixation of a mutant gene in a finite population. Genetics 61: 763–771.

Kimura, M. and T. Ohta. 1971. Protein polymorphism as a phase of molecular evolution. Nature 229: 467–469.

Kimura, M. and T. Ohta. 1972. On the stochastic model for estimation of mutational distance between homologous proteins. J. Mol. Evol. 2: 87–90.

King, J. L. and T. H. Jukes. 1969. Non–Darwinian evolution. Science 164: 788–798.

Klaer, R., S. Kühn, E. Tillmann, H. J. Fritz and P. Starlinger. 1981. The sequence of *IS4*. Mol. Gen. Genet. 181: 169–175.

Klotz, L. C., N. Komar, R. L. Blanken and R. M. Mitchell. 1979. Calculation of evolutionary trees from sequence data. Proc. Natl. Acad. Sci. USA 76: 4516–4520.

Kocher, T. D., W. K. Thomas, A. Meyer, S. V. Edwards, S. Pääbo, F. X. Villablanca and A. C. Wilson. 1989. Dynamics of mitochondrial DNA evolution in animals: Amplification and sequencing with conserved primers. Proc. Natl. Acad. Sci. USA 86: 6196–6200.

Kohne, D. E. 1970. Evolution of higher–organism DNA. Quart. Rev. Biophys. 33: 327–375.

Koop, B. F., M. Goodman, P. Xu, K. Chan and J. L. Slightom. 1986a. Primate η-globin DNA sequences and man's place among the great apes. Nature 319: 234–238.

Koop, B. F., M. M. Miyamoto, J. E. Embury, M. Goodman, J. Czelusniak and J. L. Slightom. 1986b. Nucleotide sequence and evolution of the orangutan ϵ-globin gene region and surrounding *Alu* repeats. J. Mol. Evol. 24: 94–102.

Kornberg, A. 1982. *Supplement to DNA Replication*. Freeman, San Francisco.

Kreitman, M. 1983. Nucleotide polymorphism at the alcohol dehydrogenase locus of *Drosophila melanogaster*. Nature 304: 412–417.

Kushiro, A., M. Shimizu and K.-I. Tomita. 1987. Molecular cloning and sequence determination of the *tuf* gene coding for the elongation factor Tu of *Thermus thermophilus* HB8. Eur. J. Biochem. 170: 93–98.

Lamb, B. C. and S. Helmi. 1982. The extent to which gene conversion can change allele frequencies in populations. Genet. Res. 39: 199–217.

Lampson, B. C., J. Sun, M.-Y. Hsu, J. Vallejo-Ramirez, S. Inouye and M. Inouye. 1989. Reverse transcriptase in a clinical strain of *Escherichia coli*: Production of branched RNA-linked msDNA. Science 243: 1033–1038.

Leeds, J. M., M. B. Slabourgh and C. K. Mathews. 1985. DNA precursor pools and ribonucleotide reductase activity: Distribution between the nucleus and cytoplasm of mammalian cells. Mol. Cell. Biol. 5: 3443–3450.

Lehrman, M. A., D. W. Russell, J. L. Goldsmith and M. S. Brown. 1986. Exon–*Alu* recombination deletes 5 kilobases from the low density lipoprotein receptor gene, producing a null phenotype in familial hypercholesterolemia. Proc. Natl. Acad. Sci. USA 83: 3679–3683.

Levinger, L. and A. Varshavsky. 1982. Protein D1 preferentially binds A+T-rich DNA *in vitro* and is a component of *Drosophila melanogaster* nucleosomes containing A+T-rich satellite DNA. Proc. Natl. Acad. Sci. USA 79: 7152–7156.

Levinson, G. and G. A. Gutman. 1987. Slipped-strand mispairing: A major mechanism for DNA sequence evolution. Mol. Biol. Evol. 4: 203–221.

Lewin, B. 1990. *Genes IV*. Oxford University Press, Oxford.

Lewin, R. 1981. Evolutionary history written in globin genes. Science 214: 426–429.

Li, W.-H. 1981. Simple method for constructing phylogenetic trees from distance matrices. Proc. Natl. Acad. Sci. USA 78: 1085–1089.

Li, W.-H. 1983. Evolution of duplicate genes and pseudogenes. pp. 14–37. *In* M. Nei and R. K. Koehn (eds.), *Evolution of Genes and Proteins*. Sinauer Associates, Sunderland, MA.

Li, W.-H. and M. Tanimura. 1987. The molecular clock runs more slowly in man than in apes and monkeys. Nature 326: 93–96.

Li, W.-H., C.-I. Wu and C.-C. Luo. 1984. Nonrandomness of point mutation as reflected in nucleotide substitutions in pseudogenes and its evolutionary implications. J. Mol. Evol. 21: 58–71.

Li, W.-H., C.-C. Luo and C.-I. Wu. 1985a. Evolution of DNA sequences. pp. 1–94. *In* R. J. MacIntyre (ed.), *Molecular Evolutionary Genetics*. Plenum, New York.

Li, W.-H., C.-I. Wu and C.-C. Luo. 1985b. A new method for estimating synonymous and nonsynonymous rates of nucleotide substitution considering the relative likelihood of nucleotide and codon changes. Mol. Biol. Evol. 2: 150–174.

Li, W.-H., M. Tanimura and P. M. Sharp. 1987a. An evaluation of the molecular clock hypothesis using mammalian DNA sequences. J. Mol. Evol. 25: 330–342.

Li, W.-H., K. H. Wolfe, J. Sourdis and P. M. Sharp. 1987b. Reconstruction of phylogenetic trees and estimation of divergence times under nonconstant rates of evolution. Cold Spring Harbor Symp. Quant. Biol. 52: 847–856.

Li, W.-H., M. Tanimura and P. M. Sharp. 1988. Rates and dates of divergence between AIDS virus nucleotide sequences. Mol. Biol. Evol. 5: 313–330.

Linial, M. 1987. Creation of a processed pseudogene by retroviral infection. Cell 49: 93–102.

Loomis, W. F. 1988. *Four Billion Years: An Essay on the Evolution of Genes and Organisms*. Sinauer Associates, Sunderland, MA.

Luo, C.-C., W.-H. Li and L. Chan. 1989. Structure and expression of dog apolipoprotein A-I, E, and C-I mRNAs: Implications for the evolution and functional constraints of apolipoprotein structure. J. Lipid Res. 30: 1735–1746.

Maeda, N., J. B. Bliska and O. Smithies. 1983. Recombination and balanced chromosome polymorphism suggested by DNA sequence 5′ to the human delta-globin gene. Proc. Natl. Acad. Sci. USA 80: 5012–5016.

Maeda, N., C.-I. Wu, J. Bliska and J. Reneke. 1988. Molecular evolution of intergenic DNA in higher primates: Pattern of DNA changes, molecular clock and evolution of repetitive sequences. Mol. Biol. Evol. 5: 1–20.

Margoliash, E. 1963. Primary structure and evolution of cytochrome c. Proc. Natl. Acad. Sci. USA 50: 672–679.

Margulis, L. 1981. *Symbiosis in Cell Evolution: Life and its Environment in the Early Earth*. Freeman, San Francisco.

Markert, C. L. and H. Ursprung. 1971. *Developmental Genetics*. Prentice-Hall, Englewood Cliffs, NJ.

Markert, M. L., J. J. Hutton, D. A. Wiginton, J. C. States and R. E. Kaufman. 1988. Adenosine deaminase (*ADA*) deficiency due to deletion of the *ADA* gene promoter and first exon by homologous recombination between two *Alu* elements. J. Clin. Invest. 81: 1323–1327.

Marks, J., J.-P. Shaw and C.-K. J. Shen. 1986. Sequence organization and genomic complexity of primate θ1 globin gene, a novel α-globin-like gene. Nature 321: 785–788.

Maruyama, T. and M. Kimura. 1974. A note on the speed of gene frequency changes in reverse directions in a finite population. Evolution 28: 162–163.

McCarrey J. R. and K. Thomas. 1987. Human testis-specific *PGK* gene lacks introns and possesses characteristics of a processed gene. Nature 326: 501–505.

Miklos, G. L. G. 1985 Localized highly repetitive DNA sequences in vertebrate and invertebrate genomes. pp. 241–321. *In* R. J. MacIntyre (ed.), *Molecular Evolutionary Genetics*. Plenum, New York.

Miyamoto, M. M., J. L. Slightom and M. Goodman. 1987. Phylogenetic relationships of humans and African apes from DNA sequences in the ψη-globin region. Science 238: 369–373.

Miyata, T. and T. Yasunaga. 1978. Evolution of overlapping genes. Nature 272: 532–535.

Miyata, T. and T. Yasunaga. 1980. Molecular evolution of mRNA: A method for estimating evolutionary rates of synonymous and amino acid substitutions from homologous nucleotide sequences and its application. J. Mol. Evol. 16: 23–36.

Morton, D. G. and K. U. Sprague. 1982. Silkworm 5SRNA and alanine tRNA genes share highly conserved 5′ flanking and coding sequences. Mol. Cell. Biol. 2: 1524–1531.

Muller, H. J. 1935. The origination of chromatin deficiencies as minute deletions subject to insertion elsewhere. Genetics 17: 237–252.

Mullis, K. B. 1990. The unusual origin of the polymerase chain reaction. Sci. Am. 262(4): 56–65.

Muto, A. and S. Osawa. 1987. The guanine and cytosine content of genomic DNA and bacterial evolution. Proc. Natl. Acad. Sci. USA 84: 166–169.

Muto, A., F. Yamao, H. Hori and S. Osawa. 1986. Gene organization of *Mycoplasma capricolum*. Adv. Biophys. 21: 49–56.

Muto, A., F. Yamao and S. Osawa. 1987. The genome of *Mycoplasma capricolum*. Prog. Nucleic Acids Res. Mol. Biol. 34: 29–58.

Nadal-Ginard, B. and C. L. Markert. 1975 Use of affinity chromatography for purification of lactate dehydrogenase and for assessing the homology and function of the A and B subunits. pp. 45–67. *In* C. L. Markert (ed.), *Isozymes II: Physiological Function*. Academic Press, New York.

Nagylaki, T. 1984. The evolution of multigene families under intrachromosomal gene conversion. Genetics 106: 529–548.

Nagylaki, T. and T. D. Petes. 1982. Intrachromosomal gene conversion and the maintenance of sequence homogeneity among repeated genes. Genetics 100: 315–337.

Nakamura, Y., M. Leppert, P. O'Connell, R. Wolff, T. Holm, M. Culver, C. Martin, E. Fujimoto, M. Hoff, E. Kumlin and R. White. 1987. Variable number of tandem repeat (VNTR) markers for human genetic mapping. Science 235: 1616–1622.

Nathans, J., D. Thomas and D. S. Hogness. 1986. Molecular genetics of human color vision: The genes encoding blue, green, and red pigments. Science 232: 193–202.

Needleman, S. B. and C. D. Wunsch. 1970. A general method applicable to the search of similarities in the amino acid sequence of two proteins. J. Mol. Biol. 48: 443–453.

Nei, M. 1969. Gene duplication and nucleotide substitution in evolution. Nature 221: 40–42.

Nei, M. 1975. *Molecular Population Genetics and Evolution*. North-Holland, Amsterdam.

Nei, M. 1987. *Molecular Evolutionary Genetics*. Columbia University Press, New York.

Nei, M. and T. Gojobori. 1986. Simple methods for estimating the number of synonymous and nonsynonymous nucleotide substitutions. Mol. Biol. Evol. 3: 418–426.

Nei, M. and Y. Imaizumi. 1966. Genetic structure of human populations. II. Differentiation of blood group gene frequencies among isolated populations. Heredity 21: 183–190.

Nei, M. and W.-H. Li. 1979. Mathematical model for studying genetic variation in terms of restriction endonucleases. Proc. Natl. Acad. Sci. USA 76: 5269–5273.

Nei, M., P. A. Fuerst and R. Chakraborty. 1978. Subunit molecular weight and genetic variability of proteins in natural populations. Proc. Natl. Acad. Sci. USA 75: 3359–3362.

Nuttall, G. H. F. 1904. *Blood Immunity and Blood Relationship*. Cambridge University Press, Cambridge.

Ohno, S. 1970. *Evolution by Gene Duplication*. Springer-Verlag, Berlin.

Ohno, S. 1972. So much "junk" DNA in our genome. Brookhaven Symp. Biol. 23: 366–370.

Ohta, T. 1980. *Evolution and Variation of Multigene Families*. Springer-Verlag, Berlin.

Ohta, T. 1983. On the evolution of multigene families. Theor. Pop. Biol. 23: 216–240.

Ohta, T. 1984. Some models of gene conversion for treating the evolution of multigene families. Genetics 106: 517–528.

Oparin, A. I. 1957. *The Origin of Life on Earth*. Academic Press, New York.

Orgel, L. E. and F. H. C. Crick. 1980. Selfish DNA: The ultimate parasite. Nature 284: 604–607.

Östergren, G. 1945. Parasitic nature of extra fragment chromosomes. Bot. Notiser 2: 157–163.

Ouenzar, B., B. Agoutin, F. Reinisch, D. Weill, F. Perin, G. Keith and T. Heyman. 1988. Distribution of isoaccepting tRNAs and codons for proline and glycine in collageneous and noncollageneous chicken tissues. Biochem. Biophys. Res. Commun. 150: 148–155.

Palmer, J. D. 1985 Evolution of chloroplast and mitochondrial DNA in plants and algae. pp. 131–240. *In* R. J. MacIntyre (ed.), *Molecular Evolutionary Genetics*. Plenum, New York.

Palmer, J. D. and L. A. Hebron. 1987. Unicircular structure of the *Brassica hirta* mitochondrial genome. Curr. Genet. 11: 565–570.

Patthy, L. 1985. Evolution of the proteases of blood coagulation and fibrinolysis by assembly from modules. Cell 41: 657–663.

Perlman, P. S. and R. A. Butow. 1989. Mobile introns and intron-encoded proteins. Science 246: 1106–1109.

Piatigorsky, J. and Wistow, G. J. 1989. Enzyme/crystallins: Gene sharing as an evolutionary strategy. Cell 57: 197–199.

Piatigorsky, J., W. E. O'Brien, B. L. Norman, K. Kalumuck, G. J. Wistow, T. Borras, J. M. Nickerson and E. F. Wawrousek. 1988. Gene sharing by δ-crystallin and argininosuccinate lyase. Proc. Natl. Acad. Sci. USA 85: 3479–3483.

Post, L. E., G. D. Strycharz, M. Nomura, H. Lewis and P. P. Dennis. 1979. Nucleotide sequence of the ribosomal protein gene cluster adjacent to the gene for RNA polymerase subunit β in *Escherichia coli*. Proc. Natl. Acad. Sci. USA 76: 1697–1701.

Quentin, Y. 1988. The *Alu* family developed through successive waves of fixation closely connected with primate lineage history. J. Mol. Evol. 27: 194–202.

Razin, A. and A. D. Riggs. 1980. DNA methylation and gene function. Science 210: 604–610.

Rees, H. and R. N. Jones. 1972. The origin of the wide species variation in nuclear DNA content. Int. Rev. Cytol. 32: 53–92.

Ritossa, F. M., K. C. Atwood, D. L. Lindsley and S. Spiegelman. 1966. On the chromosomal distribution of DNA complementary to ribosomal and soluble RNA. Natl. Cancer Inst. Monogr. 23: 449–472.

Rubin, G. 1983. Dispersed repetitive DNAs in *Drosophila*. pp. 329–361. *In* J. A. Shapiro (ed.), *Mobile Genetic Elements*. Academic Press, New York.

Saiki, R. K., S. J. Scharf, F. Faloona, K. B. Mullins, G. T. Horn, H. A. Erlich and N. Arnheim. 1985. Enzymatic amplification of β-globin genomic sequences and restriction site analysis for diagnosis of sickle cell anemia. Science 230: 1350–1354.

Saiki, R. K., D. H. Gelfand, S. Stoffel, R. Higuchi, G. T. Horn, K. B. Mullis and H. A. Erlich. 1988. Primer-directed enzymatic amplification of DNA with a thermostable DNA polymerase. Science 239: 487–491.

Saitou, N. and M. Nei. 1986. The number of nucleotides required to determine the branching order of three species with special reference to the human–chimpanzee–gorilla divergence. J. Mol. Evol. 24: 189–204.

Saitou, N. and M. Nei. 1987. The neighbor-joining method: A new method for reconstructing phylogenetic trees. Mol. Biol. Evol. 4: 406–425.

Samson, M.-L. and M. Wegnez. 1988. Bipartite structure of the 5S ribosomal gene family in a *Drosophila melanogaster* strain and its evolutionary implications. Genetics 118: 685–691.

Sarich, V. M. and A. C. Wilson. 1967. Immunological time scale for hominid evolution. Science 158: 1200–1203.

Sarich, V. M. and A. C. Wilson. 1973. Generation time and genomic evolution in primates. Science 179: 1144–1147.

Sattath, S. and A. Tversky. 1977. Additive similarity trees. Psychometrika 42: 319–345.

Sawyer, S. A., D. E. Dykhuizen, R. F. DuBose, L. Green, T. Mutangadura-Mhlanga, D. F. Wolczyk and D. L. Hartl. 1987. Distribution and abundance of insertion sequences among natural isolates of *Escherichia coli*. Genetics 115: 51–63.

Scharf, S. J., G. T. Horn and H. A. Erlich.

1986. Direct cloning and sequence analysis of enzymatically amplified genomic sequences. Science 233: 1076–1078.

Schmid, C. W. and P. L. Deininger. 1975. Sequence organization of the human genome. Cell 6: 345–358.

Schultz, A. H. 1963. *Classification and Human Evolution*. Aldine, Chicago.

Schuster, W. and A. Brennicke. 1987. Plastid, nuclear and reverse transcriptase sequences in the mitochondrial genome of *Oenothera*: Is genetic information transferred between organelles via RNA? EMBO J. 6: 2857–2863.

Schwarz, Z. and H. Kössel. 1980. The primary structure of 16S rDNA from *Zea mays* chloroplast is homologous to *E. coli* 16S rRNA. Nature 283: 739–742.

Scott, A. F., P. Heath, S. Trusko, S. H. Boyer, W. Prass, M. Goodman, J. Czelusniak, L.-Y. E. Chang and J. L. Slightom. 1984. The sequence of the gorilla fetal globin genes: Evidence for multiple gene conversions in human evolution. Mol. Biol. Evol. 1: 371–389.

Sellers, P. H. 1974. On the theory and computation of evolutionary distances. SIAM J. Appl. Math. 26: 787–793.

Sharp, P. M. and W.-H. Li. 1986. An evolutionary perspective on synonymous codon usage in unicellular organisms. J. Mol. Evol. 24: 28–38.

Sharp, P. M., E. Cowe, D. G. Higgins, D. Shields, K. H. Wolfe and F. Wright. 1988. Codon usage patterns in *Escherichia coli, Bacillus subtilis, Saccharomyces cerevisiae, Schizosaccharomyces pombe, Drosophila melanogaster*, and *Homo sapiens*: A review of the considerable within-species diversity. Nucleic Acids Res. 16: 8207–8211.

Shaw, J.-P., J. Marks and C.-K. J. Shen. 1987. Evidence that the recently discovered θ1-globin gene is functional in higher primates. Nature 326: 717–720.

Shen, S.-H., J. L. Slightom and O. Smithies. 1981. A history of the human fetal globin gene duplication. Cell 26: 191–203.

Shinozaki, K., M. Ohme, M. Tanaka, T. Wakasugi, N. Hayashida, T. Matsubayashi, N. Zaita, Chunwongse, J, J. Obokata, K. Yamaguchi-Shinozaki, C. Ohta, K. Torazawa, B. Y. Meng, M. Sugita, H. Deno, T. Kamogashira, K. Yamada, J. Kusuda, F. Takaiwa, A. Kato, N. Tohdoh, H. Shimada and M. Sugiura. 1986. The complete nucleotide sequence of the tobacco chloroplast genome: Its gene organization and expression. EMBO J. 5: 2043–2049.

Sibley, C. G. and J. E. Ahlquist. 1984. The phylogeny of the hominoid primates, as indicated by DNA–DNA hybridization. J. Mol. Evol. 20: 2–15.

Simpson, G. G. 1961. *Principles of Animal Taxonomy*. Columbia University Press, New York.

Singer, C. E. and B. N. Ames. 1970. Sunlight ultraviolet and bacterial DNA base ratios. Science 170: 822–826.

Singer, M. F. 1982. SINEs and LINEs: Highly repeated short and long interspersed sequences in mammalian genomes. Cell 28: 433–434.

Smith, T. F., M. S. Waterman, and W. M. Fitch. 1981. Comparative biosequence metrics. J. Mol. Evol. 18: 38–46.

Sneath, P. H. A. and R. R. Sokal. 1973. *Numerical Taxonomy*. Freeman, San Francisco.

Soares, M. B., E. Schon, A. Henderson, S. K. Karathanasis, R. Cate, S. Zeitlin, J. Chirgwin and A. Efstratiadis. 1985. RNA-mediated gene duplication: The rat preproinsulin I gene is a functional retroposon. Mol. Cell. Biol. 5: 2090–2103.

Sogin, M. L., H. J. Elwood and J. H. Gunderson. 1986. Evolutionary diversity of eukaryotic small-subunit rRNA genes. Proc. Natl. Acad. Sci. USA 83: 1383–1387.

Sogin, M. L., J. H. Gunderson, H. J. Elwood, R. A. Alonso and D. A. Peattie. 1989. Phylogenetic meaning of the kingdom concept: An unusual ribosomal RNA from *Giardia lamblia*. Science 243: 75–77.

Sokal, R. R. and C. D. Michener. 1958. A statistical method for evaluating systematic relationships. Univ. Kansas Sci. Bull. 28: 1409–1438.

Sparrow, A. H. and A. F. Nauman. 1976. Evolution of genome size by DNA doublings. Science 192: 524–529.

Sparrow, A. H., H. J. Price and A. G. Underbrink. 1972. A survey of DNA content per cell and per chromosome of prokaryotic and eukaryotic organisms: Some evolutionary considerations. Brookhaven Symp. Biol. 23: 451–494.

Stebbins, G. L. 1981. Coevolution of grasses and herbivores. Ann. Missouri Bot. Garden 68: 75–86.

Stein, J. P., J. F. Catterall, P. Kristo, A. R. Means and B. W. O'Malley. 1980. Ovomucoid intervening sequences specify functional domains and generate protein polymorphism. Cell 21: 681–687.

Stewart, C.-B. and A. C. Wilson. 1987. Sequence convergence and functional adaptation of stomach lysozymes from foregut fermenters. Cold Spring Harbor Symp. Quant. Biol. 52: 891–899.

Stryer, L. 1988. *Biochemistry*, 3rd Ed. Freeman, New York.

Sueoka, N. 1964. On the evolution of informational macromolecules. pp. 479–496. *In* V. Bryson and H. J. Vogel (eds.), *Evolving Genes and Proteins*. Academic Press, New York.

Suzuki, D. T., A. J. F. Griffiths, J. H. Milller and R. C. Lewontin. 1989. *An Introduction to Genetic Analysis*, 4th Ed. Freeman, New York.

Sved, J. A. 1976. Hybrid dysgenesis in *Drosophila melanogaster*: A possible explanation in terms of spatial organization of chromosomes. Austral. J. Biol. Sci. 29: 375–388.

Tanaka, T. and M. Nei. 1989. Positive Darwinian selection observed at the variable region genes of immunoglobulins. Mol. Biol. Evol. 6: 447–459.

Temin, H. M. 1989. Retrons in bacteria. Nature 339: 254–255.

Thomas, R. H., W. Schaffner, A. C. Wilson and S. Pääbo. 1989. DNA phylogeny of the extinct marsupial wolf. Nature 340: 465–467.

Ticher, A. and D. Graur. 1989. Nucleic acid composition, codon usage, and the rate of synonymous substitution in protein-coding genes. J. Mol. Evol. 28: 286–298.

Ullu, E. and C. Tschudi. 1984. *Alu* sequences are processed 7SL RNA genes. Nature 312: 171–172.

Upholt, W. B. 1977. Estimation of DNA sequence divergence from comparison of restriction endonuclease digests. Nucleic Acids Res. 4: 1257–1265.

Wallace, D. C. and H. J. Morowitz. 1973. Genome size and evolution. Chromosoma 40: 121–126.

Walsh, J. B. 1985. Interaction of selection and biased gene conversion in a multigene family. Proc. Natl. Acad. Sci. USA 82: 153–157.

Walsh, J. B. 1987. Persistence of tandem arrays: Implication for satellite and simple-sequence DNAs. Genetics 115: 553–567.

Watson, J. D., N. H. Hopkins, J. W. Roberts, J. A. Steitz and A. M. Weiner. 1987. *Molecular Biology of the Gene*, 4th Ed. Benjamin/Cummings, Menlo Park, CA.

Weatherall, D. J. and J. B. Clegg. 1979. Recent developments in the molecular genetics of human hemoglobin. Cell 16: 467–479.

Weiner, A. M., P. L. Deininger and A. Efstratiadis. 1986. Nonviral retroposons: Genes, pseudogenes and transposable elements generated by the reverse flow of genetic information. Annu. Rev. Biochem. 55: 631–661.

Weiss, E. H., A. Mellor, L. Golden, K. Fahrner, E. Simpson, J. Hurst and R. A. Flavell. 1983. The structure of a mutant *H-2* gene suggests that the generation of polymorphism in *H-2* genes may occur by gene conversion-like events. Nature 301: 671–674.

Widegren, B., U. Arnason and G. Akusjarvi. 1985. Characteristics of conserved 1,579-bp highly repetitive component in the killer whale, *Orcinus orca*. Mol. Biol. Evol. 2: 411–419.

Willard, C., H. T. Nguyen and C. W. Schmid. 1987. Existence of at least three distinct *Alu* subfamilies. J. Mol. Evol. 26: 180–186.

Williams, S. A. and M. Goodman. 1989. A

statistical test that supports a human/chimpanzee clade based on noncoding DNA sequence data. Mol. Biol. Evol. 6: 325–330.

Wilson, A. C., S. S. Carlson and T. J. White. 1977. Biochemical evolution. Annu. Rev. Biochem. 46: 573–639.

Wistow, G. J., J. W. M. Mulders and W. W. De Jong. 1987. The enzyme lactate dehydrogenase as a structural protein in avian and crocodilian lenses. Nature 326: 622–624.

Woese, C. R. 1987. Bacterial evolution. Microbiol. Rev. 51: 221–271.

Wolfe, K. H., W.-H. Li and P. M. Sharp. 1987. Rates of nucleotide substitution vary greatly among plant mitochondrial, chloroplast, and nuclear DNAs. Proc. Natl. Acad. Sci. USA 84: 9054–9058.

Wolfe, K. H., P. M. Sharp and W.-H. Li. 1989a. Mutation rates differ among regions of the mammalian genome. Nature 337: 283–285.

Wolfe, K. H., P. M. Sharp and W.-H. Li. 1989b. Rates of synonymous substitution in plant nuclear genes. J. Mol. Evol. 29: 208–211.

Wood, W. G., J. B. Clegg and D. J. Weatherall. 1977. Developmental biology of human hemoglobins. pp. 43–90. *In* X. E. B. Brown (ed.), *Progress in Hematology*. Grune and Stratton, New York.

Wu, C.-I. and W.-H. Li. 1985. Evidence for higher rates of nucleotide substitution in rodents than in man. Proc. Natl. Acad. Sci. USA 82: 1741–1745.

Wu, C.-I. and N. Maeda. 1987. Inequality in mutation rates of the two strands of DNA. Nature 327: 169–170.

Wu, C.-I., T. W. Lyttle, M. -L. Wu and G. -F. Lin. 1988. Association between a satellite DNA sequence and the *Responder* of *Segregation Distorter* in *Drosophila melanogaster*. Cell 54: 179–189.

Wu, C.-I., J. R. True and N. Johnson. 1989. Fitness reduction associated with the deletion of a satellite DNA array. Nature 341: 248–251.

Yoshitake, S., B. G. Schach, D. C. Foster, E. W. Davie and K. Kurachi. 1985. Nucleotide sequence of the gene for human factor IX (antihemophilic factor B). Biochem. 24: 3736–3750.

Zimmer, E. A., S. L. Martin, S. M. Beverley, Y. W. Kan and A. C. Wilson. 1980. Rapid duplication and loss of genes coding for the α chains of hemoglobin. Proc. Natl. Acad. Sci. USA 77: 2158–2162.

Zuckerkandl, E. 1976. Gene control in eukaryotes and the C-value paradox: "Excess" DNA as an impediment to transcription of coding sequences. J. Mol. Evol. 9: 73–104.

Zuckerkandl, E. and L Pauling. 1962. Molecular disease, evolution and genic heterogeneity. pp. 189–225. *In* M. Kash and B. Pullman (eds.), *Horizons in Biochemistry*. Academic Press, New York.

Zuckerkandl, E. and L. Pauling. 1965. Evolutionary divergence and convergence in proteins. pp. 97–166. *In* V. Bryson and H. J. Vogel (eds.), *Evolving Genes and Proteins*. Academic Press, New York.

Zuckerkandl, E., J. Derancourt and H. Vogel. 1971. Mutational trends and random processes in the evolution of informational macromolecules. J. Mol. Biol. 59: 473–490.

INDEX

ABOUT THE BOOK

Book Design: Joseph J. Vesely
Cover Design: Rodelinde Graphic Design
Editor: Andrew D. Sinauer
Copy Editor: J. David Baldwin
Editorial Coordinator: Carol J. Wigg
Artist: Fredric Schoenborn
Production Manager: Joseph J. Vesely
Art and Page Layout: Janice Holabird
Composition: DEKR Corporation
Cover Manufacture: New England Book Components
Book Manufacture: R. R. Donnelley & Sons